The Science Beneath Organic Production

The Science Beneath Organic Production

Edited by

David Atkinson
SRUC
Aberdeen, Scotland
UK

Christine A. Watson
SRUC
Aberdeen, Scotland
UK

WILEY Blackwell

This edition first published 2020
© 2020 John Wiley & Sons Ltd

The right of David Atkinson and Christine A. Watson to be identified as the authors of the editorial material in this work has been asserted in accordance with law.

Registered Offices
John Wiley & Sons, Inc., 111 River Street, Hoboken, NJ 07030, USA
John Wiley & Sons Ltd, The Atrium, Southern Gate, Chichester, West Sussex, PO19 8SQ, UK

Editorial Office
The Atrium, Southern Gate, Chichester, West Sussex, PO19 8SQ, UK

For details of our global editorial offices, customer services, and more information about Wiley products visit us at www.wiley.com.

Wiley also publishes its books in a variety of electronic formats and by print-on-demand. Some content that appears in standard print versions of this book may not be available in other formats.

Library of Congress Cataloging-in-Publication data has been applied for:

9780470023938 [paperback]

Cover Design: Wiley
Cover Images: © Firstsignal/iStock.com, © Eraxion/iStock.com, © Rasica/iStock.com

Set in 10/12pt Warnock by SPi Global, Pondicherry, India

Printed and bound by CPI Group (UK) Ltd, Croydon, CR0 4YY

10 9 8 7 6 5 4 3 2 1

Contents

List of Contributors

David Atkinson
SRUC
Aberdeen, Scotland, UK

Thomas F. Döring
Department of Agroecology and Organic
Farming
Institute of Crop Science and Resource
Conservation
University of Bonn
Bonn, Germany

Tony C. Edwards
SRUC
Aberdeen, Scotland, UK

Ruth E. Feber
Wildlife Conservation Research Unit
Recanati-Kaplan Centre
Department of Zoology
University of Oxford
Oxford, UK

Paul J. Johnson
Wildlife Conservation Research Unit
Recanati-Kaplan Centre
Department of Zoology
University of Oxford, Oxford, UK

David W. Macdonald
Wildlife Conservation Research Unit
Recanati-Kaplan Centre
Department of Zoology
University of Oxford, Oxford, UK

Bruce D. Pearce
Organic Research Centre
Newbury, UK

Pete Ritchie
Nourish Scotland and organic farmer
Edinburgh
Scotland, UK

Elizabeth A. Stockdale
NIAB
Cambridge, UK

Robin L. Walker
SRUC
Aberdeen, Scotland, UK

Christine A. Watson
SRUC
Aberdeen, Scotland, UK
and
Swedish University of Agricultural Sciences
(SLU), Uppsala, Sweden

Martin S. Wolfe
Organic Research Centre
Newbury, UK
and
Wakelyns Agroforestry
Eye, UK

Lawrence Woodward
Elm Farm Research Centre
Newbury, UK

Preface

There are many books dealing with organic farming so why another one? Agriculture is a major business in most countries of the world and a total global activity. Agricultural products are traded globally and so they need to be defined in relation to provenance, identity, composition and quality. This has resulted in many rules and regulations which influence what can be sold in what market and the prices of the various commodities. This all makes agriculture very political and means that the science base which informs how crops and farm animals are grown is shaped by issues beyond mere science. The ways in which crops are produced inevitably interact with all these political issues, leading to acrimony between those who produce in different ways.

While the production of any given crop can differ in a range of ways, much recent tension has arisen between those who farm in a way that allows their produce to carry an organic certified label and those who use a range of chemical inputs such as inorganic fertilisers and pesticides. The introduction of genetically modified crops increased the level of disagreement between producers in a debate which at times became an argument between what were being stereotyped as traditional farming and scientific farming. The former was characterised as being a relic of the past while the latter was seen as being more progressively logical and based on clear science. The advent of gene editing and the use of CRISPR/Cas9 and related technologies have reignited this debate.

Discussion around the place of these approaches commonly and rapidly moves to a consideration of the relative crop yields produced per unit area of land. The ability of the approaches to feed an expanding global population is then linked to such yields as if this were the sole criterion for something as economically and socially complex as world food production. In no other part of global endeavour is a simple interpretation of production the basis of how a major enterprise should be effected. Profitability and its social setting, part of social science, are always important issues. Food production involves much of the world's population and therefore it has many cultural links. Behind all of this lies an inherent suggestion that, unlike modern forms of agriculture, organic production is centred on folk lore and is a continuation of outdated methods. Agriculture is a major global source of greenhouse gases and so viewing different approaches to agriculture is important as part of current discussions about the amelioration of the impact of global climate change.

So is organic production scientific? In the sense that the question is most commonly asked, the answer is an unequivocal 'yes'. When the debate over food production methods in the UK reignited at the end of World War II, the organic movement was at pains to point out the distinguished scientific qualifications of its pioneers like Sir Albert Howard and Sir Robert McCarrison. These people had worked from a starting point in agricultural or the physical sciences to the conclusion that an ecological approach to food production was likely to be most sustainable. Organic production has thus always had a strong basis in science. However, the science which informs organic production is different from that underpinning current western food production. This difference means that it has benefited to only a small extent from much of the agricultural research done over the past half century. The emphasis of much of that research has been on the use of fertilisers to supply mineral nutrients and on chemicals and genetic modification to provide crop protection. Development of the explicitly organic farming science base has been limited.

Much of the science base is, however, shared by all forms of agriculture. For growth, all crops need to intercept light and absorb nutrients from the soil. Agricultural systems need to provide these basic resources and how effectively that is done inexorably affects yields. Data on how agricultural practice influences these aspects of production are of value to all producers, including those who produce to organic standards. There are, of course, significant differences in how these basic needs are met, which inevitably link back to why organic crops are being produced in a way which is distinct from other approaches to food production. The scientific information which is used in any form of production at any time depends on its underlying production aims and its perceived values. This results in the science important to organic production being selected by reference to a different sociological and economic model to that which informs other types of farming.

There is a distinctive element in the science base of organic production. Does this mean that information from other forms of agriculture cannot be used in evaluating the working of organic systems? No! Basic information is valid in a range of situations although the selection of what is most relevant may differ. In this volume, we detail science that is shared with other forms of production and that which is different and central to organic production.

The chapters in this volume have a number of authors. Farming is complex and few have a total overview of all elements. In respect of organic production, elements of social science are critical to the aims and objectives of producers. History is important to an understanding of how we reached this point and so we review the development of organic farming as a distinctive approach. History and social science determine the parts of the total science base which are pertinent. In editing this volume, our aim has been not only to detail key elements of the science which lies behind crop production but also to give an insight into why this approach to the use of science is important. Why producers elect to produce in accord with organic standards is a key question. Its answer lies as much in the cultural context as in mere economics. In a world of global climate change, global resource use, shifting populations and a digital revolution which systematically aims to reduce the need for human labour, such questions require a broad study of science.

Inevitably, this takes us beyond mere considerations of which approach can produce the largest yield per unit area. It also goes beyond how many could be fed were the approach to be replicated on a global scale. It asks questions about sustainability, preservation of our restricted reserve of soil, use and management of genetic resources and our use of people as both producers and consumers. It asks questions about the cultural context of food production on a global scale. Scientific facts may be absolute but their use and interpretation are culturally driven. The acquisition of new knowledge is a cultural phenomenon and commonly driven by economic forces. Where there is an absence of an industrial link, new research information is restricted. This has been an issue for organic production and so the authors of chapters in this volume, as well as summarising what is known about the various elements which make up organic crop production, have attempted to identify current gaps in understanding and future research needs.

While we have dealt with the production of crops and of forage, we have not discussed organic animal production other than in terms of the return of animal wastes to the soil system. Organic animal husbandry involves issues centred on animal health, care and husbandry which require a treatment as substantial as those covered here for crops and so we have chosen to leave that for others to discuss. Nevertheless, we believe that here we have covered many of the key issues linked to the use of science by organic producers and have identified a distinctive way of looking at the production of food and the science upon which it depends.

Looking ahead requires consideration not just of gaps in our information base but also of how that information needs to be obtained. All of agriculture is dominated by interactions which is why approaches which work well in models or in laboratory or controlled environment settings may either fail to work under field conditions or more commonly may fail to work consistently or reliably across a very diverse globe. Approaches devised to study the impact of inputs may not work for complex situations in which it is necessary to define the nature of such variation and, more importantly, what can be done to manage it. Here, we discuss how research related to organic production might best be carried out in future.

This volume has had a long gestation. During the period over which it was written, the ways in which people both obtain and use information have changed more profoundly than over any comparable period since the invention of printing. The digital revolution and the use of international databases such as Google have changed fundamentally how we work. Initially, we had aimed to produce an encyclopaedia. The increase in new publications (a single conference run by the International Society for Horticultural Science in 2010 gave rise to 89 new papers) made this approach impossible even when being selective. We had then hoped to pull together key themes in the available literature. Finally we concluded that to be of real and lasting value, what was needed in the current climate was a volume which would provide enough information to help its readers interact with the internet and to be able to use Google and similar information systems in a critical way, so enhancing knowledge and understanding.

Asking questions of Google is easy. Asking questions formulated in such a way that real information is produced is much harder. In this book, we have aimed to help readers ask better questions. However, there is also a need to understand what types of answers are likely to be helpful. We give examples of the type of answer which might be helpful in response to such key questions. While the internet is a wonderful source of information, it is important to remember that the digitisation of science only began seriously in the 1980s. For some subjects, such as molecular biology where most information has appeared over the past two decades, this is not a problem. Discussion of organic agriculture and related science is, however, of a much older vintage. Key issues were actively debated in the immediate postwar period, many long-term trials were established and this was the era when government funding for science was at its maximum. Here, we aim to provide a view of these discussions as a basis for the results of more recent studies. We see this as critical to an understanding of the division in approach to food production which dated from that time. What we have aimed for is thus something which will answer many of the most important questions but which will allow the reader to better interact with both scientific information and today's issues. This volume is the product of discussions over many years with colleagues with whom we have worked at East Malling Research Station (now EMR), The Macaulay Institute for Soil Research (now JHI) and the Scottish Agricultural College (nowSRUC). We thank them and many other colleagues interested in Organic Farming.

David Atkinson
Christine A. Watson

1

Science and Organic Agriculture

An Introduction

David Atkinson and Christine A. Watson

SRUC, Aberdeen, Scotland, UK

1.1 What is the Role of this Volume?

Our objective is to demonstrate that a substantial body of science underpins organic food production. Much of this is shared with other systems of production but there are real and major differences in terms of the parts of the science base which are mission critical. The reliance of organic systems on natural soil processes means that it has much in common with the science base of natural ecosystems. Information derived from research on such systems has a greater significance for organic production than it has for some other systems of production.

Here we do not aim to be encyclopaedic. We have not aimed to cover all aspects of organic agriculture. It would not be possible to cite every published paper relevant to those subjects selected for discussion. The cited papers have been selected as being illustrative and representative. By adopting this approach, we have been able to make use of literature from a number of decades rather than just the latest papers. We hope that this has given a timeless perspective and may even help today's readers gain access to work from before the digital era. We have tried to approach the subject through asking questions and by allowing the authors of chapters to bring their individual approaches to the discussion. In most ways, this is more a book about 'why' rather than 'how' – that is, why would you want to follow that approach? – rather being a practical organic farming textbook.

Our principal aim is to demonstrate that organic farming has a science base and that that science base is to be located in the social sciences as well as in the more traditional areas of physics, chemistry and biology. Hence, we begin by examining the societal context of organic production, which argues for this form of agriculture as both a source of employment and as the basis of societal relationships with food and food producers. We then deal with the basic science which underpins this form of production through chapters exploring the restrictions imposed by land capability, the functioning of rotations, the processes relating to the availability and storage of nutrients such as N and P in soils and the mechanisms which allow crops to access these nutrients. The culmination of this process is the production of food in the form of

The Science Beneath Organic Production, First Edition. Edited by David Atkinson and Christine A. Watson.
© 2020 John Wiley & Sons Ltd. Published 2020 by John Wiley & Sons Ltd.

either plant material or animal products. Claims have been made for the value of organic production in relation to food quality and so we examine these claims in relation to what is known of the impact of methods of production on food quality. In organic production, the genetics of the crop is important in relation to both its production features and its ability to resist disease and so we examine this as part of organic production's science base. If the link with producer communities is one reason for the adoption of this approach to agriculture, then the other is its environmental impact. We assess the basis of the interaction of this form of land use with environmental impact.

Finally, on the basis of all this we look ahead to where there is need for more information and to where the approach which organic production represents can help to solve some of the world's current problems related to food production and environmental impact.

In compiling this volume, we have not felt constrained only to include information derived from studies carried out on organic experiments but have used information relevant to organic systems, from the full range of sources. We have focused on the issues related to crop and forage production. There is a significant science base related to animal production but covering this would have either made this volume very long or resulted in a coverage which was inadequate to the importance of this issue. We leave it for others to fill that niche. We begin by reviewing what we mean when we use the term 'organic'.

1.2 What is Organic Agriculture?

The term *organic agriculture* came into being around 1940 following ongoing discussions in the 1920s and 1930s. It was a means of describing farms which were managed in such a way that the farm could be a self-contained unit with few external inputs. When the term was originated, the use of mineral fertilisers and pesticides by other types of agriculture was relatively modest. The difference between organic and other systems was small. A defining moment, in respect of identity, came during World War II when there was need to gain exemption for the Haughly experiment, which was begun in 1939, from wartime regulations which required the use of fertilisers so as to maximise crop production (Balfour 1976). From this point in time, 'organic' began to have a distinctive ethos even though in practice differences were often small.

One of the earliest statements about the distinctive nature of the organic approach came with the publication of Lady Eve Balfour's book *The Living Soil: Evidence of the importance to human health of soil vitality, with special reference to post war planning* in October 1943. Reviewing this book, Lawrence Woodward (Chapter 2) commented that:

> *The Living Soil* is not a book about farming. It is about citizenship and community – the community of all living things; it is about health – the health of individuals and the health of the communities they make up; and it is about life itself – the vitality that flows through all living things from the soil through plants, animals and man.

This approach, which is amplified in Chapter 2, indicates that any evaluation of organic farming, its practices and its science will have a wider context with the science being set against goals significantly beyond mere food production. This, it can be argued, is true of all farming and food production but perhaps the major difference is in the intentionality. Organic production by its very nature incorporates all of these elements and they all are key deliverables.

Tinker (2000) reviewed the ways in which organic production differed from other approaches and some of the many legislatively driven definitions. He concluded that it had characteristics which allowed it to be easily differentiated from other types of agriculture, which he termed 'conventional' but that it was hard to define it precisely. Official definitions tend to emphasise this difficulty. For example, the Food and Agriculture Organization/World Health Organization (FAO/WHO) defines organic agriculture as follows.

> Organic agriculture is a holistic production management system, which promotes and enhances agri-ecosystem health, including biodiversity, biological cycles and soil biological activity. It emphasises the use of management practices in preference to the use of off-farm inputs, taking into account that regional conditions require locally adapted systems. This is accomplished by using, where possible, agronomic, biological and mechanical methods as opposed to using synthetic materials to fulfil any specific function within the system.

With time, differences seem to have become greater although organic production is still most commonly distinguished more by the things it does not permit, such as fertilisers, pesticides and genetically modified organisms (GMOs), than its positive virtues.

The overall aims of organic agriculture were set out by Howard (1945) in his foundation text.

1) The birthright of all living things is health.
2) This is true for soil, plant, animal and humanity. The health of all of these is a connected chain.
3) Any weakness in the health of an earlier link in the chain is carried on through the chain.
4) The pests and diseases of agriculture are evidence of failure in the second and third links.
5) Impaired human health is also a consequence of this failure.
6) General failure in the final three links can be attributed to a failure in the first link, the soil; the undernourishment of the soil is at the root of all
7) Going back is not difficult as long as we are mindful of nature's requirements for (i) the return of wastes to the land, (ii) the mixture of animal and plant systems, (iii) maintaining an adequate system to feed the plant, i.e. we must not interrupt the mycorrhizal association.

These overlap with but have a very different emphasis from the features of organic systems identified by Tinker (2000).

1) To avoid the use of synthetic, highly toxic or soluble chemicals on crops.
2) To ensure that the soil and its biota are healthy.
3) To use sound husbandry methods so that crops and stock are healthy.
4) To ensure the welfare of farm animals.
5) To use biological natural cycles rather than distorting them.

Tinker (2000) did not produce a parallel list for what he termed conventional agriculture although the key issues he suggested are as follows.

1) Its main drive is towards productivity and efficiency as in most industries.
2) It covers a wide range of practices and can include the use of chemicals combined with nature conservation measures.
3) It requires the use of science and technology to the fullest extent and so permits the use of any chemical found to be beneficial and without serious disadvantages.
4) Safety is dependent on careful regulation by appropriate authorities.

The contrast between these definitions identifies that soil health and its microbial populations will always be critical to successful organic production and may outweigh the emphasis on production which is at the heart of current business models. Distinguishing between approaches to food production is helped by Tinker's (2000) analysis of the deficiencies of earlier systems

> The main constraints on productivity were shortages of the essential nutrient elements in the soil, pests and diseases.

This helps us to focus on the areas where the science important to organic production may be different from that in other systems. It also indicates areas where there will not be differences in relevance or importance. This helps to set the context for this volume where we aim to confirm areas of science which are common to all forms of agriculture but to highlight areas where the different approaches to nutrient supply and crop protection lead to a different use of the science base and some of the consequences of this choice.

The stress placed on the role of science in 'conventional agriculture' and its apparent omission from some of the key issues associated with organic production is both an attitude of mind and recognition of the limited amount of research carried out in support of organic production. It recalls the comments of the Secretary of the British Board of Agriculture in an earlier era:

> I cannot conceive of the circumstances in which the Board will be at all interested in scientific work (quoted in Plumb 1998).

Too often this view has been expressed about organic production. In the same paper, Plumb considered the energy needs of conventional agriculture and commented that new products:

> increase our dependence on fossil fuels for production and processing. We must begin to think more seriously of how we use the whole range of our natural resources.

This comment made by Plumb in 1973 has a very contemporary climate change significance, an issue discussed at the end of this chapter.

The need to characterise the distinctiveness of organic from other ways of production has in recent years been driven as much by commercial considerations and marketing as by ethical deliberations. Organic farming and the food it produced had to be defined so that they could be distinguished, at the point of sale, from food produced in other ways. Organic agriculture is the only form of agriculture defined in this way although current definitions are not helpful in understanding the science. The simplest definition would be that organic agriculture is the form of agriculture defined as such in national and international legislation. The development of organic agriculture in the UK has been chronicled by Conford (2001).

The intensification of agriculture during World War II led to an industrialised view of the future of agriculture, which was continued after the end of the war and enshrined in the 1947 Agriculture Act. This emphasised productivity per man and led to the replacement of mixed farming by specialist enterprises. Barker (2010) detailed the steps which led to the enactment of parallel legislation in the USA. US legislation was linked to legislation in the EU and elsewhere to provide a basis for international trade. Produce currently classed as organic was initially described in ways such as 'ecologically grown'. Expansion of organic agriculture in the 1970s as an alternative to food produced with increasing use of agrochemicals led to a need to certify provenance and for bodies to carry out this certification. In the USA, this lead to the Organic Foods Production Act 1990 and a national Organic Programme which was to be responsible for standard setting and an Organic Materials Review Institute which was to define acceptable products. There were parallel developments in UK and EU.

Organic farming as described in legislation can be inflexible. This is a consequence of the need to be able to define it in the market place. As its core attributes are the ways in which it has been produced and a series of quality attributes, which can be hard to demonstrate at the point of sale, clear definition of such features is important and so have the elements of a brand. The difficulties of using less rigorous descriptors have been emphasised by decisions of the UK Advertising Standards Authority (Anon 2012).

Organic agriculture is thus at the same time a brand, a marketing concept, a means of identifying food produced in a particular way and within a code but probably for most of its practitioners an ethical approach to thinking about farming and our use of the natural environment. While organic production is commonly described as production without added chemical inputs, it would be better to define it in terms of what it does do rather than what it doesn't do and to define it as agriculture which obviates the need for added chemical inputs as a result of its success in managing soil resources and ecological processes.

Both Howard (1945) and Balfour (1976) emphasised the importance of the soil and of mycorrhizal fungi as the basis of sustainable agriculture and health in crops, farm animals and the human population. Soil and especially its microbial populations affect cropping. It matters that agriculture should work with the soil as the basis of productivity. It is here we need to look for the distinctiveness of the science which underpins organic agriculture.

1.3 So What is Distinctive About its Science Base?

The features critical to the success of organic agriculture are not unique. Most are important in all systems of food production although in other systems they may be less critical to success. Tinker (2000) identified the factors limiting yields in organic systems as the supply of nutrients, especially nitrogen, and resistance to pests and diseases. The approaches adopted in conventional agriculture do not determine whether *any* crop at all is produced, how much is produced and its quality to the same extent as in organic systems. Greenland (2000) widened the discussion of where there might be differences. He commented:

> For arable production the great advantage of conventional systems using fertilisers and pesticides is that the methods are simple and economic to use, the crop is well fed and vigorous, and the effects of weed competition, insect attack and disease problems, minimised. Continued use of these methods in arable agriculture does reduce the soil organic matter content and this may cause deterioration in some soil properties. However, under UK conditions the changes usually have little if any effect on present or future productivity of the soil and the deleterious effects are mostly reversible.

Almost 20 years later, there is a much wider recognition of the importance of reduced organic matter in soils and its limiting effect on crop production. There is growing interest in returning to more mixed farming systems to benefit soil organic matter, albeit with potentially different models to the original model of crop and livestock integration on almost every farm (Moraine et al. 2017).

All of this led Greenland to the general conclusion that:

> Soil fertility and crop production should be supported by the integrated use of organic manures and inorganic fertilisers.

While distinguishing between organic and other production, it suggests that some things critical to organic production will be particularly important. This centres on issues related to nutrition. Systems can underplay the importance of soil organic matter on soil structure and on the ability of the soil both to provide the crop with water and to function as a matrix which holds and supplies nutrients, including those from fertilisers. Global climate change, with its associated effects on rainfall patterns, including the intensity of storms, makes this an increasingly important consideration. Soil structure and its impact on water holding and release seem likely to be the basis of the acknowledged greater relative performance of organic systems in hot and tropical environments.

To succeed, all crops need to develop a canopy of leaves so as to absorb the sun's radiant energy, to be able to absorb from the soil sufficient quantities of water and nutrients such as N to permit leaves to be constructed and to survive sufficiently long so as to allow the development of that part of the plant which is to be used as a food. The importance of these aspects to organic production means that in this volume we have focused on how organic production meets these challenges and the consequences of the approach.

In conventional agriculture, externally produced chemical resources achieve such aims. Herbicides eliminate competition from other vegetation, so allowing the easier development of the crop canopy and reducing total water use early in the season. Fertilisers are placed so as to allow rapid and early uptake. Insecticides, fungicides and growth regulators help to sustain the canopy and facilitate assimilation into harvested product. This approach makes little use of natural systems and processes, which become critically important if external crop protection and sources of nutrients are not available, just as they are in uncropped vegetation. This helps us to ask questions such as what can be done to both crop and soil to facilitate the uptake of nutrients from natural sources and what can be done to sustain the functioning of the leaf canopy in the absence of chemical protection? These issues influence the growth, development and survival of uncropped vegetation. Thus knowledge from uncropped system ecology will be a major resource for organic production.

1.4 The Ecological Roots of Organic Production

In 1988, as part of its 75th anniversary celebration, the British Ecological Society sought to identify the key ecological concepts that were important to an understanding of the natural world (Cherrett 1989). The 10 concepts within natural ecosystem ecology which were selected were as follows.

1) The ecosystem
2) Succession

3) Energy flow
4) Conservation of resources
5) Competition
6) Niche
7) Materials cycling
8) The community
9) Life history strategies
10) Ecosystem fragility

A comparison of these with the key elements identified by Howard (1945) and by Balfour (1976) shows a considerable degree of similarity. Howard emphasised the importance of a holistic view of the agricultural system characterised by linking the production of crops and animals. This defines the organically farmed unit as a distinct ecosystem. Both Balfour and Howard emphasised the importance of rotation – crops following each other in a reasoned sequence and with the intention that particular crops should have their needs met as a result of their place in that succession and as a consequence of the impact of earlier crops. These issues are discussed further in Chapter 6. This parallels an ecological understanding of succession.

Both Howard and Balfour emphasise the importance of optimum conditions for the functioning of arbuscular mycorrhizal fungi (AMF) as a key requirement for a healthy organic system. In natural ecosystems, AMF are important to energy flows, conservation of resources, materials cycling and the community. The organic approach to living with organisms which in conventional agriculture would be classed as weeds, pests or diseases parallels the concepts of competition and niche. The ecological concept of life history strategies has much in common with the organic farmer's need to identify the crop species and cultivars which best fit into the rotation at a particular point. The absence of ecosystem fragility, i.e. sustainability, is a key requirement. The organic ecosystem needs to be able to sustain itself without excess inputs and to resist the problems found in conventional systems when particular pathogenic fungi develop resistance to a particular fungicide or identify weaknesses in a genetically modified construct.

In 1968, the British Ecological Society held a symposium to discuss the ecological impact of mineral nutrition. In his introduction to the event, Clapham (1969) identified the three major ways in which nutrient supply could affect the performance and composition of vegetation. All of these elements fall within the workings of the above ecological principles.

1) The specification of the whole plant soil system in the context of uptake and utilisation within the range of naturally occurring conditions.
2) The identification, specification and measurement of differential behaviour both within and between species in the uptake and utilisation of ions.
3) The identification, specification and measurement of competitive effects both between and within species in respect of ion uptake.

A comparison of conventional agriculture with the above ecological constructs would find less similarity. It is harder to regard the monoculture of a single crop grown in the same spatial area for a number of years as an ecosystem. Similarly, it is hard to identify the concept of community where the use of pesticides is the means of establishing dominance over all other organisms and where any adverse impact on soil microbes would be regarded as collateral damage. In addition, the concept of energy flow and material cycling has a very different flavour when the source of nutrient input to the system is the use of significant proportions of global fossil fuel reserves.

Atkinson and Watson (2000) assessed whether the future research needs of organic production were similar to those of other systems of production and suggested that they were rather different, principally for the reasons identified above. In assessing the needs of natural ecosystem ecology, Cherrett (1989) noted that ecology lacked the equivalent of the Newtonian Laws of Physics or the chemist's Periodic Table, which resulted in each fact having to be discovered for itself and remembered in isolation. This is also true of organic farming. Balfour (1976) wrote, when detailing the thinking behind the Haughley Experiment:

> Crop and livestock in an ecological situation may differ from behaviour in a fragmented system. Unless applied research in any biological field has its roots in fundamental research it can lead to practices with disastrous effects.

We are aware of much about soil but our knowledge is inadequate for precise management. The plasticity found in organic systems means that the level of mechanistic information, such as is available for agrochemicals, will never be known to the same extent. However, by identifying what is most relevant to our needs, as Cherrett (1989) did, we will improve our ability to seek what is needed from the science base.

1.5 Key Elements in the Science Context of Organic Agriculture

While organic production may share much with natural ecosystem ecology in relation to basic processes, ultimately, as in other farming systems, a key aim is to achieve the dominance of one plant species and to ensure that species gains a disproportionate share of available resources. Our focus is on the science which underpins organic systems. This requires discussion of why science has a different context in organic systems. This does not imply a natural superiority of organic systems over all other options. The complexity of global agriculture and its wide range of intended deliverables make any such judgements unhelpfully simplistic. However, food production must be seen in a wider context and this is at the heart of organic methodology.

Scientific experimentation is usually carried out in an objective way but subjective factors are important in selecting the experiments carried out, and how the results of those experiments are applied. Thus, while science is objective, the context of past research is subjective because all research is carried out in a specific societal context. Sociological and ethical considerations rather than fact tend to dominate decisions on application. In this volume we compare different systems of agriculture. Some of the comparisons identify where there are differences in values. These will have different impacts as a result of what is most important to society. These objectives influence the choices made in both the use of natural resources and the values placed on human capital (Atkinson 2009).

It is possible, however, to see all of the systems currently being employed in the agriculture of developed countries as gradations in two distinct sets of values, ranging from those which primarily emphasise efficiency and cost control to those which place their primary emphasis on integrating food production into the biological environment and setting agricultural production into a cultural context. This is summarised in Figure 1.1.

In the management of intensive (conventional) farming systems, the primary objective is commonly the maximisation of production within limits set by the genetic make-up of the crop

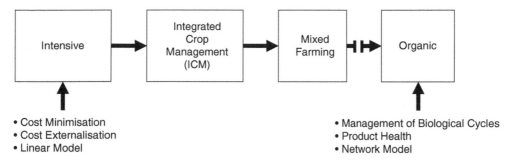

Figure 1.1 The gradation of types of farming system grouped on the basis of their dominant values. *Source:* Modified from Atkinson and Watson (2000).

or animal species and prevailing legislation on environmental impact or animal welfare. The yield obtained and the financial return are of primary importance. It is a testimony to the 1947 UK Act that over half a century later, productivity rather than environmental impact or public engagement remains dominant. This makes comparisons with the output from systems such as organic farming difficult.

Balfour (1976) argued that it is important to assess the production of organic systems over the course of a rotation during which the production of cereals or other crops is likely to represent only a part of total output. Labour is a major cost in conventional production and so efficiency is commonly understood in labour productivity terms. Simple linear models govern management. These may include linking the presence of particular weeds to specific herbicide applications or the expected attainment of yields to particular levels of fertiliser application. Minimising human involvement is seen as a necessity. Indeed, the release of people from agricultural activities was identified as a positive virtue of the increased use of herbicides for weed control (Holm 1976). This approach aims to externalise as many costs as possible, such as the costs of removing leachate emanating from added fertilisers or pesticides from water and the consequences of reducing employment in food production.

Set against the objectives of past legislation, this model has been a success. It is associated with increases in crop and animal yields and a fall in the percentage of disposable income spent on food. However, it has also been associated with significant environmental impacts through the damage caused by inputs and the amounts of fossil fuel energy used to produce inputs such as fertiliser nitrogen and pesticides. In addition, it has affected the viability of rural communities, relationship networks and available employment in food production. The approach has been associated with an increase in farm size and an increase in the importance of the providers of key inputs. Its major justification is its ability to produce high yields and feed western consumers. Those who champion this approach see it as the solution to feeding an increasing world population.

However, this assumes that the only global priority is food supply. It ignores the social context of agriculture. As a consequence, issues beyond food supply such as the importance of the environment and communities and relationships within the world's population tend to be minimised. In addition, while this form of agriculture is now the dominant approach in the western world, i.e. European and North American agriculture, it produces only around 30% of the food consumed by the global population. Most global production goes unrecorded because it is not traded. This leaves as a core question the extent to which the western approach to food production is the global way ahead. Have global priorities changed? What now should be the goals of agriculture? Solutions depend on how we view major issues such as the involvement of communities, the causes of global climate change, the wisdom of continuing with current levels of fossil fuel use and our initial exploration of the global genome of which current GM technologies represent just the start.

At the other end of the spectrum (see Figure 1.1) is organic production, as set out by the founders of the organic movement: Northbourne, Balfour and Howard in the UK, Steiner in Europe, Rodale in the USA and others in New Zealand, Germany and Japan. These systems are designed to be in harmony with the environment and to positively involve people and communities. As the primary driver is not maximised production, it is unsurprising that the yields obtained are smaller than when maximum yield is the main purpose. The higher reliance on basic soil resources means that the models which underpin organic systems are not linear. More time is required for management, which allows community involvement. Similarly, the desire that crop production should co-exist with the presence of other species on the same land unit, i.e. that species conservation should proceed in parallel with production, must reduce the proportion of total resources for the crop and so reduce yields. There is no such thing as a free lunch.

Differences in production achieved by using different systems continue to be of significant interest and there have been many studies attempting to compare the yields obtained by systems described as conventional and organic. Under western agricultural conditions, organic systems commonly produce lower yields. Given the research effort put into fertiliser and agrochemical technology, it would be surprising if this were not so. Despite this, sometimes organic and conventional yields can be similar (Lang 2014). This is most commonly when the duration of comparative studies has been long such as in the Rodale Institute trials. It takes time for distinctive soil microbial populations to establish and for the impacts of increased organic matter content in the soil to become apparent. Organic crops commonly do well under more challenging environmental conditions. The skill of the farmer is not insignificant in this (Martini et al. 2004).

The other agricultural systems shown in Figure 1.1 are intermediate between the extremes illustrated above. Many share with organic systems the willingness to trade some production or efficiency of management for a reduced environmental impact or higher welfare standards for livestock. In such approaches, there is a shared interest with organic production in ways of optimising production by the use of natural resources and in accepting that maximum total production is not the sole goal. Defining the products of such systems in marketing terms can, however, be difficult (Anon 2012).

1.6 Some Areas of Different Science

Crop genetics are important to the ability of crops to function in any system. The differences in the requirements of varieties intended for organic systems and for systems dependent on external inputs illustrate how a similar science base may be differently applied. Newer varieties have delivered increased yields in recent years. The basis of much of this increase has been an increase in the proportion of crop resources which are partitioned into edible product, and the ability to utilise high levels of nitrogen. The objectives underpinning the breeding of varieties for use in organic systems have been more diverse. Organic systems do not permit GM technologies simply because the targeted nature of such approaches, which recommends them to those who use non-organic systems, lacks the flexibility and resilience seen as important to varieties used in organic systems of production. In addition, issues related to the ownership of the genes used in GM varieties by multinational organisations and consequently the resulting inability to use saved seed and freely share seeds are difficult to reconcile with the philosophical basis of organic production.

In the crop sector, importance is placed on the use of mixtures of genotypes. Breeding for the components of a mixture differs in a number of ways from breeding for a monoculture system. The ability to cope with a nutrient supply which is more complex in terms of its speciation and more variable in the timing of its availability changes the needs of organic systems. In addition, breeding for co-existence with potentially disease-causing micro-organisms and insects requires a different approach from classic breeding for disease resistance or tolerance to applied pesticides. It seems unlikely that GM technologies will be able to meet such complex needs although the innovation of genome editing with systems such as CRISPR/Cas9 may increase possibilities, especially where the basis of genetic resistance is relatively simple.

A key element in the organic approach to agriculture is the rotation-based system and the ability to use different qualities of land constructively (Chapters 5 and 6). The importance of rotations was described and recommended in a series of government publications first issued in 1925 and consolidated on the basis of experience in World War II (Saunders 1944). Rotations were practised in Roman times as a necessary consequence of settled land use. The need to remain on a specific permanent land unit made it important to avoid exhaustion of the land and so to vary the crops on that land. The design of rotations still depends heavily on early studies of the impact of the Norfolk four-course rotation from the time of its introduction in the eighteenth century. On the basis of such evaluations, it was concluded that:

> To think of rotations as outmoded restrictions on enterprises is entirely wrong. Modern science permits; and modern wars demand, deviation from established practice but the perils which beset the innovator are all the greater because of the slowness with which they manifest themselves.

Saunders (1944) identified the principles underpinning the rotation as the following.

1) The farm must ever be regarded as one organic whole. Only after fundamental needs have been met should cash crops be considered.
2) The rotation must maintain soil in condition. The rotation brings to the field an orderly succession of crops. This results in an appropriate series of tillage operations, maintains nutrients and humus, controls weeds, pests and diseases, and requires crops to be matched to the climate and soil of the individual farm.

These issues have recently been revisited by Stockdale and Watson (2011a, 2011b) who concluded:

> the changing climate, rising prices of inputs and changes in global markets are all driving changes in farming systems. It is therefore appropriate to consider whether crop rota-tions could re-emerge as the corner stone of farming systems because of their role in

providing vegetation and habitat diversity on farm as well as facilitating in situ delivery of pest disease and weed management and soil fertility management through biological interactions.

Rotations thus endeavour to use the effects on soil structure, chemistry and microbiology of one crop to optimise conditions for the following crops. The supply of nutrients is a critical issue in all systems of crop production. In non-organic systems, nutrient needs are met by the combination of fertiliser applications and an estimate of what might be released from the soil. The nutrient additions aim to ensure that nutrients are available in large quantities at the times when they are needed for maximum crop growth. Much of the increases in yields which have been achieved by such systems in recent years has come about as a result of the timely application and availability, early in the season, of nitrogenous and phosphatic fertilisers.

In organic systems, all of the nutrients required for crop growth must arise as a result of the interactions of soil chemical and microbiological processes. Soil nutrient supply is a resultant of soil microbial processes and of basic chemical cycles, which are influenced by the design of the rotation and heavily affected by soil temperature. Management of organic systems depends on the ability to use this information to gear crop growth to soil cycles.

All agricultural systems result in nutrients being removed from the system as food products. In non-organic systems, losses are made good by nutrient additions given as part of the fertiliser programme for the next production system. Organic systems also need to replace depleted elements such as phosphorus but this is effected by feeding nutrients which are not immediately available into appropriate soil cycles. This requires increased knowledge of some elements of soil chemistry, which are less important to other systems of production. The key aspects of soil chemistry for organic production are discussed in Chapter 4.

The energy balance of such approaches has become of increased interest as a result of current concerns over climate change. Calculating exact energy balances can be difficult, as Schafer and Blanke (2012) illustrated. The boundaries of the system, that is, the allowances for the processes which produce inputs such as N fertiliser or machinery, and the differential weight to be given to the different products of energy use, such as CO_2, N_2O, CH_4, etc., are key issues. Energy balances are important in relation to input/output ratios, which is the amount of energy which needs to be input to generate a particular calorific content of output. They are also important in relation to the total energy inputs to organic systems compared to other systems and to life cycle analysis. The loss of yield which is commonly experienced in organic production is balanced by reductions in resource use, especially of energy. The loss of yield is also compensated by the greater social integration of the system of production into the local community and its encouragement of the production of food on small farms.

In addition to questions related to the environmental and sociological impact of systems of food production (discussed in Chapter 3), there are also a series of questions regarding whether organic food is better for you (discussed in Chapter 12). At the heart of this is a discussion about the ways in which food from organic systems differs from that produced in other ways and the significance of such differences. This is not a simple issue around human diets; diet selection and its impact are influenced by lifestyle. What is a matter of fact is that different crop varieties and the ways in which varieties are grown produce food products that have different nutrient contents and also that how farm animals are raised influences the resultant food composition. (Heimler et al. 2012; Kalinova and Vrchotova 2011; Konvalina et al. 2011; Murphy et al. 2011). The significance of such changes has to be seen in the context of a total diet and activity.

There are, however, deeper questions about food quality than just those which can be detected by chemical or physical analysis. Food has a strong sociological footprint. Social science deals with people in the food chain. Whether food has been produced and traded in a fair way matters to consumers. This leads to concepts such as embedded value (Atkinson et al. 2012) which seek to expand our understanding of food quality into areas beyond conventional analysis.

1.7 Production Systems Compared

The basic production of plant-based foods or of forage for use in animal production in all systems can be summarised as follows (Figure 1.2).

Light and water facilitate the production and development of leaves, which photosynthesise and, in association with nutrients absorbed from the soil, allow the development of the plant mass, including most of the crop leaves (see Figure 1.2). This plant mass is then partitioned into crop, the part of the plant which becomes food, and other plant material such as roots and stems which are not eaten but which may be important in moving carbon from the atmosphere into the soil or into other stored forms such as woody tissues. In conventional production systems (Figure 1.3), inputs are used to maintain photosynthesis at the highest rates and to maximise partitioning into edible product.

In an organic system, the basic processes are similar but optimisation into crop production is achieved with different forms of inputs but essentially for the same basic purposes (Figure 1.4).

External inputs are fewer, and depend less on fossil fuel carbon and more on knowledge of basic processes. This asks questions about our understanding of crop, soil and animal biology and their interactions. In most crops, photosynthesis captures only a small proportion of the energy which is available to it from the sun. Total net primary production is commonly around

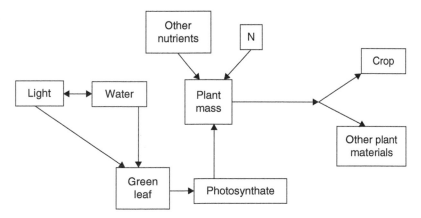

Figure 1.2 The basic process underpinning all crop production. *Source:* Reproduced from Atkinson (2002).

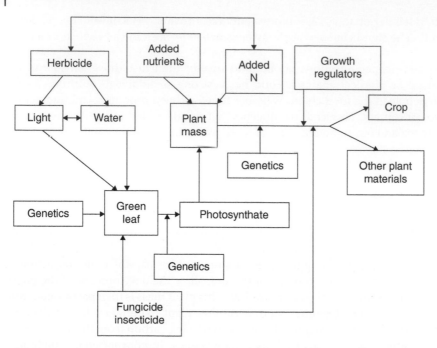

Figure 1.3 Crop production in a conventional system. *Source:* Reproduced from Atkinson (2002).

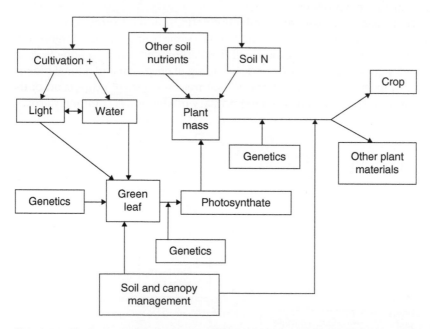

Figure 1.4 The basic model of crop production using an organic approach. *Source:* Reproduced from Atkinson (2002).

0.1% of incoming total solar radiation and food production is around 0.5% of net primary production (Newman 2000). This may relate to the proportion of the year when a substantial canopy is in place or to the efficiency of the photosynthetic process. Both genetics and management will influence this measure of efficiency. In conventional systems, nitrogen fertilisers are added to promote canopy development, herbicides to maximise the proportion of radiant energy intercepted by the crop, and fungicides and insecticides to maintain the photosynthetic efficiency of the canopy. In organic systems, management of soil resources, especially of soil microbial processes, must supply the nutrients required for canopy development and growth. The genetics of the crop must allow it to co-exist with other plant species and with non-symbiotic micro-organisms and to cope with insect-mediated reductions in effective leaf area (Atkinson 2002).

Successful organic systems must have soils which provide nutrients with the appropriate chemical speciation, in the right amounts and at the correct time. Soil microbial processes are responsible for nutrient transformations and for maintaining the supply of nitrogen. This occurs through nitrogen fixation in legumes and via the series of microbial transformations which supply nitrogen to crops in accessible forms. They are also responsible for the infection of crop roots with symbionts such as mycorrhizal fungi, which influence the uptake of relatively immobile nutrients such as phosphorus (Atkinson 2006). The microbial ecology of the soil also influences crop susceptibility to soil-borne diseases. Much of the later coverage in this volume deals with these issues.

1.8 A Science Base for All Production

The impact of the means of production was one of the key issues addressed by Atkinson and Watson (2000). The objectives of organic farming, as discussed above, are more complex than those of other systems of production. In considering systems as a whole, it is helpful to go back to the science literature of 50 years ago and prior to the current differentiation in farming systems.

At a basic level, energy capture from the sun and the fixation of CO_2 into organic compounds use similar mechanisms regardless of how crops are grown (Heath 1969). Beyond this, crop management can have a significant impact because the relationship between light intensity and photosynthesis is non-linear, in part because of the response of the crop leaf area to received light intensity. The relationship is also influenced by soil water potential with fixation falling with reducing soil water potentials. As a result, mean net assimilation rates vary within and between crops; for example, for *Lolium multiflorum* net assimilation rate (NAR) can be as high as $0.82 \, g/dm^2/week$ or as low as 0.32. During the main part of the growing season, NAR varies between crops, from 0.68 in sugar beet to 0.32 in *L. multiflorum*. Basic physiology may be the same but it can be modified by crop species and by management. These issues are discussed further in Chapter 8.

How this fixed carbon is partitioned within the crop also matters because that which ends in either the crop root system or in soil microbes has the potential to increase the organic matter content of the soil and so to remove carbon from the atmosphere. The significance of this in respect of climate change is discussed in the next section.

Table 1.1 Changes in soil properties in soil managed as grassed or bare for five years

Treatment	% Soil organic matter	Bulk density g/mL	% of soil volume as pores <30 μm
Grass	3.03	1.34	35.3
Bare	2.14	1.54	28.2

Source: Data from Atkinson and White (1981) and Atkinson et al. (1978).

Organic matter also changes the ability of the soil to retain and release water and so has the ability to change the crop water budget. Soils are enabled to hold water by pores of different sizes in the soil fabric and organic matter has a key role in stabilising these pores. Soils made from fine particles such as clay minerals will both hold more water in total and release a greater proportion of that water at lower soil water potentials, that is, values between −0.1 and −1.5 MPa, than those made of coarser particles such as sands (Slatyer 1967). The bulk density of most soils increases with a decrease in soil organic matter (Hall et al. 1977). A comparison of soil from grassed or bare soil (Atkinson and White 1981; Atkinson et al. 1978) showed a reduced organic matter content, a lower soil bulk density and an increased proportion of soil water held in the diameter of pores likely to be important to crop water supply, that is, those holding water against a suction of greater than −10 kPa (Table 1.1).

Again, the basic relationship between crop production and water supply from soils is similar for all crops but the ability to supply water is affected by management.

The ability of the crop plant to extract complex nutrients from the soil is a more important element in organic systems than in conventional production where the addition of fertilisers leads to high concentrations of nutrients in defined areas of the soil which only requires the generation of a sufficient root length to facilitate uptake (Nye and Tinker 1977). Consideration of the factors responsible for facilitating the uptake of nutrients from soil by plants and its impact on assessments of the ability of a soil to provide nutrients points us to the different approaches taken by plant physiologists and soil chemists to issues of optimum nutrient supply. In considering this, Nye and Tinker (1977) commented:

> The agricultural chemists and the plant physiologists scarcely seemed to communicate with each other. The former concentrated on finding a suitable extractant for available nutrients in the soil … It led agricultural chemists to be satisfied with correlations and regressions between fertiliser responses and chemical extracts and inhibited the search for more detailed explanations of their results. Meanwhile plant physiologists eschewing such a complex medium as soil in favour of nutrient culture solution established the essential nutrient elements. Unfortunately their experiments were usually made at concentrations very much higher than the soil solution so that quantitative aspects of their work are all too often irrelevant to soil conditions.

While the issue identified by Nye and Tinker (1977) remains relevant for all agricultural practice, it is especially important to organic agriculture as much of our understanding of crop nutrient supply relates to tests designed to assess fertiliser requirements. In organic systems, the root system needs to be more widely spread and with activity linked to resource and microbial cycles

(Atkinson 1990). However, the basic uptake processes of the root system are essentially similar in all farming systems. Jennings (1963) commented:

> There is no reason to believe that an absorption mechanism for a specific solute need necessarily differ from one plant cell to another; however, within any one plant cell the same mechanism need not be responsible for the absorption of all ions of like sign or for molecules of similar chemical composition

It is in this later element that uptake of nutrients in organically managed soils may differ. As important as root activity is to assessing the potential of a crop to make use of soil nutrients, one of the major changes to assessments of soil came through the realisation that it was home to a large number of micro-organisms, and that there was a particularly active population in that part of the soil immediately around the root. In organic soils, effective root systems will have arbuscular mycorrhizas (Atkinson 2006) and so all consideration about the performance of the crop root system must take account of the compound root system. In reviewing basic information about the role of mycorrhizas, Harley and Smith (1983) commented:

> Mycorrhizas are indeed the chief organs involved in the nutrient uptake of most land plants. The presence of the fungal associate in the root region and surrounding soil ensures that it influences the absorption of soil-derived substances. In this respect the mycorrhizal fungus is a specialised member of the rhizosphere microorganisms. Mycorrhizal infection usually increases the efficiency of nutrient absorption.

In addition, we are now aware that after infection, many plant genes responsible for nutrient uptake are downregulated so as to enhance the potential of the fungal component of the association (Ferrol and Perez-Tienda 2009). Developing sustainable crop production systems requires inputs from a variety of sources and the same is true of organically farmed systems (Atkinson and McKinlay 1995).

Nitrogen has always been a special case. Nitrogen in the soil has four main sources (McKee 1962).

1) Nitrogen released from the parent rock.
2) Nitrate and ammonia in rainfall.
3) Organic nitrogen, which is transformed by soil micro-organisms.
4) Gaseous nitrogen fixed by free-living and symbiotic micro-organisms.

All these mechanisms function to some extent in all systems but in the absence of fertiliser additions, they become critical to organic systems. Understanding how these processes might be managed to the best advantage of the crop is key to the development of organic production. The management of soil structure is important to this (Walters 1970). Organic additions and mineral additions in relatively insoluble forms have a greater residual value (Anon 1998) but this is part of a shared understanding of the soil's ability to supply nutrients from different chemical forms.

Russell (1977) reviewed what was known of crop root systems and their interactions with the soil. He divided what was known about roots into chapters dealing with root and shoot relationships, the growth of the root system, its ability to absorb water and nutrients, its ability to interact with soil micro-organisms and chapters detailing how the root system responded to specific sets of soil conditions such as compact soils and to being grown in minimum tillage systems. Much of the basic information on root functioning and on the uptake process had been obtained from studies of plants grown in solution or in sand culture and with nutrients supplied as simple chemical forms because this is the form in which nutrient uptake occurs in all systems and because it is easiest to characterise basic process by such a reductionist approach. As Read (2002) has pointed out:

> The response of many experimentalists has been to adopt reductionist approaches. Experimental designs simplified to the extent that their ecological relevance was compromised.

This basic information has relevance whatever its source. A major strand in agricultural research over recent years has been to assess the impact upon crops of varying growth conditions. This allows the plasticity, the scope for adaptation, to be assessed and its potential contribution to be evaluated. In a study on the impact of planting density on the form of the apple root system, Atkinson (1978) illustrated the inherent plasticity within a crop which would usually have a root system with one set of characteristics but which, when conditions require, can produce a system of a very different kind. These issues are discussed further in Chapters 9 and 10.

1.9 The Changing Context of Farming

The debate about appropriate systems of production has in the past been dominated by issues related to productivity, environmental impact and food quality (especially the significance of pesticide residues). Climate change has introduced a different set of considerations. This is briefly reviewed here to illustrate the impact of wider societal concerns on farming and food production and the science needed. A major issue has become the need to mitigate the adverse consequences of climate change. There is the need to consider the relationship between systems of food production and the release of greenhouse gases such as CO_2, CH_4 and N_2O. Maximising crop yields seems unlikely to be the only output expected of crop systems in the future.

Agriculture is a major producer of greenhouse gases (GHGs). The major external elements in any climate affect assessment of agriculture related to the energy required to manufacture nitrogen fertilisers and pesticides. Within the production system, CO_2 released as a consequence of the oxidation of soil organic matter following cultivation, the fuel required to power machinery and the emission of methane by ruminant livestock and nitrous oxide following the use of nitrogen fertilisers have been identified as major elements. It has been suggested that numbers of ruminant animals should be reduced and such reductions used as an indicator of progress in meeting climate change objectives (Bailey 2014). The global livestock sector is currently responsible for producing around 15% of GHGs largely as a result of the methane produced by ruminant digestion and the nitrous oxide from the fertilisers used to grow the crops which feed intensively reared livestock. This could be the application of a linear model to non-linear organic systems. The logic which underpins the idea of reducing ruminant numbers would thus be problematic for most practitioners of organic agriculture as it would preclude the easy transfer of resources from a grass or grass-clover phase of a rotation to the cereal and crop phase. In describing the rationale of the Haughley Experiment, Balfour (1976) wrote:

> An ecological approach to soil fertility precludes monoculture of livestock equally with monoculture of crops. Maximum diversity within practical possibilities is the objective.

Grassland systems cropped by ruminant animals represent a major land use on a worldwide basis. Grazed grassland is thus important both for assessing the contribution of all types of agriculture to the release of GHGs and as a basis for comparing the relative contributions of organic agriculture and other systems of production (Figure 1.5). To do so requires information about the ability of agricultural systems to sequester carbon both in plant parts (the major contribution of forest systems) and in soils (the major contribution of grassland). However, agricultural systems differ from most other activities in that part of the total balance is C sequestration, which is not an element in climate change assessments of other activities. In agriculture, it is not appropriate just to ask what are the outputs of GHGs and what can be done to reduce or offset these losses. There is a need to appreciate that sequestration in the soil may well be reduced without ruminant grazing and that emissions of methane from ruminants may be the price to pay for carbon storage in soil. What is known is that processes such as crop root growth are responsible for the transfer of significant amounts of carbon to soil organic matter and that this varies greatly between species and between different climatic areas (Table 1.2).

An additional element in the debate is whether carbon fixation is or is not significantly greater in ruminant-grazed grassland than in ungrazed grassland or in other potential land

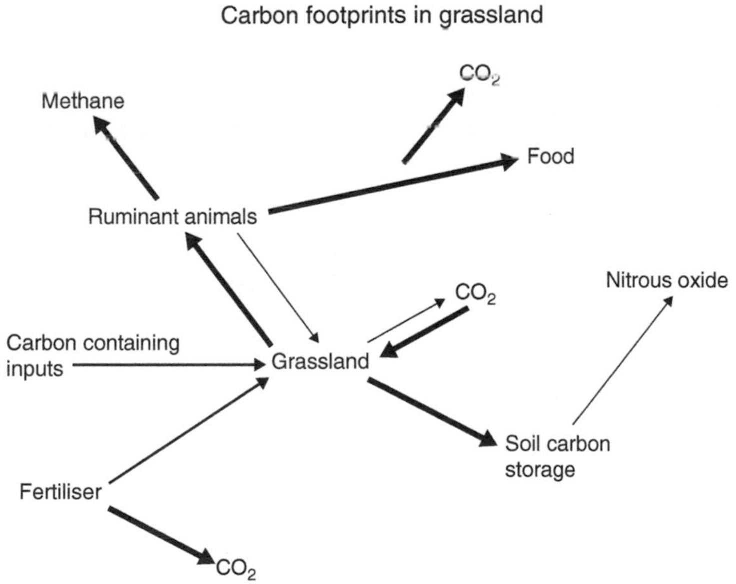

Figure 1.5 Schematic description of the components leading to the release and the fixation of CO_2 in grassland grazed by ruminant animals. The thickness of the line shows the probable importance of the process.

Table 1.2 The magnitude of the standing crop and the turnover of grassland in two geographically contrasting climates

Species	Location	Standing crop	Root turnover
L. perenne	UK	19.1	120
	Italy	13.1	150
T. repens	UK	1.9	11
	Italy	2.5	27

Source: Data from Black (1997).
The units are tonnes per hectare.

Table 1.3 Energy use in different farming systems

Trial	Rodale	Glenlea, Manitoba	Therwil, Switzerland	DEFRA 2000 study	FAO study
Type of enterprise	27 year, corn and soy bean	11 year wheat, pea, flax and wheat alfalfa flax	20 year crop and grass	Wheat, vegetables, dairy, beef and sheep	Range
Comparative energy use	Organic corn used 31% less energy and soy 17% less	Organic wheat, pea rotation used 65% less energy and wheat, alfalfa 55% less	Organic crops used 20–56% less energy than conventional	Organic crops and dairy, beef and sheep all used less, commonly >50%, energy than conventional crops	Organic typically used 30–50% less energy

Source: Data from Hill (2009).

uses. The complexity of these issues is such that we have not sought to add numbers to Figure 1.5 (www.un.org/press/en/1999/19990706.SAG44.html). Some of these issues were reviewed in a volume edited by Atkinson (1993).

When the contribution of agriculture of all types is considered against climate change reduction strategies, it becomes important to look at the energy efficiency of different means of production. Pimentel et al. (1983) compared the energy use of different types of farming systems and found that in general, organic systems were 29–70% more efficient than conventional systems although conventional apple production was more efficient than organic. Hill (2009) reviewed published data from a number of studies which had assessed the relative energy uses of conventional and organic production systems. Hill identified that there were large differences in energy use between what might be regarded as either conventional or organic and that any analysis was complicated when postproduction practices such as processing, packaging, storage and distribution were taken into account. In conventional systems, the energy required to produce and apply fertilisers was always a major element. Energy use is summarised in Table 1.3.

Soil carbon storage is important to the maintenance of soil fertility. Processes linked to this are commonly poorly quantified although values available indicate the significance of the area. Whitman et al. (1998) reviewed data on the amounts of global carbon held in living prokaryotes and concluded that around 350–550 Pg of carbon was held in this form, an amount equal to between 60% and 100% of the amount present as plants. In addition, prokaryotes held 85–130 Pg of N and 9–14 Pg of P, both of which are around 10 times the amounts held in plants. In soils, there are around $0.25–2.5 \times 10^{30}$ organisms with an average turnover time of 2.5 years. The greatest numbers of organisms are found in tropical and temperate grasslands, desert shrubland and cultivated soils. Outwith the soil, significant numbers of organisms are found in the guts of ruminant animals, especially cattle. Better quantification matters to climate change strategies and to the development of organic production.

Organic production does depend on science and in the remainder of this volume we attempt to examine what this is and what is needed to develop this approach further.

References

Anon 1998 Scottish Standing Committee for the Calculation of Residual Values of Fertilisers and Feeding Stuffs, Fiftieth Annual Report. HMSO, Edinburgh.

Anon 2012 "Pork not porkies" advert banned. Independent, 8077, 29th August, p. 13.

Atkinson, D. (1978). The use of soil resources in high density planting systems. *Acta Horticulturae* 65: 79–89.

Atkinson, D. (1990). Influence of root system morphology and development on the need for fertilisers and the efficiency of use. In: *Crops as Enhancers of Nutrient Use* (ed. V.C. Baligar and R.R. Duncan), 411–451. San Diego: Academic Press.

Atkinson, D. (1993). *Global Climate Change: Its Implications for Crop Protection*. Farnham: BCPC.

Atkinson, D. (2002). Future farming systems. In: *Crop Protection*, 2020. Farnham: BCPC.

Atkinson, D. (2006). Arbuscular mycorrhizal fungi and the form and functioning of the root system. In: *Microbial Activity in the Rhizosphere* (ed. K.G. Mukerji, C. Manoharachary and J. Singe), 199–222. Berlin: Springer.

Atkinson, D. (2009). Soil microbial resources and agricultural policies. In: *Mycorrhizas: Functional Processes and Ecological Impact* (ed. C. Azcon-Aguilar, J.M. Barea, S. Gianinazzi and V. Gianinazzi-Pearson), 1–16. Berlin: Springer.

Atkinson, D., Harvey, W., Leech, C. et al. (2012). Food security: a churches together approach. *Rural Theology* 10: 27–42.

Atkinson, D. and McKinlay, R.G. (1995). Crop protection in sustainable farming systems. In: *Integrated Crop Protection; Towards sustainability?* (ed. D. Atkinson and R.G. McKinlay), 483–488. Farnham: BCPC.

Atkinson, D., Swain, R.W., and Fricker, D. (1978). The effect on soil condition of herbicide use in fruit plantations. In: *Agrochemicals in Soils* (ed. A. Banin and U. Kafkafi). Oxford: Pergamon.

Atkinson D and Watson CA (2000) The research needs of organic farming: distinct or just the same as other agricultural research? Proceedings of the BCPC Conference on Pests and Diseases, pp. 151–159.

Atkinson, D. and White, G.C. (1981). The effects of weeds and weed control on temperate fruit orchards and their environment. In: *Pests Pathogens and Vegetation* (ed. J.M. Thresh), 415–428. London: Pitman.

Bailey R (2014) Impact of livestock on climate change cannot be ignored. Chatham House. Available at: www.chathamhouse.org/expert/comment.

Balfour, E.B. (1976). *The Living Soil and the Haughley Experiment*. New York: Universe Books.

Barker, A.V. (2010). *Science and Technology of Organic Farming*. Boca Raton: CRC Press.

Black KE (1997) Root longevity as affected by biotic and abiotic factors. PhD thesis, University of Aberdeen.

Cherrett, J.M. (1989). *Ecological Concepts: The Contribution of Ecology to an Understanding of the Natural World*. Oxford: Blackwell.

Clapham, A.R. (1969). Introduction. In: *Ecological Aspects of the Mineral Nutrition of Plants* (ed. I.H. Rorison), xv–xxi. Oxford: Blackwell.

Conford, P. (2001). *The Origins of the Organic Movement*. Edinburgh: Floris Books.

FAO/WHO (1999) Codex Alimentarius Commission approves guidelines for organic food. Available at: www.un.org/press/en/1999/19990706.SAG44.html.

Ferrol, N. and Perez-Tienda, J. (2009). Coordinated nutrient exchange in arbuscular mycorrhiza. In: *Mycorrhizas-Functional Processes and Ecological Impact* (ed. C. Azcon-Aguilar, J.M. Barea, S. Gianinazzi, et al.), 73–87. Berlin: Springer Verlag.

Greenland, D. (2000). Effects on soils and plant nutrition. In: *Shades of Green* (ed. P.B. Tinker), 6–20. Stoneleigh: RASE.

Hall DGM, Reeve MJ, Thomasson AJ, Wright VF (1977) Water retention, porosity and density of field soils. Technical Monograph 9. Soil Survey, Harpenden.

Harley, J.L. and Smith, S.E. (1983). *Mycorrhizal Symbiosis*. London: Academic Press.

Heath, O.V.S. (1969). *The Physiological Aspects of Photosynthesis*. London: Heinemann.

Heimler, D., Vignolini, P., Arfaioli, P. et al. (2012). Conventional, organic and biodynamic farming: differences in polyphenol content and anti oxidant activity of Batavia lettuce. *Journal of the Science of Food and Agriculture* 92: 551–556.

Hill H (2009) Comparing energy use in conventional and organic cropping systems. National Sustainable Agriculture Information Service. Available at: file:///C:/Users/Owner/Downloads/croppingsystems.pdf

Holm L 1976 The importance of weeds in world food production. Proceedings of the BCPC Conference on Weeds, pp. 1–16, BCPC, Farnham.

Howard, A. (1945). *Farming and Gardening for Health or Disease*. London: Faber.

Jennings, D.H. (1963). *The Absorption of Solutes by Plant Cells*. Edinburgh: Oliver and Boyd.

Kalinova, J. and Vrchotova, N. (2011). The influence of organic and conventional crop management, variety and year on the yield and flavonoid level in common buckwheat groats. *Food Chemistry* 127: 602–608.

Konvalina, P., Capouchova, I., Stehno, Z. et al. (2011). Composition of essential amino acids in emmer wheat landraces and old and modern varieties of bread wheat. *Journal of Food, Agriculture and Environment* 9: 193–197.

Lang SS (2014) Organic farms produce the same yields as conventional farms. Cornell Chronicle. Available at: www.sciencedaily.com/releases/2005/07/050714004407.htm.

Martini, E.A., Buyer, J.S., Bryant, D.C. et al. (2004). Yield increases during the organic transition: improving soil quality or increasing experience? *Field Crops Research* 86: 255–266.

McKee, H.S. (1962). *Nitrogen Metabolism in Plants*. Oxford: Clarendon Press.

Moraine, M., Melac, P., Ryschawy, J. et al. (2017). A participatory method for the design and integrated assessment of crop-livestock systems in farmers' groups. *Ecological Indicators* 72: 340–351.

Murphy, K.M., Hoagland, L.A., Yan, L. et al. (2011). Genotype X environment interactions for mineral concentration in grain of organically grown spring wheat. *Agronomy Journal* 103: 1734–1741.

Newman, E.I. (2000). *Applied Ecology and Environmental Management*. Oxford: Blackwell.

Nye, P.H. and Tinker, P.B. (1977). *Solute Movement in the Soil-Root System*. Oxford.: Blackwell.

Pimentel, D., Beradi, G., and Fest, S. (1983). Energy efficiency of farming systems. *Agriculture, Ecosystems and Environment* 9: 359–372.

Plumb, H. (1998). The future of agriculture in the EEC. In: *The Bawden Memorial Lectures*, 1–8. Farnham: BCPC.

Read, D.J. (2002). Towards ecological relevance – progress and pitfalls in the path towards an understanding of mycorrhizal functions in nature. In: *Mycorrhizal pp.* (ed. M.G.A. van der Heijden and I.R. Sanders), 3–29. Berlin: Springer.

Russell, R.S. (1977). *Plant Root Systems: Their Function and Interaction with the Soil*. London: McGraw Hill.

Saunders HG (1944) Rotations. Bulletin 85, Ministry of Agriculture and Fisheries. HMSO, London.

Schafer, F. and Blanke, M. (2012). Farming and marketing system affects carbon and water foot print- a case study using Hokaido pumpkin. *Journal of Cleaner Production* 28: 113–119.

Slatyer, R.O. (1967). *Plant Water Relationships*. San Diego: Academic Press.

Stockdale EA, Watson CA 2011a Making crop rotations fit for the future. Aspects of Applied Biology 113. AAB Warwick.

Stockdale, E.A. and Watson, C.A. (2011b). Can we make crop rotations fit for a multifunctional future. *Aspects of Applied Biology* 113: 119–126.

Tinker, P.B. (2000). Introduction. In: *Shades of Green* (ed. P.B. Tinker), 1–5. Stoneleigh: RASE.

Walters, H. (1970). Nitrate in soil, plants and animals. *Journal of the Soil Association* 16: 1–22.

Whitman, W.B., Coleman, D.C., and Wieb, W.J. (1998). Prokaryotes: the unseen majority. *Proceedings of the National Academy of Sciences USA* 95: 6578–6583.

2

Science, Research and Organic Farming

Lawrence Woodward

Former director of Organic Research Centre – Elm Farm, Newbury, Berkshire, UK

2.1 Introduction

Organic farming has been dismissed as unscientific. This perception is often based on a lack of understanding of underlying concepts. Researchers in this field accept the scientific discipline while seeking new methodologies to explore the holistic character of these concepts. Of central importance is the concept that health, food quality and the nature of the agricultural system are inextricably linked. This chapter argues that a holistic approach to scientific investigation is valid even if at present it is more of an aspiration than a reality.

2.2 The Roots of the Approach

In 1943, Lady Eve Balfour was one of the leading protagonists of the ideas that lead to the creation of organic agriculture. Lady Eve did not have a scientific background but many of her colleagues, supporters and fellow protagonists had formal training and careers in such areas as medicine, nutrition, chemistry, biology, soil science and agriculture. What is notable is their insistence that their ideas should be developed through research as well as practice and that the scientific method be adhered to.

This scientific and research tradition has been a critical part of the organic movement ever since its beginning and remains so today. It is a matter of fact that ever since the 1920s, private and publicly funded research into organic food and farming has been carried out in some way in many countries. The vast majority of this work has been undertaken using mainstream reductionist methods and most of it by researchers with mainstream qualifications from mainstream universities and institutions. There is now a large body of literature reporting on this work in the form of peer-reviewed papers, government reports and conference proceedings as well as a vast amount of so-called 'grey' literature which has been one of the best routes for technology transfer.

The Science Beneath Organic Production, First Edition. Edited by David Atkinson and Christine A. Watson.
© 2020 John Wiley & Sons Ltd. Published 2020 by John Wiley & Sons Ltd.

But it is also a matter of fact that those researchers investigating the underlying conceptual framework as opposed to the techniques of organic agriculture have often been frustrated by the limitations of reductionist methodology and statistical analysis when applied to whole biological and ecological systems, especially when the concepts concerned are based on the theory of whole organisms or 'living entities' which are not 'entirely resolvable into parts'.

Yet the charge that organic agriculture is unscientific still finds credence in some quarters. Why this is so needs to be addressed, and there are three areas that are worthy of comment.

2.2.1 Is it a philosophical or political movement and cannot therefore be scientific?

Yes, the organic movement is in part a philosophical movement and in part in a broad sense a political movement. It is based on a 'worldview' that is relatively distinct, if not always coherent. Not all the researchers working in this field share that worldview. Some do but that does not mean they are not able to employ the scientific method in an objective and truthful way. The organic movement's scientific tradition needs to assert itself more against the tendency to cliché, superficiality and hype that has poured from supporters and opponents alike in recent years.

Far too often, organic agriculture is defined or described by what it is not. This is detailed in Chapter 1. The most common example is the notion that 'organic farming is farming without chemicals'. Whilst this description has the advantage of being clear and concise, it is unfortunately untrue and misses out on several characteristics which are of fundamental importance. However, organic farming is a system that seeks to avoid the direct and/or routine use of readily soluble chemicals and all biocides, including those that are naturally occurring or nature identical. Where it is necessary to use such materials or substances, then the least disruptive at both micro and macro levels, that is to say from bacteria through plants and animals to man's wider environment, are used.

Another common misconception is that organic farming merely involves using so-called 'organic' inputs instead of 'agrochemical' ones. A straight substitution of mineral fertiliser by manure is likely to have the same – probably adverse – effect on plant quality, disease susceptibility and environmental pollution. This neo-conventional approach is not organic farming. The misuse of organic materials either by excess, inappropriate timing, or both, will effectively short circuit or curtail the development of the biological cycles that the organic system seeks to build upon.

Also mistaken is the notion that organic farming is a return to farming as it was some time in the past. It is true that some traditional farming characteristics remain features of the organic system, including rotations, mixed farms and mechanical methods of weed control. However, organic farming seeks to build upon the increased understanding of such things as mycorrhizal associations, the rhizosphere, the turnover of organic matter and other areas of soil life, crop and animal husbandry that have come from modern science. Although the organic movement might reject some technologies such as aspects of genetic engineering, it does seek to use scientific knowledge and information – for example, from genomics – because the appropriate application of science is essential for its further development. These issues are explored in more detail in other chapters of this volume.

2.2.2   Does it lack coherence except as a marketing exercise?

In the last two decades there have been many documents published which set out the details of organic standards and regulations. However, they all have much that is shared. Arguably, still the most accessible definition of organic agriculture is the one framed by the United States Department of Agriculture (USDA) in 1981. It provides a handy description of the key practices as they are found on the farm.

> Organic farming is a production system which avoids or largely excludes the use of synthetically compounded fertilisers, pesticides, growth regulators and livestock feed additives. To the maximum extent feasible, organic systems rely on crop rotations, crop residues, animal manures, legumes, green manures, off farm organic wastes and aspects of biological pest control to maintain soil productivity and tilth, to supply plant nutrients and to control insects, weeds and other pests.
>   The concept of the soil as a living system ... that develops ... the activities of beneficial organisms is central to this definition.

Here we can see what organic farmers do not do, what positive things they do instead and the context in which they work, that is, the living soil. Here is the key to understanding what organic farm management looks like – or should look like – wherever it is. It concentrates primarily on adjustments within the farm and farming system, in particular rotations and appropriate manure management and cultivations, to achieve an acceptable level of output. External inputs are generally adjunct or supplements to this management of internal features. This is the common basis of organic agriculture wherever it is found in the world – practical, clear and coherent enough for all but the dullest or most obstructive.

That is not to say that organic farms are carbon copies of each other. Far from it, as the USDA recognised:

> Many organic farmers have developed unique and productive systems of farming ... It is also likely that these systems are highly complex and involve unknown or poorly understood chemical and microbiological interactions.

The USDA concluded that a new research approach is necessary.

> A holistic research approach, which may involve the development of new methodologies, is needed to thoroughly investigate these interactions.

A number of guidelines have been established by researchers experienced in this field.

1) Research should be conducted within a viable farming system.
2) Research methods such as pot trials should only be a subsidiary component of any organic research programme and not the sole method of investigation.
3) All researchers involved should have a thorough grounding in and understanding of organic agriculture.
4) Researchers should work closely with and be prepared to learn from established organic farmers.

Despite the increased interest in organic research and the number of projects, we have barely progressed our understanding of how an organic system works and of the validity of its underlying concepts (although reasonable progress has been made on specific technical or organisational applications). William Lockeretz, for more than 10 years editor of the *American Journal of Alternative Agriculture*, summarised the situation as follows:

> Organic research is hardly different from conventional research. Few papers in the field can honestly be called holistic or system orientated and very few are even slightly multi-disciplinary. The main difference compared with conventional research is what gets studied – some aspects of organic farming – not how one studies it ... Neither in the kinds of questions posed nor in the way they are answered do I find any clear systematic distinction between research or organic and on conventional systems.

There are many names worldwide for what we in the UK call organic agriculture and there is some variation between them However, government-backed regulations in many countries, including the European Union, now exist to provide a legal framework and floor which affords reasonable equivalence for the purpose of consumer protection.

2.2.3   Is it inherently unscientific because it is based on concepts that are not explicable in rational scientific terms?

In conceptual terms, the genesis of organic agriculture is arguably found in three schools of thought, which first originated in the first three decades of the twentieth century.

1) The biodynamic or anthroposophical school of Steiner.
2) The organic-biological school of Muller and Rusch.
3) The organic school of Howard and Balfour.

Also important is the work of Schuphan and Voisin. The conceptual basis of organic farming is also discussed in Chapter 1 while the approaches of the pioneers are detailed by Conford (2001). But there is an essential core of agreement which (although battered and shabby in parts) remains at the heart of the organic movement. This has four aspects.

1) The concept of the farm as a living organism, tending towards a closed system in respect to nutrient flows but responsive and adapted to its own environment.
2) The concept of soil fertility through a 'living soil' which has the capacity to influence and transmit health through the food chain to plants, animals and humans.
3) The notion that these linkages constitute a whole system within which there is a dynamic yet to be understood.
4) The belief in science and an insistence that while these ideas might be challenging orthodox scientific thinking, they could be explored, developed and eventually explained through appropriate scientific analysis.

These pioneers sought to examine, test and develop their ideas scientifically. However, from a modern perspective, a good deal of their research seems inadequate. Some experiments are poorly reported, some are badly designed and suffer from the limitations of the equipment and methodology of the time. It is not surprising that the science of these pioneers is found wanting. To truly examine these concepts requires not only large sums of money but also a large intellectual leap, possibly into a black hole but also, many have argued, into holism.

There is a growing body of work regarding the production system, the idea of biological food quality and the link to the health of animals and humans. This includes the observations and experiments of Sir Robert McCarrison as reported by Balfour (1943). McCarrison systematically observed many peoples and many diets and realised that there was a quality in the diets of the healthiest peoples which was absent from the least healthy:

> ... that food in all these diets is, for the most part, fresh from its source, little altered by preparation and complete; and that, in the case of foods based on agriculture, the natural cycle is complete. Animal and vegetable waste – soil – plant – food – animal – man; no chemical or substitution stage intervenes.

It also includes the Haughley Experiment (Balfour 1976), where over 11 years a completely closed organic system was compared with a rotational conventional system and a non-rotational stockless system (discussed further in Chapter 6). Even over this short period of time, differences were noted in food quality and the productivity, longevity and general health of livestock, in favour of the closed system.

A relatively large amount of research has been completed on the quality of both crops and livestock. There is a clear trend that in appropriate crops, organic produce contains more desirable components (vitamins, dry matter, protein) and fewer undesirable substances (pesticide residues, nitrates, sodium) than conventional produce. In livestock trials, animals fed on organically grown feed generally show greater fertility and longevity than those fed on conventionally produced feed.

Most of this work has been carried out using mainstream methods and statistical analysis although some so-called novel methods of analysis have been used in some trials. Picture developing methods, copper chloride crystallisation, measurement of low-level luminescence and forced storage tests have been used to measure factors that are not revealed by chemical analysis.

As the science of ecology has developed in recent decades, it may now be that the concept of the farm as a living organism might not sound as preposterous to conventional scientists as perhaps it once did. But the idea that the way food is produced affects the health of those who consume it and that food may contain qualities that are not accessible to conventional analysis flies in the face of orthodox science. The conventional view of nutrition is that it is enough to ensure that a diet contains adequate proteins, carbohydrates, fats, vitamins, mineral salts, trace elements, etc., and that how they got there is of no relevance and the presence of pesticide residues *per se* is of no consequence to the health of the consumer. A recent medical conference looking at the health implications for research and food policy of the mapping of the human genome considered that:

> The discovery of only 35 000 genes instead of the expected 150 000 means that the nutritional environment played a pivotal role in human evolution ... We now know that the key to evolution and the behaviour of human disease is in the environment interacting with the gene and not the other way around.

2.3   Agricultural Science: Some Reflections

As argued in Chapter 1 and developed in Chapter 3, science always has a societal context. It matters in assessing the science underlying organic production that we examine the role of individual science practitioners in the development of thinking about agricultural science. We can ask whether agriculture trusts science or science agriculture or whether agricultural science trusts the twenty-first century or vice versa. There is a supplementary question: if there is continuing agricultural science, can anyone trust it or trust scientists? The possibilities for gene editing using CRISPR/Cas9 technology make these questions of continuing importance.

This is a manifestly important issue for any society that aspires to be civilised and for every citizen – scientist or not – in such a society. It is equally clear that in our society there has been some erosion of trust between some citizens and some scientists. Note I did not say science – although the validity of some scientific methods in some circumstances may be questioned; it is not clear to me that there is a serious erosion of trust in science *per se*. The problem is most commonly to do with scientists, not science.

It is an interesting question as to why this has come about but I believe that there is a readjustment, even a redefining, of the relationship which people who are labelled 'scientists' have with other citizens. The process is dynamic and, certainly from my perspective, has some regrettable negative aspects but it also has some beneficial and positive aspects. It is important that the scientific community engages in this process and does not arrogantly rail against it.

There are within this community people capable of breath-taking naivety – putting it politely – and trust will not be encouraged if they are leading. The public debate about the proposed uses of genetically modified organisms (GMOs) in the closing years of the twentieth century is a case in point (Bruce and Bruce 1998). A letter sent to the Prime Minister as part of the GM debate in November 2003 by a number of leading scientists and their subsequent media interviews are an example of this folly. Whatever merit there was in their argument – and that's moot – was lost in hopeless timing and poor exposition. It came across to many as a piece of whingeing which ignored the fact that pro-GM scientists have dominated the make-up of all the various advisory and steering bodies on GM; that in all the debates the pro case was politely listened to – if then rejected as not being convincing; that there was ample opportunity for pro-GM scientists to organise meetings, debates and generally engage but the fact is that generally they did not. To react as this group did presents scientists as arrogant, elitist, undemocratic and contemptuous of their fellow citizens. It suggests that parts of the scientific community have learned little from the experiences of bovine spongiform encephalopathy (BSE) and foot and mouth disease (FMD). Fortunately, there are others in the scientific community who give cause for optimism.

David Shannon, then Chief Scientist of MAFF/DEFRA, wrote in the aftermath of FMD:

> The public has had its eyes opened to science through the experience of the BSE epidemic. It learned that science is mainly evidence, not fact, that it can change over time; that there's frequently not just one scientific view – there may be conflicting views – and that scientific evidence is subject to the weaknesses of all other sorts of evidence: such as the motivation of the provider, funding issues and even competition between disciplines. Some scientists want their discipline to be the one to solve the problem. Unless we solve

it with this technique we haven't solved it. We should seek to inform the public and let it arrive at its own view and experts must listen to the public: that's one of the important things we must do much more of. Understand where the public is coming from.

To me, this view of science and scientists is not only refreshing, it is an essential starting point for rebuilding trust between citizens and scientists of all disciplines and should be prominently displayed in all scientific institutions. It is especially important for agricultural science because agricultural science is – or should be – concerned with the most basic issues of food, nutrition, health and the maintenance of the biological and ecological resources of our fragile environment. The questions surrounding these issues cannot be addressed by technology and science alone. They go right to the heart of how we organise our society and to who we are, all of us – citizens or merely consumers – and that question has exercised the organic movement since its beginning.

Does agricultural science – of any type – have anything other than the most peripheral role in the modern world? Food technology, the creation of junk food and the like seems to have a role. Creating new products for multinational companies in the global economy has a role but does agricultural science have a role in dealing with things that really matter? One reads of claims made by scientists that their particular discipline is set to make a significant contribution to, for example, feeding the hungry world. This claim is heard a lot in the debates about GM and related crops and it is fatuous. Money, economic structures, distribution and power are the most significant factors in equitable food production (issues explored further in Chapter 3), not agricultural technology of whatever type. And Agri-Culture in its fullest sense, with a capital A and a capital C, is peripheral at best in the global economy.

A former EU Agriculture Commissioner, Franz Fischler, said:

> 'Organic agriculture epitomises the aims of CAP, it is environmentally beneficial, it considers quality production second nature and sustainability is at its core.'

DEFRA tends to follow a similar line though in a less effusive manner. They do so because they hope that by tapping into a particular market organic farmers can demonstrate that sustainability is compatible with economic competitiveness and trade liberation.

I do not believe that it is. I believe that there is a fundamental conflict here that may be irreconcilable But even if this is not the case and these notions are ultimately compatible, a great deal of thought, effort and change in farming and food systems policy and practice would need to occur for that reconciliation to be achieved. The muddled thinking, marketing and dissembling of the past decade that have allowed us to believe that we can 'have our cake and eat it' – we can have sustainability and competitiveness – without making more than relatively minor modifications to our farming and food systems have significantly contributed to the crisis agriculture is now facing. I do not believe that the fundamental problem that our society will face in the next two decades – the fallout from climate change, problems with finite resources such as soil and water and the effect on economic activity due to the problems of energy – have been adequately considered within agricultural policy. There is a pervading and fatalistic World Trade Organization (WTO) 'business as usual' drive, which in my view is folly.

Anyone can draw up scenarios but who decides on the actions? That should not be left to the WTO bureaucrats, multinational companies or the short-termism of politicians. These decisions are – or should be – the business of all of us informed citizens. And here is the role of agricultural science and of scientists; informing, engaging, participating as scientists and as citizens in a great unprecedented act of citizenship in placing Agri-Culture, capital A and C, not agri-business, not commodity capitalism, at the heart of a sustainable society.

2.4 Conclusion

Discussion of the science base of organic production thus requires us to both assess the underlying ethos of this approach to food production and ask questions about the development of the whole agri-science base and the various pressures which facilitate some forms of science and neglect other forms. This chapter thus sets out organic agriculture's underpinning concepts – scientific or philosophical – that the health of soil, plant, animal and human is indivisible brought right up to date by the most modern science. This science base – social, physical, chemical and biological – is discussed in the remainder of this volume.

References

Balfour, E. (1943). *The Living Soil*. London: Faber and Faber.
Balfour, E. (1976). *The Living Soil and the Haughley Experiment*. New York: Universe Books.
Bruce, D. and Bruce, A. (1998). *Engineering Genesis*. London: Earthscan Books.
Conford, P. (2001). *The Origins of the Organic Movement*. Edinburgh: Floris Books.

3

Framing and Farming

Putting Organics in a Societal Context

Pete Ritchie

Nourish Scotland, Edinburgh, Scotland, UK

3.1   Introduction

This chapter aims to place organic agriculture in its wider social, economic and philosophical context. While much of this book is concerned with the scientific basis of organic agriculture, essentially the science which makes organic approaches 'work', this chapter considers some of the major 'why' issues – essentially *why organic*? Here we suggest the importance of seeing what are suggested as science facts or conclusions as part of a social construct and the need to see food production in a wider societal, environmental and ethical context.

In an era when food can be produced cheaply and in abundance using fertilisers and chemicals to provide the precise nutrients the crop needs while protecting it from competition and predators, why would anyone argue for an approach which turns its back on these aids to production? When livestock can be reared quickly and efficiently in concentrated and controlled conditions, why would anyone keep farming them extensively and outside?

This chapter examines three arguments – an economic argument, an environmental argument and an ethical argument. The economic argument reflects the rapid growth in organic agriculture and sets out why farming organically can make sense in purely financial terms for many farmers. The environmental argument puts the costs and benefits of organic and conventional farming in a wider frame than simply farmgate yields. Finally, the ethical argument looks at the values inherent in and associated with an organic approach. Before laying out these arguments, we introduce the organic approach by taking a historical perspective: what was the problem organics was designed to solve?

The Science Beneath Organic Production, First Edition. Edited by David Atkinson and Christine A. Watson.
© 2020 John Wiley & Sons Ltd. Published 2020 by John Wiley & Sons Ltd.

3.2 The Origin of Organics

Organic agriculture did not develop primarily as a production technique (like intensive poultry production or aquaponics). Instead, it emerged as a family of approaches in response to broader concerns about human health and the changing environment. For most of human history, farming has used organic methods. Soil fertility was maintained through rotations, fallows and the thrifty use of animal and human excreta. Livestock breeding and natural selection adapted animals to their local environment. Diversity, intercropping, plant-based chemicals and human attention were used to protect crops from disease and predators. Farmers still invented and innovated, and the pace of diffusion increased where new market opportunities arose. Better ploughs, seed drills, livestock breeding, wheat breeding all had an impact on agricultural productivity. The upheavals of the enclosures and the clearances in the UK privileged production over people but until the middle of the nineteenth century, agriculture would have been recognisable to earlier farmers. Working with Nature was still the default, energy provided by humans and animals, fertility by manure.

The organic movement arose in the early twentieth century specifically in response to the new science of chemistry, including soil chemistry. Scientists had isolated the chemicals essential to plant growth. Guano was being imported in vast quantities from bird and bat colonies, and nitrates were being harvested from coke production and the saltpetre reserves in Chile.

The Haber–Bosch process developed in 1909 and commercialised before the First World War – as much for munitions as for fertiliser – was not just a triumph of chemistry (Hager 2008). It accelerated the shift from agriculture as a 'circular economy', based on recycling nutrients mostly at farm or near-farm level, to a 'linear economy', described in terms of inputs and outputs. This difference in basic model still remains as perhaps the most profound difference between the two approaches.

William Crookes, President of the British Academy of Sciences, had argued in 1898, in true Malthusian fashion, that mass starvation would soon happen with rising populations and all the world's land already in use. He declared, 'It is the chemist who must come to the rescue' to stop food supplies running out – and this at a time when world population was around 1.6 billion. This fundamental shift towards an industrial model – farm as factory with inputs and outputs – led to the articulation of a countervailing 'organic' argument. The proponents of organic farming criticised conventional farming as reductionist – reducing agriculture to the application of mineral nutrients.

Many different strands coalesced in the early organic movement (Conford 2001). The empirical approach of Albert Howard's work on the value of compost and real-life on-farm experiments went alongside a much broader critique, not just of chemical techniques in agriculture but also of contemporary society. Whether expressed in terms of a more 'natural' state of society, either in the past or in a far-off country, or in terms of the need to conform to 'Nature's laws', there was a sense from many of the founders of the organic movement that society needed to get back on track (Balfour 1943). These two elements – an alternative approach to agricultural practice and an alternative set of values about how we should live – continue to characterise the organic movement. These elements are tightly coupled in the biodynamic approach where specific techniques are prescribed based on an esoteric cosmology.

> **Box 3.1 The IFOAM Principles**
>
> **Principle of Health** Organic agriculture should sustain and enhance the health of soil, plant, animal, human and planet as one and indivisible.
>
> **Principle of Ecology** Organic agriculture should be based on living ecological systems and cycles, work with them, emulate them and help sustain them.
>
> **Principle of Fairness** Organic agriculture should build on relationships that ensure fairness with regard to the common environment and life opportunities.
>
> **Principle of Care** Organic agriculture should be managed in a precautionary and responsible manner to protect the health and well-being of current and future generations and the environment.

In the mainstream organic sector, the broad principles (Box 3.1) which underpin an organic approach are set out by the International Federation of Organic Agriculture Movements (IFOAM) while the detailed and prescriptive organic standards seek to reflect these in agricultural practice. The IFOAM principles are about what should happen, how the world should be: organic agriculture is not simply one way to produce food but a contributor to a better world. We explore these in more depth later in this chapter.

The alternative set of values associated with organics in the UK has been less consistent. Many of the founders of the organic movement were well-connected landowners, and there was a strong Christian connection. Some influential supporters were reactionary, with distaste for modern urban life and wanting a land-based traditional 'Anglo-Saxon' society.

Later on, from the 1970s to the 1990s, organics acquired a stronger commercial base but also became associated with liberal rather than traditional values. Organics was less about harking back to a better previous era, more about creating a better and more progressive alternative and confronting corporate control, particularly in relation to genetic modification (GM).

As the organic sector has grown and become part of the mainstream food market, some mainstream organic bodies have become less vocal about wider political issues, at least in the UK. While there is more self-confidence and depth in the scientific challenge to contemporary agricultural practice, the wider challenge to power and values in the food system is no longer led by the UK Soil Association. Instead, a broader range of organisations and campaigns has evolved to challenge current food and farming practices on ethical as well as environmental grounds. The distinctive cluster of values associated with organics and agro-ecology as a contemporary social movement forms the third argument for organics.

First, we consider the economic and environmental case.

3.3 The Argument from Economics: Is More Better?

Debates in everyday conversations or newspapers about organic and conventional farming tend to gravitate quickly to the question of relative yields and various expressions of productivity. That conventional farms are more productive on a crop per unit area basis than organic ones seems self-evident. In the UK and European contexts, good conventional yields per hectare are higher than good organic yields. Recent estimates suggest an overall yield gap of just under 20%. It's often a short step from there to concluding, as William Crookes did over a century ago, that 'feeding the world' can only be done by using science to increase production. Proponents of GM such as Owen Paterson restate Crookes' argument with a contemporary spin:

> We should all keep one fact at the front of our minds. At this very moment there are one billion people on this planet who are chronically hungry. Are we really going to look them in the eye and say, 'We have the proven technology to help, but the issue's just too difficult to deal with, it's just too controversial'? It won't be long until the population moves from seven billion to nine billion and we'll have even fewer resources to feed them. It is our duty to explore technologies like GM because they may hold the answers to the very serious challenges ahead.

Alongside the argument that more food production is needed in absolute terms, there's an argument that producing more intensively on a smaller area of land allows more 'room for Nature' on the land thereby 'spared'. The case for higher yields is presented as requiring no further explanation. Of course more is better. Why would anyone want to produce less than the maximum possible?

At a global level, there is clearly no economic argument for maximising global yield (although there are good economic and social arguments for increasing smallholder yields to support local food security). Hunger is not caused by a global shortage of food, but by local shortages (usually caused by humans rather than Nature) and by poverty. The challenge is one of distribution, both of food and of the means of food production. Growing more food to put into storage is not the answer.

In 1898, William Crookes could not have foreseen that despite a world population almost five times larger than when he spoke, the twenty-first century food system would be able to use 33% of cropland for animal feed and 4% for biofuels. More than a quarter of the USA's cropland grows corn, overwhelmingly used for animal feed, biofuels and other industrial uses rather than feeding people. At the same time, an estimated one-third of total food production is wasted; reducing this would be much more resource efficient than simply producing more. There is also land in reserve: an estimated 1.4 bn hectares of rainfed land suitable for agriculture is not currently used for production (Alexandratos and Bruinsma 2013) because the costs of using that land to bring food to market are not justified by the current level of effective demand.

So there is no absolute global imperative to maximise rather than optimise yields – but what about at farm level? Depending on the price of land, the price of inputs and the price of the crop, many farmers make rational decisions to grow less per acre than they could. Farmers in the US produce far less wheat per acre than farmers in Europe, because the extra costs of water and fertiliser would often exceed the value of the extra crop. So purely in narrow economic terms, organic agriculture may be a rational decision at farm level: lower yields with lower inputs and higher prices make financial sense. Financial performance on organic and conventional farms is comparable in many EU states, (ec.europa.eu/agriculture) with the growth in organic production and land area in Europe driven by consumer demand. Finally, there is growing evidence that organic methods can achieve yields on a par with conventional methods in different climatic and soil conditions, for example in Africa (Adamtey et al. 2016).

Even where yield from conventional systems is higher, choosing to farm organically is a rational choice in current market conditions. (Crowder and Reganold 2015) This depends in part on organic produce attracting a premium, whether that's organic milk for school meals in Scotland or organic pineapples in Uganda for export to Germany. And that premium depends in part on the people who buy organic food seeing yield in broader terms than tonnes per acre.

3.4 The Argument from the Environment: Externalities Matter

A linear model of inputs and outputs focuses attention on farmgate commodity outputs (and income associated with those outputs). Other social and environmental outputs and outcomes tend to be seen as secondary. However, the ideas of yield and productivity should be expanded to include all the proximate impacts of farming. The environmental impacts are often referred to as 'externalities' – consequences of a commercial activity that impact on third parties. Briefly, the consequences of mainstream farming practice, in comparison to organic farming practice include:

- lower biodiversity (fibl.org), as a result of specialisation, landscape simplification and the use of pesticides
- greater use of antibiotics and anthelmintics, and consequently more selection pressure for resistance to emerge
- due to higher nitrogen inputs, greater leakage of reactive nitrogen to water (Benoit et al. 2014) and air, with consequences for human health estimated to exceed the value added to agricultural production (Sutton et al. 2011)
- higher leaching of pesticides to the wider environment, and higher pesticide residues in food.

Organic farming also tends to require more labour – which can be seen either as an input to be minimised or as a useful benefit from farming in a different way. Later chapters set these out in more detail.

The key point is that all these environmental impacts are part of the farming 'yield'; they happen as a direct result of farming activity but do not appear on any profit and loss account or balance sheet. Subsidies for organic conversion and maintenance are designed to reflect the public benefits of the differential impact of organic farming.

3.5 The Argument from Ethics: There's Something Wrong with the System

In the two sections above, the economic and environmental case for organics was set out. We can have economic and environmental arguments about this, within a common frame of reference. Maybe the economic case for organic farming will weaken once a threshold of 10% or 20% of the market is reached; maybe conventional farming will clean up its act and reduce or eliminate pesticides, and new laws raise the bar for animal welfare to narrow the gap between conventional and organic. This section deals with values and beliefs – what matters most. We can articulate and discuss different conceptions of what we are and should be doing on this planet, or what a better world looks like but no amount of evidence can clinch the argument.

The values, concerns and beliefs associated with the wider organic (or agro-ecological) social movement cluster around three main questions.

1) What is the place of humans in nature?
2) What is farming for, and what makes for good farming?
3) How can the way we produce food promote social justice?

This section briefly explores each of these huge questions.

3.5.1 What is the Place of Humans in Nature?

Elon Musk recently described two paths for humanity. We could, he said, 'stay on earth forever' and suggested that sooner or later there would be a doomsday event, or we could become an 'interplanetary species'. For Elon Musk, it's a no-brainer: of course we want to build a colony on a barren cold planet like Mars. Staying at home is clearly for losers. Inherent in this idea of colonising space is a sense of humans as outsiders, detachable from the rest of Nature on this planet. But what if we found another planet, not lifeless like Mars or toxic like Venus but teeming with life like Earth – what would we do there? What would our job be there? We would probably express it along the lines of the IFOAM principles: 'to sustain and enhance the health of ecosystems and organisms'. We would try to appreciate Earth 2, understand how it works, and not mess it up.

Back on Earth 1, an organic approach is about seeing ourselves in, not apart from, Nature; as biological beings in an ecosystem, which we have come to dominate before we've understood or appreciated it. If we see ourselves as part of the system, we have a part to play in relation to sustaining the whole. If we see ourselves outside the system, we are free to take as much as we can get away with. This holistic perspective – seeing Nature as a whole and ourselves as part of that whole – is a way of seeing and thinking, feeling and appreciating, rather than a hypothesis to be proved or disproved. Astronauts who see the Earth from space get it, but there has been a strong tendency in western thought to emphasise both the separateness of individuals and a clear difference between 'man' and Nature. If we see ourselves as part of a whole, our task is to be in right relationship with that whole, to play our part. From this perspective, right action is to contribute, to serve, to create harmony. A very different concept of right action drives modern capitalism: here we praise people who win, people who do it their way, play their own tune.

The focus on the whole rather than the part is reflected in the organic understanding of health. Again from the IFOAM principles:

> Health is the wholeness and integrity of living systems. It is not simply the absence of illness, but the maintenance of physical, mental, social and ecological well being. Immunity, resilience and regeneration are key characteristics of health.

Seeing health as a characteristic of dynamic systems in constant interaction with their environment encourages a much broader process of building and maintaining health. For Steiner, these interactions reach out through the solar system and include astral forces: however, more secular organic approaches also look at plant, animal and soil health in broader terms than simply the absence of pathology. If plants are unhealthy, an organic approach is first of all to look for immunity and resilience and regeneration: what is it about the plant's interaction with its environment which is undermining its health?

Similarly, an organic approach sees co-operation, exchange, balance and stability as key characteristics of healthy ecosystems. Unlike late-stage capitalism, ecosystems don't operate on a 'winner takes all' system but on the basis of 'everyone gets along'. The practice of stewardship is integral to an organic approach. Importantly, this ascribes intrinsic value to the ecosystem, not merely use value. What makes the ecosystem valuable is not just the things it does for us (and what they can be traded for in monetary terms) but the sheer beauty and complexity of it, the absurd abundance of beetles. And, of course, the fact that until Elon Musk gets us to Earth 2, it's the only one we have.

Appreciating the beauty and complexity of the ecosystem should inspire a proper sense of awe: how can it be that plants use the fungal network to warn each other about predators? How can it be that sheep recognise human faces from two-dimensional photographs? How did ants evolve to use triage on their wounded companions and only bring home and nurse the ones likely to survive? How did dung beetles get started with using the moon, sun and stars to navigate? It should also inspire a cautious attitude towards messing about in it with cascading interventions. So organic philosophy is not antiscience or anti-innovation but it's inclined to caution; as the IFOAM principles state, 'Practitioners of organic agriculture can enhance efficiency and increase productivity, but this should not be at the expense of jeopardising health and well-being'.

Take pesticides, for example. Over the last 70 years, we have released tens of millions of tonnes of pesticides – synthetic chemicals designed to kill living organisms – into an open system. These pesticides are now ubiquitous, several links in the chain away from their introduction to the system, with DDT accumulating in polar bears, glyphosate and its derivatives being commonplace in human urine, and neonicotinoids found in honey on every continent. This has been a 'suck it and see' experiment on a global scale, without truly informed public debate or an ethics committee. No amount of laboratory testing on animals such as rats or laboratory microcosms could predict the wider impact on entire ecosystems. We simply do not know whether or not there have been or will be long-term public health consequences from the widespread ingestion of a complex cocktail of different pesticides, each tested separately.

It would be very inconvenient for agribusiness and for the current business model if any of these wider causal links were established – even more inconvenient than it was for the motor industry when, after decades of evidence about the harm caused by lead, concerns finally resulted in actions to ban its inclusion in petrol. However, because the stakes are high, keeping an open mind on the possibility of harm is often claimed by advocates of conventional farming as antiscience. Discussions are often conducted in terms of absolute safety rather than of a balance of risks. It is too often assumed that 'the public' is unable to understand and engage with discussions of risk in relation to food. So it is difficult to have a sensible conversation about possible low-level but widespread health impacts of pesticides in human food – even though it is agreed that many foods contain detectable levels of pesticides, that around 2% of samples in the UK contain more than the maximum residual limit set by government, and that some of these compounds may be carcinogens, endocrine disruptors, etc.

This raises a number of important and very challenging questions. Would a small, but measurable, rise in childhood autism be a reasonable price to pay for the higher production enabled by pesticide use in conventional agriculture? What about a rise in the incidence of certain cancers, or allergies? Would the people or the health services affected have a claim on the people who sold the food, those who used the pesticide or those who made it? And how could they establish liability for their specific condition as resulting from a particular food, pesticide or combination of pesticides?

Organic thinking wants to open up these conversations, to look at risks versus benefits. Reductionist thinking wants to shut them down – not through scientific argument but through excluding them from the frame. The assertion that pesticides or GMOs are 'proven to be completely safe' is not in the conventional or logical sense a scientific statement but tends to be a bid to control the agenda for discussion. It easily becomes a declaration of confidence in a course of action which has already been chosen. The higher the stakes, the more confidence is needed but we would be doing better science if we spent more time testing our hypotheses. The absence of doubt in contemporary advocates of Big Food has no true scientific basis

3.5.2   What is Farming For, and What Makes for Good Farming?

The purpose of farming in organic systems is to maintain the health of soils, plants and animals so that the farm can nourish people. This differs from a typical modern account of farming in a number of ways. First, we tend to talk about 'farm businesses' rather than farming, and these farm businesses may produce no food at all – instead, for example, growing crops to feed an anaerobic digester. The purpose of farming is sometimes defined not in terms of producing food but producing 'raw materials for the food and drink industry' (Scottish Government motion for a debate on Scotland's Food and Drink Industry 2017).

However, the conventional farming perspective would see the core purpose of farming as producing and selling food commodities in order to generate an income.

The organic approach extends this purpose in both directions. At one end, it focuses on stewardship of the farm's natural resources as the core of the work: food production happens as a result of focusing on the health of the farm ecosystem. At the other end, it brings in people: organic farming at heart is about food for people, not food as a commodity. It is a service business. This is not to say that organic farms always have a direct relationship with the people who eat their food. Organic farms export food, and supply commodities into the wholesale market. Nevertheless, an organic or agro-ecological approach retains a focus on the people who eat the food, not just the merchant who pays the farmer. The point of farming is to nourish people, not simply to sell a commodity.

This is not the place to argue about the quality of food produced from organic systems. In Chapter 12 the concept of 'inherent value' (Atkinson et al. 2012) is introduced and explained. This suggests that food has attributes which go beyond those which can easily be described by scientific analysis but which are important to consumers and which are part of the purchasing decision. The organic concept can be part of this type of decision making. The organic label indicates that food has been produced in certain ways and commonly results in producers being rewarded for food being produced in this way. It may also result in the food having some differences in composition and quality, issues that are also discussed in Chapter 12.

Maybe some food produced in this way has added 'vitality'. Science measures health only by the number of nutrients present in a foodstuff but that is not the whole story. In biodynamics, health is much more than mere nutrients; it's about vitality, which in turn is the holistic indicator of quality (Biodynamic Association website). Maybe food produced in this way just has more nutrients and fewer 'nasties' (Barański et al. 2014). The key point is that organic farming aims not just to produce food but to produce the highest quality food it can, by focusing on the overall health of the system. Farmers should be able to stand behind their food with confidence – whether literally in the farm shop or metaphorically.

Organics also has a clear account of what makes for good farming. It values human-scale farming, and attaches no virtue to size. It sees the farm as a living entity, both physical and social. It places a value on generating and using resources (such as seed, feed and fertility) from within or close to the farm rather than simply providing a platform which imports materials from elsewhere to produce food. It values connections between the farm and the community around it. A healthy farm, on this account, like any other healthy organism, has immunity, resilience and capacity for regeneration. The focus on animal welfare in organic farming is not simply about health, ensuring that animals thrive by providing good living conditions. It's a more proactive approach, respecting the integrity and sentience of farm animals.

Good farming means extending the constituency of care – to customers and suppliers, to other species on the farm, to future generations. As part of a web connected across both space and time, good farming is about co-operation and right relationship. This account of good farming is not unique to organics. It reflects traditional farming values but these values have been eroded where farmers have been pushed into practices which deaden, isolate or neglect the farm.

3.5.3   How can the way we produce food promote social justice?

IFOAM argues, 'Organic agriculture should … contribute to food sovereignty and reduction of poverty'. We seem to be moving a long way away from this book's focus on the science base of organic agriculture; how can a set of farming techniques have a bearing on the huge social challenges of inequality and hunger in the midst of plenty? Or on how power in the food system is held in so few hands? These are questions to ask about how we understand science and why we have information about some aspects of agriculture, such as the transformations of added nutrients in soil, but much less on the microbial basis of organic matter decomposition.

Is the answer to 'why organics?' really 'because it promotes food justice'? Isn't organic food increasing unfairness because it's priced beyond the reach of many? It would be difficult to argue that social justice was at the heart of the organic movement from the outset. While there was a focus on the relationship between 'man' and Nature, there was less interest in fairness between 'men', and little consideration of the role of 'women'. Of course, the certification system enables smallholder producers across the world to sell certified organic products into western markets – and this undoubtedly raises incomes for producers, as well as reducing producers' exposure to toxic chemicals. (On the other hand, it's important that this production for export doesn't work against local food security.)

In the late 1990s, the UK Soil Association campaigned strongly against GM trials. This has been their most prolonged and vigorous campaign, and is ongoing. More recently, 'good food for all' forms part of the organisation's strategy. The UK Soil Association in general has not led the debate on wider food justice issues – however, the organic movement is now much broader. Many other organisations in the UK advocate an agro-ecological approach, defined in Coventry University's *Mainstreaming Agroecology* as follows:

> The discipline of agro ecology has emerged as a different paradigm to address the problems of the global industrial agriculture and food system. It is based on the principles of sustainability, integrity, equity, productivity and stability.

The People's Food Policy, developed and supported by over 100 civil society organisations in 2017, sets out a food policy for England where:

> Everybody, regardless of income, status or background, has secure access to enough good food at all times, without compromising on the wellbeing of people, the health of the environment and the ability of future generations to provide for themselves.

This vision of food justice is linked not just to food worker rights and land reform but also to a wholesale shift to an agro-ecological approach. Agro-ecology as a social movement is closely linked to food sovereignty and issues of how the food system is governed. The introduction to this volume emphasises the ecological basis of organic production and of ecological principles being at the heart of the science of organic production.

In the UK, discussions of food justice sometimes skirt round the 'o' word because of its elitist connotations, while in other countries organic production is seen as part of a social, environmental and economic struggle for a fairer food system. For example, Nayakrishi Andolon in Bangladesh has adopted a biodynamic/organic approach grounded in local spiritual traditions to engage 300 000 families in a transformation of agricultural practice and community regeneration. Its farming practice, however, is closely allied with a political challenge:

> We must interrogate the dominant discourses related to food, nutrition and agriculture and stand against the global processes that replace and eliminate farming communities in order to install transnational companies to take over the food production and control the global food chain. (http://ubinig.org)

In Brazil, MST – the movement of landless rural workers – has opened a warehouse in São Paolo to sell organic produce from smallholders.

3.6 Aligning Organics with Social Justice

There is work to be done. Since the rise of organic certification and the commercial success of organic products, it has been too easy to reduce the challenging organic principles set out by IFOAM into a set of techniques and standards. We have not done enough to articulate organics as a coherent and far-reaching philosophical approach, and this theoretical work is now more likely to be framed in terms of agro-ecology. In a sense, this doesn't matter. What's important is that we connect food justice issues with the issues discussed in this chapter – our relationship with the natural world and the role of farming. We need people who are passionate about ecosystems to realise that a fairer, more equal and more democratic society is better placed to respect and protect Nature, and people who are passionate about food security and health to be interested in how food is produced, not just how it is distributed. And of course, we need to keep learning, through experience and research, how to put our principles for better agriculture into practice. We need to develop the science base of organic production within this wider context.

3.7 Conclusion

This all takes us far from 8 t or 10 t a hectare, to questions about who we are and what we are doing in the world. Such questions are just as much part of the organic conversation as discussion of yields and feeding the world. The task of organic philosophy is to set out a whole-system picture of what better farming looks like, and to use this as a guide to achievable and desirable change.

References

Adamtey, N., Musyoka, M.W., Zundel, C. et al. (2016). Productivity, profitability and partial nutrient balance in maize-based conventional and organic farming systems in Kenya. *Agriculture, Ecosystems & Environment* 235: 61–79.

Alexandratos, N., & Bruinsma, J. (2013). The 2012 Revision World agriculture towards 2030/2050: the 2012 revision. Retrieved from: www.fao.org/economic/esa.

Atkinson, D., Harvey, W., and Leech, C. (2012). Food security. A Churches Together approach. *Rural Theology* 12: 27. 42.

Balfour, E.B. (1943). *The Living Soil*. New York: Universe.

Barański, M., Średnicka-Tober, D., Volakakis, N. et al. (2014). Higher antioxidant and lower cadmium concentrations and lower incidence of pesticide residues in organically grown crops: a systematic literature review and meta-analyses. *British Journal of Nutrition* 112 (5): 794–811.

Benoit, M., Garnier, J., Anglade, J., and Billen, G. (2014). Nitrate leaching from organic and conventional arable crop farms in the Seine Basin (France). *Nutrient Cycling in Agroecosystems* 100 (3): 285–299.

Conford, P. (2001). *The Origins of the Organic Movement*. Edinburgh: Floris Books.

Crowder, D.W. and Reganold, J.P. (2015). Financial competitiveness of organic agriculture on a global scale. *Proceedings of the National Academy of Sciences USA* 112 (24): 7611–7616.

Hager, T. (2008). *The Alchemy of Air: A Jewish Genius, a Doomed Tycoon, and the Scientific Discovery that Fed the World but Fuelled the Rise of Hitler*. New York: Harmony Books.

Sutton, M., Howard, C., Erisman, J. et al. (2011). *The European Nitrogen Assessment. Sources, Effects and Policy Perspectives*. Cambridge: Cambridge University Press.

Further Reading

EuropaAgriculture. http://ec.europa.eu/agriculture/rica/pdf/FEB4_Organic_farming_final_web.pdf

FAO 2012 www.fao.org/docrep/018/ar591e/ar591e.pdf

FAO www.fao.org/3/a-i3991e.pdf

Howard A, Yeshwant D (1931) The waste products of agriculture. OUP/IFOAM, Oxford.

Rulli, M.C., Bellomi, D., Cazzoli, A. et al. (2016). The water-land-food nexus of first-generation biofuels. *Scientific Reports* 6: 22521.

4

Soil Health and Its Management for Organic Farming

Elizabeth A. Stockdale, Tony C. Edwards and Christine A. Watson

SRUC, Aberdeen, Scotland, UK

4.1   Introduction

Soil is arguably the most important single contributor to food production. As a result the pioneers of organic production placed it at the heart of their thinking. It is equally important to all approaches to food production. Although organic and 'conventional' production differ in many ways, much of what we understand about soils is common to both approaches and equally important. A basic organic principle is that appropriately managed soil can provide the nutrients for crops and, as a result, for livestock and humans and also support crop, animal and human health without the need for mineral fertilisers or regular planned applications of crop protection chemicals. These basic principles are discussed in detail in Chapters 1 and 2.

The role of soil is complex and information on soil mineral nutrients and the consequences of soil fertility are detailed in the chapters on land capability (5), rotations (6), nutrient balances (9), root systems (10) and microbiology (11). With detail provided elsewhere in the volume, the aim of this chapter is thus to give basic information on soils, their importance to crop production and to provide a framework for the approaches discussed in later chapters. In this chapter we discuss what characterises soil and why those properties are so important. We then use this understanding to explore the ways in which soil management affects the properties of soils but we leave it for other chapters to assess the ways in which soils are managed in a distinctive way in organic production.

Soil is the basis of most agricultural and horticultural production. However, soil processes are important in providing a wide range of biophysical necessities for human life and/or making other contributions towards human welfare. For example, the Food and Agriculture Organization of the United Nations (FAO) (2015) considers that soil provides 11 key functions.

1) Provision of food, fibre and fuel.
2) Carbon sequestration.
3) Water purification and soil contaminant reduction.
4) Climate regulation.
5) Nutrient cycling.

The Science Beneath Organic Production, First Edition. Edited by David Atkinson and Christine A. Watson.
© 2020 John Wiley & Sons Ltd. Published 2020 by John Wiley & Sons Ltd.

6) Habitat for organisms.
7) Flood regulation.
8) Source of pharmaceuticals and genetic resources.
9) Foundation for human infrastructure.
10) Provision of construction material.
11) Cultural heritage.

Soil functions result from the interaction and/or integration of a number of processes which may be linked. The three most important activities supported by agricultural soils are as follows.

1) **Food and fibre production.**
Food and fibre crops require soils to be maintained in a suitable state that provides good soil structure, water retention and nutrient availability. Inappropriate soil management can lead to erosion, a loss of soil organic matter and/or degradation of soil structure. These changes can result in a decline of productive capacity.

2) **Environmental interaction (between soils, air and water)**
Soil provides the essential link between the components that make up our environment. These components include the atmosphere, surface and ground waters, above-ground habitats and human activities. Managing and protecting soil is therefore an essential part of protecting the environment as a whole. Soil forms these links through:

- the exchange of gases, such as carbon dioxide, with the atmosphere
- its role in regulating the flow of water and rainfall in the water cycle
- its role in the degradation and storage of organic matter
- the storing, degradation and transforming of solid materials, such as nutrients, organic materials and contaminants that are applied through animal and human activities or deposited by flood waters and aerial deposition.

Protecting the capacity of soils to store, transform and regulate these processes is critical to environmental sustainability.

3) **Support of ecological habitats and biodiversity**
Soils contain a very diverse biota and soil biodiversity is vitally important in maintaining soil functions and sustainable systems as many of the key processes underpinning these functions are mediated by the soil biota. Fungi, bacteria and larger organisms, particularly earthworms, play a crucial role in the generation and stabilisation of soil structure that influences rooting, aeration and drainage.

Soil quality can be considered as the capacity of the soil to deliver the range of desired functions. For agricultural soils, Larson and Pierce (1994) have suggested that this should be measured by a range of physical, chemical and biological properties of the soil within its particular environment that together indicate the capacity of the soil to provide a medium for plant growth and biological activity, regulate and partition water flow and storage in the environment, and, serve as a buffer in the formation and destruction of environmental hazardous compounds. The related concept of soil health lays slightly greater stress on the soil as a living system, that is:

> the continued capacity of a soil to function as a vital living system, within ecosystem and land-use boundaries, to sustain biological productivity, promote the quality of air and water environments, and maintain plant, animal and human health. (Pankhurst et al. 1997)

However, these two concepts of soil quality and soil health are commonly, but perhaps unhelpfully, used synonymously. The term *soil health* is often preferred by organic farmers where 'the health of soil, plant, animal and man is one and indivisible' (Balfour 1943). Organic farmers aim to maintain and increase the long-term fertility of soils while minimising nutrient losses according to the principles of the International Federation of Organic Agriculture Movements (IFOAM). The emphasis of organic farming is to encourage the efficient cycling and recycling of nutrients rather than relying on fertilisers and purchased manure (Lampkin 1990).

In this chapter, we will provide an overview of key soil properties and processes which are important for all farming but which have a critical significance for organic farming systems. We will first explore soil components, and then consider how these interact to drive some of the key soil processes for agricultural systems. This understanding will then allow a critical consideration of the impacts of agricultural practices on soil function and health within organic farming systems.

4.2   Soil Components

4.2.1   Soil Parent Material and Profile Formation

Soils form as a result of the physical and chemical alteration (weathering) of parent materials (solid rocks and drift deposits). Biological activity is a critical component of soil formation; soil is distinguished from weathered rocks as a result of the biological cycles of growth and decay leading to the incorporation of organic matter (OM). Hence, soils in any location are the unique result of the specific local interactions of climate, geology, hydrology and management. Typically, a vertical section of a soil (soil profile) will exhibit a number of distinctive horizontal layers which differ in appearance and in chemical composition. The basal layers will resemble the parent rock material while other layers are the result of physical and chemical weathering and of the interaction of this weathered material with biological processes. Soils are thus a product of the accumulation of biological material in the more superficial layers, leaching of the profile and the deposition of leached products. Soil profiles are the result of a series of complex interactions between climate, geology, topography and biological activity and where these factors are stable over a long period, very distinctive horizons can form (see Chapter 10 for the detail of a profile).

Where soils are relatively undisturbed by man, the soil surface is often characterised by a layer of plant litter with OM then incorporated into lower mineral horizons through the activity of a variety of soil organisms which decompose the organic materials and mix/move the components of the soil. OM content usually declines rapidly down the profile. Under agricultural management, soils are never in equilibrium. They are regulated by a series of human-managed perturbations (Odum 1969). Soils under agricultural management are most often distinguished by the character of their well-mixed surface horizons from adjacent soils in semi-natural systems.

Soil parent material controls a range of intrinsic soil properties including soil depth, stoniness, mineralogy and texture. Soil texture, that is the relative proportions of sand, silt and clay minerals in soil, plays a large part in determining the physical and chemical properties of soil. Farmers recognise soil textures in relation to their ease of cultivation and management, but these emergent properties also result from clay mineralogy and biologically mediated interactions between soil minerals and OM. Consequently, a good understanding of the soil profile and soil-forming factors and their variation within and between fields on a farm is important to developing effective management practices.

The principal minerals found in the soil exert great control over both inherent soil fertility and the buffering capacity of the soil to acidity, as they contain varying amounts of basic cations (Ca^{2+}, K^+, Mg^{2+}, Na^+). Negatively charged sites on the surface of clay minerals and OM in soil act as a nutrient sink for cations (known as cation exchange capacity, CEC), where total CEC depends on both edge and isomorphous replacement charges. That part of CEC which is due to clay minerals varies with the balance of different clay minerals present; for example, montmorillonite has around 30 times the capacity of kaolinite. In temperate soils, the organic fraction of mineral soils normally constitutes 30–65% of the total CEC. Differences between soils in CEC are strongly linked to texture, and effective CEC is also affected by pH. A good understanding of the inherent CEC and its variability is of importance in crop production because of its role in regulating the release of plant essential nutrients such as K, Mg and Ca and restricting their loss by leaching from soil. Understanding and working within the constraints set by these soil components is more important to organic production than to other production approaches where deficiencies in soil fertility may be compensated by chemical additions. Some of these issues are discussed further in Chapter 5.

4.2.2 Soil Organic Matter

Soil organic matter (SOM) consists of plant, animal and microbial residues in various stages of decay and provides the energy (carbon) resource for many soil organisms. Across a range of climates and systems, there is a strong correlation between total OM content in soil and the number of soil organisms; Wardle (1992) showed such a correlation between soil OM and the size of the soil microbial biomass population. Overall, the net primary production of the plant community, at any site, is a key driver of the SOM inputs; soil properties then interact to determine the relative rates of decomposition and stabilisation of fresh SOM additions. A range of site factors (climate, soil depth, stoniness, mineralogy, texture) have been identified as controlling the maximum potential SOM content (Ingram and Fernandes 2001; Dick and Gregorich 2004). SOM declines with depth in soil but even within the topsoil, SOM is not distributed uniformly. Buried litter, decaying roots and other concentrations of OM in soil have been described as resource islands (Wardle 1995). Van Noordwijk et al. (1993) observed increases in protozoa concentrations associated with buried stubble and other old organic matter. In contrast, bulk soil is a relatively resource-poor habitat. SOM may remain undecomposed as a result of its spatial location and a range of other physical and chemical stabilisation mechanisms (Powlson 1980; Six et al. 2002). As a result of these interacting mechanisms, clay soils will tend to have higher equilibrium OM contents than sandy soils; care is needed to put measures of soil OM for farms/fields in the context of soil texture and climate when interpreting them for farmers, as these factors can be as important as management in determining the OM content of soils and the potential for further carbon sequestration (Verheijen et al. 2005).

4.2.3 Soil Organisms

Plant root systems are a dynamic and varied component of below-ground ecology. Plant roots are themselves a key habitat component for a number of soil organisms, including symbiotic bacteria, mycorrhizal fungi and root pathogens and herbivores (Brussaard 1998). The rhizosphere and interactions of roots and soil organisms are considered more fully in Chapters 10 and 11.

Soil organisms not only occupy soil; they are a living part of it and, as a result of their interacting activities, also change it (Killham 1994). Soils contain a very high diversity of organisms, many of which remain unknown or, at least, little studied (Brussaard et al. 1997).

Bacteria and **Archaea** are distinct kingdoms of single-celled prokaryotes. Archaea are much less numerous than bacteria but thought to account for around 1% of prokaryotic activity (Buckley et al. 1998). The large majority of bacteria existing in soil (>95%) are not culturable and so it is difficult to characterise or study their function.

Much of our current understanding of the roles of bacteria in soil derives from studies of culturable bacteria and/or approaches which treat bacterial communities in soil as an undifferentiated unit (Stockdale and Brookes 2006). Using molecular methods, which extract DNA from soil and analyse patterns formed by DNA fragments, the vast diversity of bacterial ribotypes (i.e. subtypes of bacterial strains identified by their ribosomal nucleic acid composition) is being revealed. Molecular approaches quantify diversity based on assumptions about the degree of nucleic acid sequence similarity that can be used to differentiate taxa (Fitter 2005). In arable ecosystems, proteobacteria have been shown to dominate the ribotypes determined, with over 30% of the ribotypes attributed to this bacterial group (Smit et al. 2001; Sun et al. 2004). Acidobacteria are also very common (Smit et al. 2001; Sun et al. 2004) and found in very wide range of environments (Barns et al. 1999). Actinomycetes are predominantly decomposers of complex organic compounds and make up a small proportion of the soil community. However, they have been studied relatively intensively as they exude antibiotics as secondary metabolites. As might be expected, different patterns of dominance are beginning to emerge from studies of grassland soils (e.g. Borneman et al. 1996). Nicol et al. (2003) have shown that upland grassland archaeal communities are dominated by Crenarchaeota with some evidence that management practices influence the nature of this subcommunity. However, a considerable proportion of the prokaryotic community in soils remains unidentified (Sun et al. 2004).

Plasticity and capacity for change are very important characteristics of all soil prokaryotic populations because of their capacity for rapid growth through binary fission. The other key defining characteristic of the soil prokaryotic population is its ability to 'slow down' metabolic activity and maintain activity in a dormant state, even under conditions of very low energy and nutrient availability. At any time, a high proportion of the population is dormant and can remain this way for a long time until conditions for growth improve.

Fungi are eukaryotic and have a mycelial morphology with a mass of hyphal tubes enclosing multinucleated cytoplasm. Fungal hyphae are usually 2–10 µm in diameter, but can extend to m/km in length. Fungi are involved in a large number of interactions and processes in soil and are part of many complex relationships with other soil organisms. Mutualistic relationships between arbuscular mycorrhizal fungi (AM fungi) and crops are widespread, with only a very few crop species not forming such associations (Brassicae and Chenopodiaceae). The benefits of AM fungal associations to crops are well documented (Harrier and Watson 2003; Leake et al. 2004; Gosling et al. 2006) and these relationships are discussed further in Chapter 11. Soils also can host pathogenic fungi whose impact is discussed in Chapter 12.

Micro- and mesofaunal organisms are common in soil. Many are aquatic organisms occupying water-filled pores and water films in soil and are capable of encystment to enable survival in low moisture conditions.

Protozoa are unicellular organisms which are predators of the microbial (mainly bacterial) populations in soil. Changes in species balance and biomass within protozoan populations have been related to soil conditions and the impact of soil management practices (Foissner 1997).

Nematodes are microscopic roundworms with a diameter of <50 μm. Nematodes occupy central and diverse trophic positions within the soil food web, with at least three different functional groups identifiable: (i) plant feeding/root herbivore species are primary consumers; (ii) bacterial and fungal feeding nematodes, which are secondary consumers; (iii) predatory and omnivorous species (tertiary consumers) are also common. Species balance can therefore indicate changes in below-ground ecological relationships (Bongers and Bongers 1998; Mulder et al. 2003) and has been suggested as an ecological soil quality indicator. In agricultural systems, bacterial feeding species often dominate, with about 15 times more bacterial feeders in agricultural grasslands in the Netherlands than fungal feeders (Mulder et al. 2005). Plant-feeding nematodes can have adverse effects on crop growth both directly and through the transmission of plant viruses. These effects are discussed further in Chapter 12.

Enchytraeids are related to earthworms (class Oligochaeta) and morphologically, they look like small, white or transparent earthworms. Functionally, they are detritivores and microbial feeders and are therefore an important component of the decomposition system in soils. There is some evidence that specific soil types are inhabited by specific enchytraeid communities and that these respond to changes in management (Didden 1993).

Arthropods may use the soil for the egg or pupal stages of their life cycle; a wide range of insect species inhabit the soil for all or part of their life cycle. Larvae of beetles, flies and ants are common; in addition, woodlice, centipedes and millipedes are found in all life stages in soil. A number of these species are root herbivores and thus affect a range of above-ground plant processes (Wardle 2002). It has also been shown that root herbivores have a critical role in facilitating the rapid interchange of fixed N between legumes and associated species, such as in a mixed species sward (Murray and Hatch 1994).

Mites are the smallest (usually less than 1 mm), and also the most diverse group of arthropods in soil and therefore show a very wide range of feeding habits and life history strategies. Prostigmatid and oribatid mites have been relatively well studied in agricultural soils (Crossley et al. 1992; Siepel 1995). On average, the presence of these microarthropods increases decomposition rates across a range of environments (Seastedt 1984). Collembola, also known as springtails, also have a central role in soil food webs and affect decomposition processes. They are small (less than 6 mm in length) wingless insects in the subclass Apterygota. Different collembola species are specialised for different microhabitats in soil and litter and are quite susceptible to desiccation unless they remain in a moist environment. Some arthropods are a major pest of particular crops, an aspect discussed in Chapter 11.

Earthworms show differences between species in both burrowing and feeding activities. Earthworm species are most commonly grouped into:

- epigeic species that feed and inhabit the litter layer. Very few species in this group are found in UK agricultural systems
- anecic species, which feed on fresh organic material pulled down from the litter layer and form deep and permanent burrows. These species play a crucial role in initiating the contact

between inorganic and organic components in the soil. Anecic species make up about 70% of species present in UK agricultural systems. *Lumbricus terrestris* can reach population sizes up to a biomass of 2.5 t/ha in grassland systems (Killham 1994)

• endogeic species live and feed on OM from within the soil.

Earthworms have a very important role in the decomposition of organic matter in soils mainly as a result of the mixing of organic and mineral components and the incorporation of litter into deeper soil layers (Wolters 2000; Lavelle et al. 2001). Hogben (1983) showed that organic inputs had a significant impact upon earthworm numbers but the numbers of earthworms had little effect on aggregate stability, which was very variable. The structure of the earthworm community, as well as their abundance and biomass, has been suggested as an ecological soil quality indicator and these measures have been shown to indicate the influence of different anthropogenic land uses (Römbke et al. 2006).

4.3 Key Soil Processes in Agricultural Systems

4.3.1 Decomposition

Decomposition is a central process and results from the intermeshing vital processes of many soil organisms (Figure 4.1).

The carbon dioxide (CO_2) efflux from soils is the net effect of all heterotrophic aerobic decomposition processes. CO_2 consumption by soils is measurable but has only a small impact on the overall carbon (C) balance (Miltner et al. 2005). Carbon assimilation by soil organisms is also a net result of decomposition. Where the total amount of C input (in organic materials) to soil is increased then both the total CO_2 efflux and C assimilated are likely to increase (e.g. Jacinthe et al. 2002). However, the quality of the C input and a range of other factors affect the partitioning between CO_2 production and C assimilated to the living biomass and decomposition by-products (often called humus). Microbial efficiency has been shown to be affected by soil texture (Schimel 1986), mineralising substrate/residue type (Hart et al. 1994; Mueller et al. 1997) and temperature (Henriksen and Breland 1999). It is likely that a range of similar factors will affect the assimilation efficiency of organisms at higher trophic levels in the soil food web.

The decomposition interaction web developed for soils often shows two distinct routes by which OM is broken down, with one decomposition chain based on fungi and one based on bacteria (see Figure 4.1). This has been postulated to be linked to the difference in the capacity of fungi and bacteria to degrade organic materials; fungi are often considered to be more effective degraders of lignin, cellulose and hemicellulose (de Boer et al. 2005; Meidute et al. 2008). However, the role of fungi and bacteria may be as much a consequence of their physical location within the soil (see next section) and the availability of nutrients in the soil as the quality of the soil organic matter or added organic materials (Strickland and Rousk 2010).

The complex interaction between soil communities, carbon inputs and the physicochemical dynamics of the soil habitat means that causes and consequences can be difficult to disentangle; does the soil have higher organic matter because of the fungal dominance of the decomposer food web, or is the food web dominated by fungi because of the large inputs of recalcitrant litter? While there is some evidence of difference in the ratio of the size of soil fungal to bacterial population size in different tillage or fertilisation systems, such differences have been shown to

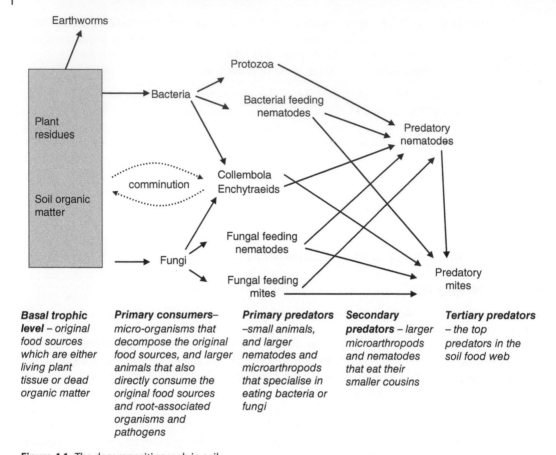

Figure 4.1 The decomposition web in soil.

have no clear pattern of impact on soil function (Strickland and Rousk 2010). Manipulation of the decomposer community to improve soil health through application of targeted inoculants or stimulants has been proposed, but there remains almost no evidence of impact (whether positive or negative) at field scale of such practices in independent trials.

The physical environment is often considered as a template on which organisms and ecological systems operate but it is clear that organisms' response to the physical environment may exhibit patterns that vary between species and are constrained by the geometry of the environment (Williams et al. 2002). Temperature has a direct effect on the rates of biological reactions; plant root growth and the activity of soil organisms increase with increasing temperature. The importance of this relationship is discussed in Chapter 8. In temperate climates, it is the interaction of temperature and moisture that largely controls the rate of biological processes and hence N cycling (Nishio and Fujimoto 1989; Recous et al. 1999). Soil animals are sensitive to overheating. Temperature may also affect the proportion of the organic matter that is decomposable (Dalias et al. 2003) and/or microbial efficiency (Henriksen and Breland 1999).

Seasonal variation in plant growth also leads to temporal variability in organic matter inputs to soil (see Chapter 10). Temporal variability in the activity and biomass of below-ground organisms and therefore the rates of soil processes can therefore be as significant as spatial variability. Decomposition processes are a major component within what is thought of as soil fertility and impact on the potential for C sequestration in soil. Farming is one of the few human activities which can result in both the generation and fixation of C and the extent of both processes varies between organic and other approaches to food production. This is discussed in Chapter 1.

4.4 Soil Structure Formation and Stabilisation

The amount and nature of the pore space in soil are dependent on soil texture (particle size) and chemical composition but particularly on the aggregation of the mineral particles together with SOM; that is, the formation and stabilisation of soil structure. Dry–wet cycles driven by frost, heat or water uptake by plants are the main physical processes which lead to the movement and realignment of soil particles; this effect is most marked in smectite clays where the individual clay crystals do not bind tightly together, thus allowing the clay to swell when water is absorbed and shrink when it dries.

The architecture of the soil pore network largely describes the space in soil available as habitat for organisms (Young and Ritz 2000) and controls the balance of oxygen and water available to organisms at any given soil moisture potential (SMP), as well as regulating access of soil organisms to one another and resources. Greenland (1977) grouped pores in soil by size and in relation to their function in the mediation of the balance of air and water. Transmission pores are >50 μm in diameter and in topsoil are usually filled with air; storage pores 5–50 μm in diameter are the main pores which fluctuate in air/water balance whereas residual pores 0.5–5 μm are commonly full of water, although plant roots can effectively empty pores down to 0.2 μm. Roots use pores of >100 μm as points of entry, root hairs, protozoa and fungi use pores of >10 μm, while bacteria can move in water films of 1 μm depth. Pores with neck sizes below a certain diameter may restrict entry by some organisms and hence protect smaller organisms from predation (Young et al. 1994); substrate may also be protected from decomposition by similar mechanisms (Powlson 1980). Nevertheless, extracellular enzymes can lead to apparent biological activity in smaller pores than organisms are able to inhabit (Young and Ritz 2000). In field soils, nematodes were shown to be dominantly associated with the pore class size of 30–90 μm, while bacterial biomass was correlated with pores 0.2–1.2 μm (Hassink et al. 1993).

Soil structure influences the nature and activity of soil organisms (Young and Ritz 2000), but soil organisms also have a key role in its formation and stabilisation (Tisdall and Oades 1982; Beare et al. 1995; Lavelle 2000; Wolters 2000). Plant roots also have a central role in structure development (Angers and Caron 1998) through physical penetration and water extraction. Some soil organisms are 'ecosystem engineers' which change the structure of soil by burrowing and/or transport of soil particles and hence create microhabitats for other soil organisms (Jones et al. 1994); in temperate agro-ecosystems, earthworms are very dominant within this functional group. Ninety percentage of the SOM in an upland improved grassland soil had been processed by earthworms and enchytraieds (Davidson et al. 2002). Supporting the biological processes of structure formation can also confer increased resistance to compaction arising during tillage and resilience to structural degradation (compaction); however, this is also closely linked to concomitant increases in soil OM content (Angers and Carter 1996) and the mechanisms supporting structural resilience are not yet well understood (Schlüter et al. 2011).

The geometry of the soil pore network also controls the movement of water. Water can only move freely in large pores (>300 μm diameter). Interactions between water and soil surfaces hold water in soils and prevent rapid movement in smaller pores. Plants can exert large forces to extract water from fine pores within the soil but except in the case of collapsing pores within clay domains, very small pores in soil (<0.2 μm diameter) will always be water filled. The drier the soil, the more direct routes for gas exchange will be in place; the converse is also true – the wetter the soil, the more direct routes for transfer of water, solutes and soil organisms. Because of the complex 3D framework that soil structure provides, a wet soil which is well structured and has a good mix of pore sizes will contain a mosaic of anaerobic volumes embedded in an aerobic matrix (Arah and Vinten 1995). Pore size distribution and connectivity are key factors controlling water movement; soil organisms affect both.

Biological interactions in soil usually have positive impacts on soil structural stability; these are fuelled by fresh organic materials, hence regular inputs of OM are more important than the total amount of soil OM (Loveland and Webb 2003). Therefore, biological modification of soil structure has significant effects on infiltration, water retention and drainage. The interaction between rainfall (amount, intensity) and soil surface structure (stability of a network of large transmission pores) determines partitioning between surface run-off and infiltration. The extent of ground cover by plants modifies rainfall intensity and tends to increase infiltration; the use of mulches has the same effect.

4.5 Below-Ground Ecological Interactions

Anderson (1975) put into words the paradox that the diversity of below-ground species poses for ecologists. How is it possible for such a large number of species to apparently co-exist without biotic mechanisms (e.g. competitive exclusion) reducing diversity? The usual explanation given is the extreme spatial (vertical and horizontal) and temporal heterogeneity in soil which gives rise to a wide range of surface types, pore sizes and microclimates, and a range of resources and resource partitioning in space and time. Most soil organisms have limited migration capacity (Fitter et al. 2005) and the motility of many soil species is low compared to the scale of resource patchiness (Ettema and Wardle 2002). Soil organisms often enter inactive or dormant states in unfavourable conditions, so that diversity is preserved; this is analogous to the role of soil seedbanks in preserving plant diversity (Ettema and Wardle 2002).

In above-ground ecology, the factors affecting diversity have been identified and ranked as trophic interactions between species, spatial habitat heterogeneity, temporal habitat heterogeneity, disturbance and nutrient resource availability (Torsvik et al. 2002). However, 'very little supports the notion that these relationships above ground can be simply transferred below ground' (Bardgett 2002). The extensive and critical review of Wardle (2002) highlights that competition is not the main regulator of trophic relationships below ground. This is not to say that competition plays no role – for macrofauna and some fungi there is evidence of competitive regulation (Wardle 2002). Consideration of interorganism interactions and their relation to function (Wardle and Giller 1996; Wardle 2002) can only take an understanding of ecological relations below ground so far; it is essential to integrate spatial habitat factors (Young and Ritz 1998).

The high degree of specialisation amongst soil animals provides evidence that increasing spatial heterogeneity increases soil animal diversity (Bardgett 2002). Spatial variability has been often treated as distracting 'noise' which obscures the key relationships between structure and function of below-ground biodiversity but it is likely that understanding the control over ecological systems imposed by spatial variability is the key to improving our ability to manage below-ground ecosystems (Ettema and Wardle 2002). This may provide the 'theory linking microbial population dynamic to biodiversity and function in terms of the soil microenvironment' which Young and Crawford (2004) conclude is more or less absent and presents a major interdisciplinary challenge for soil science.

Describing and modelling the interactions of below-ground ecology is often caught in the 'middle number' conundrum: there are too many individual components with too many complex interactions to deal explicitly with the individual yet the individual details affect the dynamics of the system as a whole, so general statistical properties yield an incomplete picture of what is going on (Weinberg 1975; Wu and David 2002). This problem is amplified by spatial and temporal variations and interdependencies, scale dependencies and thresholds. It is clear that all ecological processes occur in a spatial context but to date, modelling of soil processes has not been able to effectively deal with this challenge.

The wealth of information on the soil biota has been simplified by grouping species into trophic categories. Using this approach, the food web in soils can be described (e.g. Beare et al. 1992; Bloem et al. 1994) and modelled (Hunt et al. 1987; de Ruiter et al. 1993). The use of such models has allowed the relative importance of the interactions between trophic groups in controlling decomposition and other aspects of nutrient cycling to be investigated. Brussaard et al. (1996) found that soil fauna can account overall for 30–40% of net N released into plant-available forms. However, such models do not take account of non-trophic interactions such as impacts on soil structure; Brussaard (1998) outlined a number of further problems with this type of modelling approach. Wardle (2002) concluded that within the decomposition interaction web, the fungi-based channel has been shown to be resource driven (bottom-up regulation) while the bacteria-based channel has been shown to be predator controlled (top-down regulation).

The role of below-ground ecological interactions and the links to above-ground processes and environmental factors in determining the occurrence, severity and impact of crop diseases are covered more fully in Chapter 11. Van der Putten et al. (2009) highlight how modelling above-ground and below-ground plant–enemy interactions could lead to new avenues for sustainable control of pests and diseases in crops. Some soils have been shown to be disease suppressive (Menzies 1959); a wide range of ecological mechanisms (including parasitism, direct and indirect antagonism) is now thought likely to contribute to this effect (Mazzola 2002) and interactions with soil chemical and physical properties are also important (Duffy et al. 1997).

We commonly assume that when we have a range of organisms, there must be competition between them and often mechanisms whereby one species suppresses the growth/reproduction of another. Positive community interactions are also important within the soil biological community but are poorly understood. For example, there are different AMF species, combinations of AMF and bacteria working as a community to the benefit of all. Signalling between these components of a community is also important. Managing these interactions is a key opportunity for organic systems. This is discussed further in Chapter 11.

4.6 Nutrient Cycling and Management

4.6.1 Potassium (K) and Other Cations (Mg, Ca)

Most of the simple cationic forms of nutrients present in soil are in exchangeable forms associated with clay minerals and the organic fraction of the soil. These can rapidly exchange with the cations in soil solution; for example, of the 1–2% of the total soil K that is readily available to plants, approximately 90% is in exchangeable forms on soil surfaces (Sparks 2000). Options for managing K in organic systems are discussed in more detail in Chapter 9. Where SOM is increased, the capacity of the soil to retain nutrient cations, such as K, will be increased (Sparks and Huang 1985; Arden-Clarke and Hodges 1988). In the majority of farming systems, especially in temperate climates, the main limiting nutrient is N (Berry et al. 2002). N availability will thus largely control the demands for other nutrients (Fortune et al. 2004). Where plant uptake reduces the soil solution concentrations of a cation, this will lead to a simple exchange with release to soil solution from the exchangeable pool held on the soil surfaces; the reverse also occurs (McLean and Watson 1985; Barber 1995).

Plant-available cations (K, Mg, Ca) are routinely assessed from chemical extracts that measure this exchangeable pool, together with the much smaller solution pool (Bates and Richards 1993). The extractants used are intended to simulate the ability of the crop to remove nutrients. The amounts of extracted nutrients then indicate whether the soil is likely to be able to supply sufficient quantities of nutrient to sustain crop growth and production. The extractant used is commonly a compromise, as it has to be able to work with a number of soil types and often to extract more than one nutrient. The same extractant is likely to be used for K, Mg and Ca. The concentration of extracted nutrient is used to assess whether nutrient supply is likely to be sufficient or to require nutrient additions. For farming systems where fertilisers are used routinely, farmers are advised to maintain moderate levels of exchangeable soil K (often referred to as index 2 levels). However, it has been postulated that a lower K level (index of 1) is acceptable for maintaining production in organically managed grassland (Keatinge 1997). Fortune et al. (2005) have shown that available K, as measured routinely using ammonium nitrate extraction, can support the management of K in organic farming systems. So long as care is taken to prevent levels declining, these data also support the conclusion that maintenance of a low K index of 1 would be sufficient to maintain crop production in organic systems.

There are a number of approaches used to interpret soil tests in terms of soil fertility and/or health (Stockdale et al. 2013). One of the most disputed is the base cation saturation approach (Kopittke and Menzies 2007), which often results in the recommendation that for an ideal soil, 65% of the exchange complex should be occupied by Ca, 10% by Mg, 5% by K and 20% by hydrogen (Bear et al. 1945). It was considered that where this cation balance was achieved, chemical, physical and biological properties would all be in an appropriate equilibrium to support optimum plant growth.

The basic cation exchange phenomena in soil which control the availability of calcium, magnesium and potassium were outlined above; the degree of saturation of one cation can have significant effects on the plant availability of itself and other cations. However, review of the literature by Eckert (1987) and Kopittke and Menzies (2007) suggests that fairly wide variations in actual ratios are of little consequence for crop yield, quality or a wide range of soil physical and biological properties, as long as gross imbalances in nutrient availability are not created. Consequently, sole focus in the interpretation of soil fertility on a target Ca/Mg ratio often leads to recommendations for management that are extremely expensive and have little impact on soil health, crop productivity or profitability.

Typically, 90–98% of soil K is held in soil minerals, mainly K-feldspars and primary micas. Over a longer time period, K from these non-exchangeable pools will be released to top up the exchangeable pool and for slow-growing crops may be a significant source of K (Syers 1998; Johnston et al. 2001). While feldspars only weather very slowly, the micas and their weathering products are generally considered to be more important contributors to the supply of plant-available K. The annual release from temperate loams and clays is thought to be equivalent to about 50 kg/K/ha during the growing season (Goulding and Loveland 1987). Inclusion of non-exchangeable K (from both residual fertiliser and soil minerals) in estimates of plant-available K may significantly improve the correlation between soil supply and crop uptake (Syers 1998). For organic farming systems, there is an increased need to consider fully the possible K dynamics and the rate of release from less readily available pools, as well as measuring the content of available K at one particular time.

4.6.2 Nitrogen (N)

Nitrogen (N) is the primary nutrient limiting crop production in farming systems throughout the world. Matching the supply of N (via fertiliser applications and biologically mediated processes) to the amount, timing and location of crop N demand is critical to enable the development of environmentally acceptable farming systems that increase productivity through greater efficiency of N use and minimise losses of N. This is discussed further in Chapter 9.

Plants take up N from the soil solution as simple mineral forms, mainly ammonium (NH_4) and nitrate (NO_3); in some cases, simple organic N compounds are also absorbed (Nashölm et al. 2000; Jones et al. 2004). This is further discussed in Chapters 10 and 11. More than 90% of the total nitrogen found in soils occurs in high molecular weight organic polymers; N is an essential component of amino acids, cell peptides and proteins and incorporated into a wide range of other biologically essential compounds such as nucleic acids and chitin. On average, where mineral N fertilisers are applied, 50% of crop N uptake is derived from that fertiliser in the year of its application (Jarvis et al. 1996), with the remainder coming from the mineralisation of SOM. In organic systems, the nitrogen used by the crop comes either from the mineralisation of SOM or from the decomposition of organic manures. This is discussed further in Chapter 9. Consequently, the dynamics of N in soils are intimately connected with the decomposition of SOM.

Availability of biologically derived inorganic N to plants is difficult to predict and manage since it is regulated by a number of variables including:

- the quantity/quality of the residue material
- the rate of fragmentation and soil incorporation of residues by fauna
- the amount and timing of microbial N release
- microbial requirements for N
- N loss from the rooting zone.

The effect of variation in these variables is discussed in Chapter 8 and means of increasing resources within the system in Chapter 9. These processes are influenced by both environmental factors (e.g. temperature, wet–dry cycles), which can be difficult to predict, and agricultural management practices (e.g. tillage, green manures, residue management), which can be manipulated to enhance supply and demand. Other issues related to this are discussed in Chapters 6 and 9.

The N content of soil can be increased indirectly through symbiotic N-fixing processes. Forty million tonnes of N are estimated to be fixed by field crops and pasture species globally each year (Jenkinson 2001). In agricultural systems, much of the N fixed is harvested but a significant proportion will enter the SOM (via crop residues, excreta of grazing animals, etc.) and be subject to decomposition. Incorporation of crop residues, manures, etc. may increase soil N content in a particular place, but such applications usually represent transfers of N within/between farming systems, rather than imports of N. It has also been shown that a wide range of free-living soil bacteria can fix atmospheric N but the high energy requirements of the process mean that both the number of these organisms is small (Reed et al. 2011) and N fixation by this route in agricultural systems is very much lower than through symbiotic N fixation.

The gross release of N into mineral forms (NH_4^+) during decomposition (gross mineralisation) depends on the C/N ratio of the resource and the C assimilation efficiency of the decomposers (Hart et al. 1994). The net release also depends on the C/N ratio and nutritional status of the decomposer/predator, which controls the rate of assimilation (immobilisation) of N released. Mineralisation processes also occur within the guts of larger soil animals such as earthworms. These processes are carried out by diverse and distinct groups including bacteria, fungi and a wide range of invertebrate phyla. The direct and indirect effects of trophic (food webs) and non-trophic interactions (manipulation of habitat) between soil organisms and growing plants at a range of spatial scales within the soil result, at an ecosystem level, in the processes we measure as the soil N cycle and determine the net amount of soil N that is made available for crop uptake.

Organic materials applied to soil can have a wide range of C/N ratios; it has been estimated that in arable soils, the critical C/N ratio of added materials is c.20. The materials which can be used in organic systems are discussed further in Chapter 9. At larger C/N ratios, additional mineral N is needed to support decomposition. Soil organisms are able to immobilise NO_3, but NH_4 immobilisation is more energetically favourable (Recous et al. 1990). Critical C/N ratios will differ depending on whether the decomposer subsystem is bacterially or fungally dominated; fungi and bacteria have different assimilation efficiency and C/N ratios. In grassland soils, the role of cycling through NH_4, as described above, may be reduced in importance compared to the release and uptake of small soluble organic N compounds (Murphy et al. 2000).

Nitrification is the biological oxidation of nitrogen in soil. Nitrification in soils is dominated by the chemoautotrophic oxidation of NH_4 to NO_3 via nitrite. This is a two-step process mediated by *Nitrosomonas* and *Nitrobacter* respectively; rates are usually limited by the availability of NH_4. Nitrous oxide (N_2O), which is a potent greenhouse gas, is released during the process of nitrification; Stevens et al. (1997) showed that under many circumstances nitrification processes are the main source of this gas from soils. Heterotrophic nitrification, which releases NO_3 directly from organic N without NH_4 as an intermediary, is also known to occur, particularly under acid uncultivated situations (Pennington and Ellis 1993).

The soil NO_3 pool is particularly vulnerable to loss from the soil. Leaching losses of NO_3 are related to the rates of drainage from soils. In contrast, NH_4 is prevented from leaching in drainage as a result of cation exchange on the surfaces of clays and and/or stable SOM. Increasing numbers and connectivity of transmission pores can increase the rates of leaching loss; tillage often disrupts the continuity of pores from the surface into the subsoil (Young and Ritz 2000). The modification of pore size distribution through the activity of mesofauna can increase the amount of water retained in the soil at field capacity, particularly for coarse textured sandy soils, and may reduce drainage and hence leaching. Denitrification also leads to losses of NO_3 from soil. NO_3 replaces oxygen as the terminal electron acceptor in respiration in a wide range of bacterial groups that are facultative anaerobes whenever oxygen concentrations are limiting. The end-product of denitrifcation processes is dinitrogen gas, but significant amounts of N_2O are also released (Stevens et al. 1997).

Primary manipulation of the soil N cycle to increase N availability in organic farming systems is largely associated with retention or export of crop residues. The effective use and recycling of manures is a key component of nutrient management in mixed and livestock systems. Fine manipulation of the soil N cycle would be difficult to achieve under field conditions since the chemical composition of plant materials is not constant but strongly dependent on climate, soil and management factors (Gunnarsson and Marstorp 2002). While a range of timing and incorporation options have been considered for organic materials and leguminous residues in arable systems (Korsaeth et al. 2002), there is clearly less flexibility to alter the synchrony of N supply and demand in legume/grass pastures themselves. However, the management and timing of cutting and grazing, particularly in respect to the significant spatial impact of excretal returns on N cycling, can be an important tool in manipulating the balance between N fixation, soil N supply and plant N uptake (Ledgard 2001). Although we have a good mechanistic understanding of the individual factors regulating soil N supply, the capacity to design practicable farming systems that harness this N to its fullest is still questionable. These issues are discussed further in Chapter 9.

4.6.3 Phosphorus (P)

Total P in soils in the UK averages c.700 µg/g (Cooke 1958). However, concentrations in soil solution are very low (c.0.1 µg/g) and only a very small fraction of total P is available to plants. This is discussed further in Chapters 9–11.

Mineral P is found in amorphous and crystalline sesquioxides and calcareous compounds, with soil pH and the amounts of Al, Fe, Mn and Ca present determining the proportions of the different forms (PO_4^{3-}, HPO_4^{2-}, $H_2PO_4^{-}$) in a soil (Sharpley 2000). Where soluble P (e.g. in superphosphate fertiliser in 'conventional production') is added to soil, the P rapidly becomes associated with Al, Fe, Mg and Ca compounds in the soil, depending on the soil type and pH. With time, there is a further diffusion of the adsorbed P into the soil matrix, making the P unavailable for uptake in the short term (Johnston et al. 2001). Mineral P thus exists in the soil in soluble available/labile and slowly available forms. Both its chemistry and its availability are complex. The majority of P added in manures is also in an inorganic form, mainly as precipitated calcium phosphates; typical values range from 0.3% to 2.4% on a dry matter basis compared with <0.1–1% as organic P (van Faassen and van Dijk 1987).

The widespread use of P fertilisers in conventional agriculture has led to a situation where most arable soils in the UK have relatively large labile reserves of P (Tunney et al. 1997). Although originally, it was thought that P was fixed irreversibly in soils (Larsen 1967), it is now known that P can move from weak to strong and strong to weak bonding sites and that soil reserves of P can have a significant positive effect on crop yields in the long term (Johnston et al. 2001). Equilibration between the soil solution and the soil surfaces and other mineral constituents of the soil takes place as the result of simple exchanges resulting from changes in concentration, adsorption–desorption reactions and reduction–oxidation reactions. These processes are dominantly controlled by the concentration and composition of the soil solution, pH and aeration.

About 25% of soil P is held in organic forms (Wild 1988); the release of this P pool is linked to decomposition processes, as for N. Organic P is mineralised by the action of extracellular phosphatases (Gressel and McColl 1997), but soil micro-organisms may also immobilise significant quantities of P to meet their own requirements (Paul and Clark 1996). However, this immobilisation is short term and may protect the available forms of P from longer term adsorption in soil. More rapid turnover of microbial biomass P has also been measured in soils receiving regular inputs of livestock manures (Oberson et al. 1996). It has been estimated that (under conventional management) mineralisation of SOM supplies about 1–10 kg/P/ha annually in arable rotations and 15–25 kg/ha in grassland soils in the UK (Gasser 1962; Chater and Mattingly 1980). The activity of some key enzymes involved in the P cycle (acid phosphatases) has been found to be higher in organic and biodynamically managed soils than under conventional management (Oberson et al. 1996). A previous history of P fertiliser applications can reduce phosphatase activity (Spiers and McGill 1979). It has been shown that mineralisation of organic P is repressed when there is an ample supply of available P in soil but is induced by conditions where available P is limited (Smeck 1985). SOM also increases the number of low-energy bonding sites available for P, which is critical to regulate short-term P availability (Johnston et al. 2001).

Biological processes also play an important role in solubilising inorganic forms of P. Some bacteria, fungi and plant roots excrete organic acids which modify the chemical equilibria controlling P availability from calcium phosphates (including rock phosphate) by reducing pH and/or chelating calcium, thus increasing the solubilisation of P. This is discussed further in Chapter 6. Plant uptake of P is also facilitated where AM fungal associations with roots are present. H_2PO_4 concentrations in soil solution are not large enough to mean that mass flow (i.e. the movement of nutrients in solution to the root in response to transpiration demand) can meet plant requirements, and diffusion of H_2PO_4 is very slow. Hence, increased apparent root surface area mediated by root hairs and AM fungal associations gives plants access to an increased volume of soil solution and more available P. Issues related to this are discussed further in Chapter 11 Where soil becomes very dry, diffusion rates are reduced further, so changes in pore size distribution which increase water-holding capacity as a result of biological activity, particularly in sandy soils, can also improve the P availability for plants.

The potential supply of P from soil to crop has commonly been assessed by the use of chemical extractants, as detailed earlier for K. A large number of extractants have been assessed, of which Olsen and resin are perhaps the most common, each of which may work better at predicting crop response in different types of soils. This is a result of the more varied chemistry of phosphate in soils. Ultimately, the purpose of the soil test, as for K, is to predict whether there is a need to add nutrients to a crop so as to sustain or increase crop yields. The use of organic nutrient inputs and sparsely soluble inorganic inputs such as rock phosphate in organic production has meant that many conventional soil tests are inappropriate and the interpretation of data from them in the context of organic production is difficult.

Leaching losses of P have been observed on soils with high P concentrations resulting from long-term overfertilisation and/or excessive applications of animal manures where the accumulation of P exceeds the soil's sorption capacity. Where adsorption sites for H_2PO_4 along the walls of transmission pores become saturated, then leaching of P may also become a significant route for loss (Heckrath et al. 1995; Fortune et al. 2005). Soils with high indices (>4) are considered to be at risk of P loss by this route, although the critical index value varies markedly between soil types (McDowell et al. 2001). In the majority of soils, however, losses of P are likely to occur in surface run-off or in subsurface drainage through the transport of P associated with colloidal clay or organic matter (Edwards and Withers 1998). Factors affecting the balance between infiltration and run-off, including biological activity as discussed above, therefore control erosive losses of P. There is little direct information about P leaching and run-off from organic agriculture, but as budgets for organic farms rarely show a significant annual surplus of P, losses are assumed to be small (e.g. Watson et al. 2002). However, overall P losses are likely to be determined more by the cumulative development of P surpluses, the rate of soil erosion and the differences between the dominant loss pathways in livestock and arable systems.

There is no simple relationship between either P or K removed through crop off-take and the change in available P and K levels in soil. In part, this is the result of the ways in which crops themselves influence P and K availability by adaptation of root system morphology (increasing root surface area and/or root hair and mycorrhizal development); changing pH or excreting chelates in the root zone to solubilise minerals; releasing hydrolytic enzymes that enable organic forms of P to be used by the plant; and/or association with mycorrhizal fungi (see Chapters 10 and 11). Consequently, inclusion of crops within the rotation with a high affinity for P may help reduce stratification of P within the soil and increase overall P use efficiency (Braum and Helmke 1995). Likewise, crops that can increase the release of non-exchangeable K, such as cereals and grasses (Kuchenbuch 1987), or access subsoil nutrient reserves may increase P and K supply to following crops through the provision of increased P and K in crop residues. Hence, the crop rotation and the timing of P and/or K application within the rotation may have important effects on P and K availability both for individual crops and the rotation as a whole (see Chapter 5).

An increased understanding of the supply of P and K from the 'less available' pools that are not measured by standard extraction procedures would also be useful. However, it is important that as far as possible, P and K management in organic farming systems leads to balanced P and K budgets.

4.7 Impact of Agricultural Management Practices on Soil Function and Health

A range of largely unmanageable site factors (e.g. climate, slope, some soil factors) constrain land use potential for agriculture in any location. This is discussed further in Chapter 5. Agricultural systems represent integrated collections of structures and practices set within their local environmental constraints. Particular practices are more associated with certain farming systems than others and practices may be: (i) common across conventional and organic farming systems of the same type, such as grazed pastures for dairy production; (ii) similar but with different emphases, such as increasing diversity of leys in organic systems; or (iii) completely distinct, such as weed control using herbicides (Table 4.1). Differences in practices

Table 4.1 A typical range of agricultural practices and the degree to which they are permitted, restricted or prohibited within organic farming systems.

Practice	Permitted/restricted/ prohibited	Comment on occurrence
Tillage	Permitted	Common within arable, horticultural and ley-arable systems. Less common in intensive grass-based systems and restricted on permanent pastures Key tool for seedbed preparation and weed control
Rotation of a variety of crops	Permitted	Diversity in space and time encouraged as good practice Key tool for managing nutrient availability, crop disease risk and weed control
Grass-clover mixtures or other diverse swards	Permitted	Most common in ley-arable systems and in intensive grass-based systems. May be found on in-bye land in upland/extensive systems. Increasingly used as short-term cover crops/green manures with mixtures of grasses, legumes and herbs in arable and horticultural rotations
Incorporation of crop residues	Permitted	Incorporation of crop residues encouraged where tillage is practicable. Where residues are removed for use as livestock bedding, then effective use of the manures within the farm system is encouraged
Lime	Permitted	May be applied if crop nutrition and soil condition cannot be maintained through rotation and recycling on farm (e.g. manures, composts)
Mineral fertiliser	Restricted	Some materials may be applied if crop nutrition and soil condition cannot be maintained through rotation and recycling on-farm (e.g. manures, composts). Materials available for use are restricted, e.g. soluble N/P fertiliser is prohibited
Organic fertiliser	Restricted	May be applied from beyond the farmgate if the need is recognised by the inspection body; cannot exceed 170 kg/N/ha per year. Inputs must be free of genetically modified materials and usually subjected to composting or anaerobic fermentation if derived from non-farm sources
Herbicides	Prohibited	No chemical weed control permitted
Pesticides	Restricted	Use is restricted to a very narrow range of products; these are predominantly used in horticultural systems
Grazing intensity/ stocking rate	Restricted	Livestock systems must be land-linked and the number of animals on any holding must relate to the land area available so that there are no problems caused by overgrazing. Livestock unit equivalents should not exceed 170 kg/N/ha per year

Source: Information taken from European Organic Regulation EC 834/2007.

are likely to lead to quantitatively different characteristics at landscape, farm and field scales, such as higher weed incidence in organic arable and horticulture systems, hedge lengths per unit area of land (Fuller et al. 2005). The range of land management practices used will influence soil processes through direct impacts on organisms (through physiological effects on populations) and through impacts on a range of regulating soil properties.

While the impacts of separate management practices on soil organisms and processes can be distinguished (Stockdale and Watson 2009), at farming systems level, there will be interactions between different management practices and between management practices and the range of largely unmanageable site factors (e.g. climate, slope, some soil factors) at any site. Hence, it is difficult to draw conclusions about the impact of agricultural management practices on soil function and health without taking careful account of the site, overall farming system and its component practices.

4.8   Cropping Systems

Most studies comparing the effects of farming systems on soil function and health have been carried out in arable systems. In these systems, there are large differences between organic and conventional management. All systems use lime to maintain optimum pH (around 6.5). However because of the restriction/prohibition in use of fertiliser, pesticides and herbicides (see Table 4.1), rotations (see Chapter 6) are usually more diverse in organic systems and tillage may be more intensive.

Within fields, farmers choose which crop, which variety and when and where they grow it, hence determining species richness, genetic variability and organisation in space and in time of the crops grown. Crop management is also a key determinant of the associated weed populations. The presence or absence of particular plant species is critical to the survival of strongly root-associated species such as rhizobia, AM fungi (see Chapter 11) and plant pathogens (see Chapter 12), for example, the potato cyst nematode. Hence extended periods where host plants are absent can be used to reduce the populations of associated soil biota below critical levels so that no/limited infection takes place if the host is reintroduced. This strategy is important in the cultivation of many crops, such as brassicas and potatoes, where long rotations are used to manage the occurrence of soil-borne disease. The same mechanism also potentially reduces the effect of positive plant–microbe interactions. Each plant species (and often crop variety) contributes a unique root structure, amount and composition of root exudates and residues to the soil (see Chapter 10). These inputs of OM drive the soil food web. The use of monocultures, or simplified rotations, reduces both above- and below-ground biodiversity (Culman et al. 2010).

The design of both the crop rotation and the farm landscape is critical in contributing to the conservation and enhancement of soil biota (Jackson et al. 2007). For many insect species, a range of habitat types is required during the species' life cycle – loss of any habitat component could critically affect species survival even where the remainder of the habitat is in pristine condition. For soil biota (including crop pests), field margin habitats may provide an important buffer and maintain a source of organisms able to reinvade cropped land following disturbance (Blackshaw and Vernon 2006; Smith et al. 2008).

Diversified cropping systems are also associated with a range of sowing and harvest dates and an increased diversity of cultivation practices implemented at different times. Crop cover and OM inputs are usually increased and tillage type and intensity are diversified. These changes will affect a range of soil properties, including soil structure, which will have complex and interacting effects on the soil biota. Increased plant diversity and diversity of soil biota also mean increased diversity in weed and disease pressures, above-ground consumers and predators.

Wardle (1995) carried out an extensive review of the impacts of disturbance through tillage on food webs in agro-ecosystems. The conclusions of his meta-analysis show that tillage tends to reduce large soil organisms (beetles, spiders, earthworms) more than the smallest ones (bacteria, fungi). On average, some intermediate groups such as bacterial feeding nematodes, mites and enchytraeids even show small population increases. All tillage operations have direct negative impacts on the biomass of macrofauna, with the largest impacts seen for earthworms and beetles (e.g. Postma-Blaauw et al. 2010), usually as a result of exposure at the soil surface and subsequent dessication or predation. For most species groups, the effects on populations are caused by indirect effects arising from modification of soil habitats, particularly the continuity of water-filled pores and water films (Winter et al. 1990). Consequent smaller impacts of changes in tillage practice are often seen on very sandy soils (Spedding et al. 2004). Improvement in infiltration and drainage through changes in soil structure can simply lead to a trade-off between routes of N loss (with an increase in leaching and reduced denitrification) and their associated environmental impacts.

Return of crop residues (in contrast to baling and removal) has been shown in some studies to make a larger contribution to the increase in size of the soil microbial biomass than decreasing intensity of tillage; however, the relative magnitude of these effects is strongly dependent on the soil type (Spedding et al. 2004). Reductions in the microbial population density and diversity have been observed following stubble burning; this was linked to reductions in amount and availability of OM (Rasmussen and Rohde 1988). Increases in soil microbial biomass are commonly measured where residues are incorporated rather than removed or burnt (Powlson et al. 1987). Increased OM inputs require careful management. Leaving crop residues *in situ* may increase the risk of disease and/or pest transmission between crops. Increasing N supply through mineralisation of OM inputs may lead to an increased mismatch of N supply and crop demand increasing risks of N leaching and denitrification. The impacts on greenhouse gas emissions from soils, C sequestration and consequent overall benefits for greenhouse gas balance are less clear and will depend on application conditions and the alternative potential use of the OM input (Powlson et al. 2011).

Where soil tests indicate that nutrient availability is low, fertilising materials are routinely applied to meet plant nutrient demand and maintain a balanced nutrient budget (see Chapter 9). Over the long term, fertiliser applications that increase plant yields have also been shown to increase soil OM and microbial biomass (Murphy et al. 2003; Marschner et al. 2004), particularly where crop residues are returned. Sun et al. (2004) showed that arable plots which had received inorganic fertilisers (including N, P and K together and separately) for over 100 years had developed different microbial communities where inorganic fertiliser rather than no addition or manure is used. Careful nutrient management supports crop yield, quality and soil health. Care needs to be taken that import of fertilisers to a farming system does not bring with it an unwanted contaminant. For example, P fertilisers often contain trace heavy metals (Cd, Hg, Pb)

(McLaughlin et al. 2000) so where P fertilisers have been used regularly, long-term toxicity might arise. However, this is more often a problem with contaminated organic amendments (Giller et al. 1998).

A range of materials is also permitted for use as supplementary fertilising materials to supply P and K, where other fertility management practices have been optimised and these have a wide range of nutrient contents, solubility and other properties. It has been widely shown that there is no inherent reason why organic farming systems or crop rotations should be unsustainable with regard to P or K (Watson et al. 2002; Berry et al. 2003). However, these studies have also highlighted the variability in nutrient use efficiency between organic farms and the scope that those farming systems have to increase this efficiency.

An increase in plant diversity, whether in space or time, is likely to lead to an increase in the species richness of soil biota; in forestry, tree mixtures have been shown to enhance OM decomposition and tree growth (Brown and Dighton 1989). However, to date, increasing species richness in the soil biota (or a component of it) has not been strongly linked to improvement in any soil function or their resilience (Mikola et al. 2002), in part due to high functional redundancy between and within species groups. In agricultural systems, which are typically in non-equilibrium states (plagio-climax communities), it is equally important to determine whether diversity affects how biologically mediated processes respond to further disturbance such as climate change (Stockdale et al. 2006); this is much less well studied.

4.9   Intensive Grassland

Grazing, that is the above-ground defoliation of grass and forb species by herbivores, consumes up to half the annual above-ground net primary productivity. Defoliation has been shown to reduce the amount of root exudation with consequent reductions in the activity of the cultivable soil bacterial population (Macdonald et al. 2004). Other studies have shown increases in exudation following defoliation (e.g. Hamilton and Frank 2001). Other impacts of the livestock, particularly the returns of dung and urine to the soil surface, confound the direct impact of defoliation within grazing management. Supplementary feeding of livestock during the grazing period may also significantly increase inputs of C, N, P and other nutrients to the below-ground ecosystem via excreta. These combined effects therefore mean that grazing affects the amount and quality of C (and other nutrient) input to the soil in quite a complex way (Bardgett et al. 1997) and often increases the size and activity particularly of bacteria in soil. High stocking rates such as typically seen in lowland grassland have a negative impact, which probably arise due to increased compaction, poaching, disruption of sward and an increased proportion of bare ground in overstocked swards.

Veterinary medicines include a variety of nematicides, hormones and antimicrobials, which may impact on below-ground ecology as a result of deposition in grazing excreta or through application of manures. Direct application of antimicrobials and nematicides usually used as veterinary medicines to soil has a negative impact on soil microbial populations and affects below-ground food webs (Westergaard et al. 2001; Jensen et al. 2003; Svendsen et al. 2005). There is some evidence of reduced numbers and activity of dung beetles where veterinary drugs are used regularly (Hutton and Giller 2003), while retarded decomposition rates of dung are likely to have an impact on other species.

Application of livestock manures on-farm provides a way of recycling nutrients and OM temporally and spatially within the farming system. Manures from housed livestock are collected and handled both as solids (farmyard manure, FYM) and/or liquids (slurry); housing design largely determines the forms of manure produced on each farm. Application of livestock manures may be as raw (fresh) materials or following storage and sometimes treatment. Active management of livestock manure as a valuable resource on-farm is gradually replacing the treatment of manures as an inconvenient waste (Smith et al. 2001); this has been driven by the need for compliance with environmental legislation, increasing fertiliser costs for which manures can provide replacement, and the need to reduce both smell nuisance and ammonia volatilisation losses. Currently, the main approaches used on-farm for manure treatment are (i) composting systems or related technologies producing a useful solid product; (ii) biological systems for liquid manures that lead to decomposition of the organic materials; and (iii) separation systems concentrating solids which can then be composted while concentrating available nutrients in the liquid fraction (Martinez et al. 2009).

Communities of soil organisms are not specialised in decomposing the predominant plant litter type, in that organisms habituated to one type will not be less capable of decomposing a different type (St John et al. 2011). However, different quality plant litter, resulting from different types of vegetation, results in different soil conditions, in terms of their acidity, moisture conditions, nutrient availability and other factors. Broad vegetation type, encompassing semi-natural and agricultural vegetation, has been shown to affect the diversity of broad groups of soil invertebrates (Simfukwe et al. 2010) and their response to environmental changes (Emmett et al. 2010). Diversity within communities and diversity of different assemblages of soil bacteria have been shown to be strongly associated with soil pH and vegetation characteristics across a broad range of British ecosystems, suggesting that diversity is strongly influenced by the quality of plant inputs, and the influence these have on conditions (Griffiths et al. 2011).

Decomposition of OM inputs is often associated with increased cycling of nutrients occurring in organic forms (N, P, sulphur, S). OM inputs may also contain nutrients in plant-available forms (N, P, K, and micronutrients), hence stimulating plant growth and further stimulating C inputs through roots, root exudates and residues. Improvements in nutrient supply and soil structure, both resulting directly from the decomposition of OM inputs effected by the soil biota, have direct benefits for crop growth and also may reduce the indirect (fertiliser) and direct (cultivation) energy demands of agricultural systems. Improved soil structural stability has potential benefits for reducing sediment loss and for improving water regulation in agricultural catchments (Posthumus et al. 2011).

4.10 Conclusion

It is often expected that soil health will be improved in organic compared with conventional farming systems under the same climate/soil constraints. However, there has been little evaluation in practice of the management options that are recommended as more sustainable: 'For the most part sustainable agriculture projects assume that the practices they promote will improve sustainability without ever measuring the results to see if this is actually the case' (Holt-Gimenez 2002). Sandhu et al. (2008) calculated the total value of ecosystem services delivered in conventional and organic arable landscapes in New Zealand; while the overall value of ecosystem services delivered by organic systems was higher on average than for conventional systems, there was significant overlap on a field-by-field basis so that some conventional fields deliver more ecosystem services than organic fields. Stockdale (2011) found that while overall yields are reduced in organic farming systems, on average there can be significant improvements in soil health and resilience to weather variability as a result of the adoption of organic farming practices. Even within organic systems which, according to their principles, have a more integrated approach to crop production, often a narrow input substitution approach to soil management is taken (as an example, see the farmer perceptions revealed in Kaltoft 1999). Best practice options for soil management are therefore not restricted to organic farming.

Most of the recommended low-input strategies advocated for all agricultural systems, for example use of cover crops, green manures, animal manure, are already embedded in organic farming systems and lead to increases in soil organic matter levels compared to the levels typical of intensive conventional management approaches, where residues are removed/burnt, tillage is intensive and no inputs of organic materials are made (Matson et al. 1997). In degraded areas, application of composts has been shown to increase crop yields compared with typical field practice and with matched inputs through chemical fertilisers (for example, doubling of yields compared to field with no inputs and small increases compared to use of chemical fertilisers in Tigray, Ethiopia; Edwards 2007). Other minor impacts such as reduced crop pest problems and increased soil moisture retention have also been noted where compost is used (Edwards 2007). Sustainable soil health is associated with regular OM inputs to soil, together with a diversity of crop plans within the system.

Yield resilience is a very important benefit under more marginal conditions for crop growth, especially where irrigation inputs are fixed (or decreasing). Melero et al. (2006) found that increasing soil organic matter through compost application in organic systems increased yields of melon/watermelon crops compared to the comparative conventional system where compost was not applied. However, the range of reduced tillage approaches adopted in many arable farming systems are not widely used in organic farming systems as a result of the need to use tillage-based seedbed preparation approaches for weed management where herbicides are not available.

However, when considering effective soil management strategies for organic farming, it is key to recognise that in organic farming, nutrient management is also dominantly controlled by soil management. Nutrients are dominantly added to the system as organic (manures, compost, crop residues, legumes) or slow-release sources (e.g. rock phosphate). There is a focus on nutrient recycling to minimise reliance on external inputs and crop rotations are designed with regard to the fertility-building and -depleting role of the crops included. However, there is no strong evidence that the fundamental nutrient cycling processes in organically farmed soils differ significantly from those in soils from conventional farms although due to the differences in management practices, their relative importance and rates may differ (Stockdale et al. 2002). Most materials incorporated into the soil in organic systems do not contain readily soluble nutrients (K is an exception) and hence a greater reliance is placed on chemical and biological processes to release nutrients in plant-available forms in soil solution. While soil fertility is not fundamentally different in organic farming systems in terms of processes, this difference in emphasis is important for both research and practical management.

Low-input strategies require intensive use of information to design effective rotations and management strategies that ensure synchronisation of nutrient release with plant demand given the wider quality of nutrient sources available; productive and sustainable low-input systems require high levels of farmer knowledge. The decisions of land managers, driven by their underlying values, perceptions and level of understanding, are often the critical factor determining soil health.

References

Anderson, J.M. (1975). The enigma of soil animal species diversity. In: *Progress in Zoology* (ed. J. Vanek), 51–58. Prague: Academia.

Angers, D.A. and Caron, J. (1998). Plant-induced changes in soil structure: processes and feedbacks. *Biogeochemistry* 42: 55–72.

Angers, D.A. and Carter, M.R. (1996). Aggregation and organic matter storage in cool, humid agricultural soils. In: *Structure and Organic Matter Storage in Agricultural Soils* (ed. M.R. Carter and B.A. Stewart), 193–211. Boca Raton: CRC Press.

Arah, J.R.M. and Vinten, A.J.A. (1995). Simplified models of anoxia and denitrification in aggregated and simple-structured soils. *European Journal of Soil Science* 46: 507–517.

Arden-Clarke, C. and Hodges, R.D. (1988). The environmental effects of conventional and organic/biological farming systems. II. Soil ecology, soil fertility and nutrient cycles. *Biological Agriculture & Horticulture* 5: 223–287.

Balfour, E. (1943). *The Living Soil and the Haughley Experiment*. London: Faber and Faber.

Barber, S.A. (1995). *Soil Nutrient Bioavailability: A Mechanistic Approach*, 2nd edition. New York: Wiley.

Bardgett, R.D. (2002). Causes and consequences of biological diversity in soil. *Zoology* 105: 367–374.

Bardgett, R.D., Leemans, D.K., Cook, R., and Hobbs, P.J. (1997). Seasonality of the soil biota of grazed and ungrazed hill grasslands. *Soil Biology and Biochemistry* 29: 1285–1294.

Barns, S.M., Takala, S.L., and Kuske, C.R. (1999). Wide distribution and diversity of members of the bacterial kingdom Acidobacterium in the environment. *Applied and Environmental Microbiology* 65: 1731–1737.

Bates, T.E. and Richards, J.E. (1993). Available potassium. In: *Soil Sampling and Methods of Analysis* (ed. R.L. Carter), 59–64. Boca Raton: CRC Press.

Bear FE, Prince AL, Malcolm JL (1945) The potassium needs of New Jersey soils. New Jersey Agriculture Experimental Station Bulletin 721.

Beare, M.H., Coleman, D.C., Crossley, D. et al. (1995). A hierarchical approach to evaluating the significance of soil biodiversity to biogeochemical cycling. *Plant and Soil* 170: 5–22.

Beare, M.H., Parmelee, R.W., Hendrix, P.F. et al. (1992). Microbial and faunal interactions and effects on litter nitrogen and decomposition in agroecosystems. *Ecological Monographs* 62: 569–591.

Berry, P.M., Stockdale, E.A., Sylvester-Bradley, R. et al. (2003). N, P and K budgets for crop rotations on nine organic farms in the UK. *Soil Use and Management* 19: 112–118.

Berry, P.M., Sylvester-Bradley, R., Philipps, L. et al. (2002). Is productivity of organic farms restricted by the supply of available nitrogen? *Soil Use and Management* 18: 248–255.

Blackshaw, R.P. and Vernon, R.S. (2006). Spatio-temporal stability of two beetle populations in non-farmed habitats in an agricultural landscape. *Journal of Applied Ecology* 43: 680–689.

Bloem, J., Lebbink, G., Zwart, K.B. et al. (1994). Dynamics of microorganisms, microbiovores and nitrogen mineralisation under conventional and integrated management. *Agriculture, Ecosystems and Environment* 51: 129–143.

Bongers, T. and Bongers, M. (1998). Functional diversity of nematodes. *Applied Soil Ecology* 10: 239–251.

Borneman, J., Skroch, P.W., O'Sullivan, K.M. et al. (1996). Molecular microbial diversity of an agricultural soil in Wisconsin. *Applied and Environmental Microbiology* 62: 1935–1943.

Braum, S.M. and Helmke, P.A. (1995). White lupin utilises soil phosphorus unavailable to soybean. *Plant and Soil* 176: 95–100.

Brown, A.H.F. and Dighton, J. (1989). Mixtures and mycorrhizas: the manipulation of nutrient cycling in forestry. In: *Cumbrian Woodlands – Past, Present and Future*. ITE Symposium (ed. J.K. Adamson), 65–72. London: HMSO.

Brussaard, L. (1998). Soil fauna, guilds, functional groups and ecosystem processes. *Applied Soil Ecology* 9: 123–135.

Brussaard, L., Bakker, J.-P., and Ollf, H. (1996). Biodiversity of soil biota and plants in abandoned arable fields and grasslands under restoration management. *Biodiversity and Conservation* 5: 211–221.

Brussaard, L., Behan-Pelletier, V.M., Bignell, D.E. et al. (1997). Biodiversity and ecosystem functioning in soil. *Ambio* 26: 563–570.

Buckley, D.H., Graber, J.R., and Schmidt, T.M. (1998). Phylogenetic analysis of nonthermophilic members of the kingdom *Crenarchaeota* and their diversity and abundance in soil. *Applied Environmental Microbiology* 64: 4333–4339.

Chater M, Mattingly GEG (1980) Changes in organic phosphorus contents of soils from long continued experiments at Rothamsted and Saxmundham. Rothamsted Experimental Station Report 1979, Part 2, pp. 41–61.

Cooke, G.W. (1958). The nation's plant food larder. *Journal of the Science of Food and Agriculture* 9: 761–772.

Crossley, D.A. Jr., Muller, B.R., and Perdue, J.C. (1992). Biodiversity of microarthropods in agricultural soils: relations to processes. *Agriculture Ecosystems and Environment* 40: 37–46.

Culman, S.W., DuPont, S.T., Glover, J.D. et al. (2010). Long-term impacts of high-input annual cropping and unfertilized perennial grass production on soil properties and belowground food webs in Kansas, USA. *Agriculture Ecosystems and Environment* 137: 13–24.

Dalias, P., Kokkoris, G.D., and Troumbis, A.Y. (2003). Functional shift hypothesis and the relationship between temperature and soil carbon accumulation. *Biology and Fertility of Soils* 37: 90–95.

Davidson, D.A., Bruneau, P.M.C., Grieve, I.C., and Young, I.M. (2002). Impacts of fauna on an upland grassland soil as determined by micromorphological analysis. *Applied Soil Ecology* 20: 133–143.

de Boer, W., Folman, L.B., Summerbell, R.C., and Boddy, L. (2005). Living in a fungal world: impact of fungi on soil bacterial niche development. *FEMS Microbiology Reviews* 29: 795–811.

de Ruiter, P.C., Moore, J.C. et al. (1993). Simulation of nitrogen mineralization in the belowground food webs of 2 winter-wheat fields. *Journal of Applied Ecology* 30: 95–106.

Dick, W.A. and Gregorich, E.G. (2004). Developing and maintaining soil organic matter levels. In: *Managing Soil Quality: Challenges in Modern Agriculture* (ed. P. Schjønning, S. Elmholt and B.T. Christensen), 103–120. Wallingford: CABI Publishing.

Didden, W.A.M. (1993). Ecology of terrestrial Enchytraeidae. *Pedobiologica* 37: 2–29.

Duffy, B.K., Ownely, B.H., and Weller, D.M. (1997). Soil chemical and physical properties associated with suppression of take-all of wheat by *Trichodema koningii*. *Phytopathology* 87: 1118–1124.

Eckert, D.J. (1987). Soil test interpretations: basic cation saturation ratios and sufficiency levels. In: *Soil Testing: Sampling, Correlation, Calibration, and Interpretation*. Soil Science Society of America Special Publication 21 (ed. J.R. Brown), 53–64. Madison: SSSA.

Edwards, S. (2007). Role of organic agriculture in preventing and reversing land degradation. In: *Climate and Land Degradation*, Environmental Science and Engineering Book Series (ed. S. MVK and N. Ndiangui), 523–536. Berlin: Springer.

Edwards, A.C. and Withers, P.J.A. (1998). Soil phosphorus management and water quality: a UK perspective. *Soil Use and Management* 14: 123–130.

Emmett BA, Reynolds B, Chamberlain PM, et al. (2010) Soils Report from 2007. CS Technical Report No. 9/07. Centre for Ecology & Hydrology, Natural Environment Research Council, Bailrigg.

Ettema, C.H. and Wardle, D.A. (2002). Spatial soil ecology. *Trends in Ecology and Evolution* 17: 177–183.

Fitter, A.H. (2005). Darkness visible: reflections on underground ecology. *Journal of Ecology* 93: 231–243.

Fitter, A.H., Gilligan, C.A., Hollingworth, K. et al. (2005). Biodiversity and ecosystem function in soil. *Functional Ecology* 19: 369–377.

Foissner, W. (1997). Protozoa as bioindicators in agroecosystems, with emphasis on farming practices, biocides, and biodiversity. *Agriculture Ecosystems and Environment* 62: 93–103.

Food and Agriculture Organization of the United Nations (FAO) (2015) Available at: http://www.fao.org/3/a-ax374e.pdf.

Fortune, S., Hollies, J., and Stockdale, E.A. (2004). Effects of different potassium fertilizers suitable for use in organic farming systems on grass/clover yields and nutrient offtakes and interactions with nitrogen supply. *Soil Use and Management* 20: 403–409.

Fortune, S., Robinson, J.S., Watson, C.A. et al. (2005). Response of organically managed grassland to available phosphorus and potassium in the soil and supplementary fertilization: field trials using grass-clover leys cut for silage. *Soil Use and Management* 21: 370–376.

Fuller, R.J., Norton, L.R., Feber, R.E. et al. (2005). Benefits of organic farming to biodiversity vary among taxa. *Biology Letters* 1: 431–434.

Gasser JKR (1962) Mineralisation of nitrogen, sulphur and phosphorus from soils. Welsh Soils Discussion Group Report No. 3, p. 26.

Giller, K.E., Witter, E., and McGrath, S.P. (1998). Toxicity of heavy metals to microorganisms and microbial processes in agricultural soils: a review. *Soil Biology and Biochemistry* 30: 1389–1414.

Gosling, P., Hodge, A., Goodlass, G., and Bending, G.D. (2006). Arbuscular mycorrhizal fungi and organic farming. *Agriculture, Ecosystems and Environment* 113: 17–35.

Goulding, K.W.T. and Loveland, P. (1987). The classification and mapping of potassium reserves in soils of England and Wales. *Journal of Soil Science* 37: 555–565.

Greenland, D.J. (1977). Soil damage by intensive arable cultivation – temporary or permanent. *Philosophical Transactions of the Royal Society Series B* 281: 193–208.

Gressel, N. and McColl, J.G. (1997). Phosphorus mineralization and organic matter decomposition: a critical review. In: *Driven by Nature: Plant Litter Quality and Decomposition* (ed. G. Cadisch and K.E. Giller), 297–309. Wallingford: CAB International.

Griffiths, R.I., Thompson, B.C., James, P. et al. (2011). The bacterial biogeography of British soils. *Environmental Microbiology* 13: 1642–1654.

Gunnarsson, S. and Marstorp, H. (2002). Carbohydrate composition of plant materials determines N mineralisation. *Nutrient Cycling in Agroecosystems* 62: 175–183.

Hamilton, E.W. and Frank, D.A. (2001). Can plants stimulate soil microbes and their own nutrient supply? Evidence from a grazing tolerant grass. *Ecology* 82: 2397–2402.

Harrier, L.A. and Watson, C.A. (2003). The role of arbuscular mycorrhizal fungi in sustainable cropping systems. *Advances in Agronomy* 79: 185–225.

Hart, S.C., Nason, G.E., Myrold, D.D., and Perry, D.A. (1994). Dynamics of gross nitrogen transformations in an old-growth forest: the carbon connection. *Ecology* 75: 880–891.

Hassink, J., Bouwman, L.A., Zwart, K.B. et al. (1993). Relationships between soil texture, physical protection of organic-matter, soil biota, and C-mineralization and N-mineralization in grassland soils. *Geoderma* 57: 105–128.

Heckrath, G., Brookes, P.C., Poulton, P.R., and Goulding, K.W.T. (1995). Phosphorus leaching from soils containing different phosphorus concentrations in the Broadbalk experiment. *Journal of Environmental Quality* 24: 904–910.

Henriksen, T.M. and Breland, T.A. (1999). Decomposition of crop residues in the field: evaluation of a simulation model developed from microcosm studies. *Soil Biology and Biochemistry* 31: 1423–1434.

Hogben CD (1983) Some effects of orchard soil management with herbicides on earthworms. PhD Thesis, University of Lancaster, UK.

Holt-Gimenez, E. (2002). Measuring farmers' agroecological resistance after Hurricane Mitch in Nicaragua: a case study in participatory, sustainable land management impact monitoring. *Agriculture, Ecosystems and Environment* 93: 87–105.

Hunt, H.W., Coleman, D.C., Ingham, R.E. et al. (1987). The detrital food web in a shortgrass prairie. *Biology and Fertility of Soils* 3: 57–68.

Hutton, S.A. and Giller, P.S. (2003). The effects of the intensification of agriculture on northern temperate dung beetle communities. *Journal of Applied Ecology* 40: 994–1007.

Ingram, J.S.I. and Fernandes, E.C.M. (2001). Managing carbon sequestration in soils: concepts and terminology. *Agriculture, Ecosystems & Environment* 87: 111–117.

Jacinthe, P.A., Lal, R., and Kimble, J.M. (2002). Carbon budget and seasonal carbon dioxide emission from a central Ohio Luvisol as influenced by wheat residue amendment. *Soil & Tillage Research* 67: 147–157.

Jackson, L.E., Pascual, U., and Hodgkin, T. (2007). Utilizing and conserving agrobiodiversity in agricultural landscapes. *Agriculture, Ecosystems and Environment* 121: 196–210.

Jarvis, S.C., Stockdale, E.A., Shepherd, M.A., and Powlson, D.S. (1996). Nitrogen mineralization in temperate agricultural soils: processes and measurement. *Advances in Agronomy* 57: 187–235.

Jenkinson, D.S. (2001). The impact of humans on the nitrogen cycle, with focus on temperate arable agriculture. *Plant and Soil* 228: 3–15.

Jensen, J., Krogh, P.H., and Sverdrup, L.E. (2003). Effects of the antibacterial agents tiamulin, olanquindox and metronidazole and the anthelmintic ivermectin on the soil invertebrate species *Folsomia fimetaria* (Collembola) and *Enchytraeus crypticus* (Enchytraeidae). *Chemosphere* 50: 437–443.

Johnston, A.E., Poulton, P.R., and Syers, J.K. (2001). *Phosphorus, Potassium and Sulphur Cycles in Agricultural Soils*, Proceedings 465. York: International Fertiliser Society.

Jones, D.L., Hodge, A., and Kuzyakov, Y. (2004). Plant and mycorrhizal regulation of rhizodeposition. *New Phytologist* 163: 459–480.

Jones, C.G., Lawton, J.H., and Shachak, M. (1994). Organisms as ecosystem engineers. *Oikos* 69: 373–386.

Kaltoft, P. (1999). Values about nature in organic farming practice and knowledge. *Sociologia Ruralis* 39: 39–53.

Keatinge R (1997) Organic sheep and beef production in the uplands. ADAS Site Report. MAFF Project OF011. ADAS Redesdale Rochester, Newcastle upon Tyne.

Killham, K. (1994). *Soil Ecology*. Cambridge: Cambridge University Press.

Kopittke, P.M. and Menzies, N.W. (2007). A review of the use of the basic cation saturation ratio and the 'ideal' soil. *Soil Science Society of America Journal* 71: 259–265.

Korsaeth, A., Henriksen, T.M., and Bakken, L.R. (2002). Temporal changes in mineralisation and immobilisation of N during degradation of plant material: implications for the plant N supply and losses. *Soil Biology and Biochemistry* 34: 789–799.

Kuchenbuch RO (1987) Potassium dynamics in the rhizosphere and potassium availability. Proceedings of the 20th Colloquium, International Potash Institute, Bern, Switzerland, pp. 215–234.

Lampkin, N. (1990). *Organic Farming*. Ipswich: Farming Press.

Larsen, S. (1967). Soil phosphorus. *Advances in Agronomy* 19: 151–210.

Larson, W.E. and Pierce, F.J. (1994). The dynamics of soil quality as a measure of sustainable management. In: *Defining Soil Quality for a Sustainable Environment*. Soil Science Society of America Special Publication 35 (ed. J.W. Doran, D.C. Coleman, D.F. Bezdicek and B.A. Stewart), 37–51. Madison: SSSA.

Lavelle, P. (2000). Ecological challenges for soil science. *Soil Science* 165: 73–86.

Lavelle, P., Barros, E., Blanchart, E. et al. (2001). SOM management in the tropics: why feeding the soil macrofauna? *Nutrient Cycling in Agroecosystems* 61: 53–61.

Leake, J.R., Johnson, D., Donnelly, D.P. et al. (2004). Networks of power and influence: the role of mycorrhizal mycelium in controlling plant communities and agroecosystem functioning. *Canadian Journal of Botany* 82: 1016–1045.

Ledgard, S.F. (2001). Nitrogen cycling in low input legume-based agriculture, with emphasis on legume/grass pastures. *Plant and Soil* 228: 43–59.

Loveland, P. and Webb, J. (2003). Is there a critical level of organic matter in the agricultural soils of temperate regions: a review. *Soil & Tillage Research* 70: 1–18.

Macdonald, L.M., Paterson, E., Dawson, L.A., and McDonald, A.J.S. (2004). Short-term effects of defoliation on the soil microbial community associated with two contrasting Lolium perenne cultivars. *Soil Biology and Biochemistry* 36: 489–498.

Marschner, P., Crowley, D., and Yang, C.H. (2004). Development of specific rhizosphere bacterial communities in relation to plant species, nutrition and soil type. *Plant and Soil* 261: 199–208.

Martinez, J., Dabert, P., Barrington, S., and Burton, C. (2009). Livestock waste treatment systems for environmental quality, food safety, and sustainability. *Bioresource Technology* 100: 5527–5536.

Matson, P.A., Parton, W.J., Power, A.G., and Swift, M.J. (1997). Agricultural intensification and ecosystem properties. *Science* 277: 504–509.

Mazzola, M. (2002). Mechanisms of natural soil suppressiveness to soil borne diseases. *Antonie van Leeuwenhoek International Journal of General and Molecular Microbiology* 81: 557–564.

McDowell, R., Sharpley, A., Brookes, P., and Poulton, P. (2001). Relationship between soil test phosphorus and phosphorus release to solution. *Soil Science* 166: 137–147.

McLaughlin, M.J., Hamon, R.E., McLaren, R.G. et al. (2000). Review: a bioavailability based rationale for controlling metal and metalloid contamination of agricultural land in Australia and New Zealand. *Australian Journal of Soil Research* 38: 1037–1086.

McLean, E.O. and Watson, M.E. (1985). Soil measurements of plant-available potassium. In: *Potassium in Agriculture* (ed. R.D. Munson), 277–308. Madison: SSSA.

Meidute, S., Demoling, F., and Bååth, E. (2008). Antagonistic and synergistic effects of fungal and bacterial growth in soil after adding different carbon and nitrogen sources. *Soil Biology and Biochemistry* 40: 2334–2343.

Melero, S., Porras, J.C.R., Herencia, J.F., and Madejon, E. (2006). Chemical and biochemical properties in a silty loam soil under conventional and organic management. *Soil & Tillage Research* 90: 162–170.

Menzies, J.D. (1959). Occurrence and transfer of a biological factor in soil that suppresses potato scab. *Phytopathology* 49: 648–652.

Mikola, J., Bardgett, R.D., and Hedlund, K. (2002). Biodiversity, ecosystem functioning and soil decomposer food webs. In: *Biodiversity and Ecosystem Functioning: Synthesis and Perspectives* (ed. M. Loreau, S. Naeem and P. Inchausti), 169–180. Oxford: Oxford University Press.

Miltner, A., Richnow, H.H., and Kopinke, F.D. (2005). Incorporation of carbon originating from CO_2 into different compounds of soil microbial biomass and soil organic matter. *Isotopes in Environmental and Health Studies* 41: 135–140.

Mueller, T., Jensen, L.S., Magid, J., and Nielsen, N.E. (1997). Temporal variation of C and N turnover in soil after oilseed rape straw incorporation in the field: simulations with the soil-plant-atmosphere model DAISY. *Ecological Modelling* 99: 247–262.

Mulder, C., de Zwart, D., van Wijnen, H.J. et al. (2003). Observational and simulated evidence of ecological shifts within the soil nematode community of agroecosystems under conventional and organic farming. *Functional Ecology* 17: 516–525.

Mulder, C., Dijkstra, J.B., and Setälä, H. (2005). Nonparasitic nematoda provide evidence for a linear response of functionally important soil biota to increasing livestock density. *Naturwissenschaften* 92: 314–318.

Murphy, D.V., Macdonald, A.J., Stockdale, E.A. et al. (2000). Soluble organic nitrogen in agricultural soils. *Biology and Fertility of Soils* 30: 374–387.

Murphy, D.V., Stockdale, E.A., Brookes, P.C., and Goulding, K.W.T. (2003). Impact of micro-organisms on chemical transformations in soil. In: *Soil Biological Fertility – A Key to Sustainable Land Use in Agriculture* (ed. L.K. Abbott and D.V. Murphy), 139–152. Dordrecht: Kluwer Academic.

Murray, P.J. and Hatch, D.J. (1994). Sitona Weevils (Coleoptera, Curculionidae) as agents for rapid transfer of nitrogen from white clover (*Trifolium-Repens* L) to perennial ryegrass (*Lolium-Perenne* L). *Annals of Applied Biology* 125: 29–33.

Nashölm, T., Huss-Danell, K., and Hogberg, P. (2000). Uptake of organic nitrogen in the field by four agriculturally important plant species. *Ecology* 81: 1155–1161.

Nicol, G.W., Glover, L.A., and Prosser, J.I. (2003). The impact of grassland management on archaeal community structure in upland pasture rhizosphere soil. *Environmental Microbiology* 5: 152–162.

Nishio, T. and Fujimoto, T. (1989). Mineralisation of soil organic nitrogen in upland fields as determined by a ^{15}N-NH$_4^+$ isotope dilution technique and absorption of nitrogen by maize. *Soil Biology and Biochemistry* 21: 661–665.

Oberson, A., Besson, J.M., Maire, N., and Sticher, H. (1996). Microbiological transformations in soil organic phosphorus transformations in conventional and biological cropping systems. *Biology and Fertility of Soils* 21: 138–148.

Odum, E.P. (1969). The strategy of ecosystem development. *Science* 164: 262–270.

Pankhurst, C., Doube, B.M., and Gupta, V.V.S.R. (1997). *Biological Indicators of Soil Health*. Wallingford: CAB International.

Paul, E.A. and Clark, F.E. (1996). *Soil Microbiology and Biochemistry*, 2nde. San Diego: Academic Press.

Pennington, P.I. and Ellis, R.C. (1993). Autotrophic and heterotrophic nitrification in acidic forest and native grassland soils. *Soil Biology & Biochemistry* 25: 1399–1408.

Posthumus, H., Deeks, L.K., Fenn, I., and Rickson, R.J. (2011). Soil conservation in two English catchments: linking soil management with policies. *Land Degradation and Development* 22: 97–110.

Postma-Blaauw, M.B., de Goede, R.G.M., Bloem, J. et al. (2010). Soil biota community structure and abundance under agricultural intensification and extensification. *Ecology* 91: 460–473.

Powlson, D.S. (1980). The effects of grinding on microbial and non-microbial organic matter. *Journal of Soil Science* 31: 77–85.

Powlson, D.S., Brookes, P.C., and Christensen, B.T. (1987). Measurement of soil microbial biomass provides an early indication of changes in total soil organic matter due to straw incorporation. *Soil Biology and Biochemistry* 19: 159–164.

Powlson, D.S., Whitmore, A.P., and Goulding, K.W.T. (2011). Soil carbon sequestration to mitigate climate change: a critical re-examination to identify the true and the false. *European Journal of Soil Science* 62: 42–55.

Rasmussen, P.E. and Rohde, C.R. (1988). Stubble burning effects on winter-wheat yield and nitrogen-utilization under semiarid conditions. *Agronomy Journal* 80: 940–942.

Recous, S., Aita, C., and Mary, B. (1999). In situ transformations in bare soil after addition of straw. *Soil Biology and Biochemistry* 31: 119–133.

Recous, S., Mary, B., and Faurie, G. (1990). Microbial assimilation of ammonium and nitrate in soil. *Soil Biology and Biochemistry* 22: 597–602.

Reed, S.C., Cleveland, C.C., and Townsend, A.R. (2011). Functional ecology of free-living nitrogen fixation: a contemporary perspective. *Annual Review of Ecology, Evolution and Systematics* 42: 489–512.

Römbke, J.R., Höfer, H., Garcia, M.V.B., and Martius, C. (2006). Feeding activities of soil organisms at four different forest sites in Central Amazonia using the bait lamina method. *Journal of Tropical Ecology* 22: 313–320.

Sandhu, H.S., Wratten, S.D., Cullen, R., and Case, B. (2008). The future of farming: the value of ecosystem services in conventional and organic arable land. An experimental approach. *Ecological Economics* 64: 835–848.

Schimel, D.S. (1986). Carbon and nitrogen turnover in adjacent grassland and cropland ecosystems. *Biogeochemistry* 2: 345–357.

Schlüter, S., Weller, U., and Vogel, H.-J. (2011). Soil structure development including seasonal dynamics in a long-term fertilization experiment. *Journal of Plant Nutrition and Soil Science* 174: 395–403.

Seastedt, T.R. (1984). The role of microarthropods in decomposition and mineralisation processes. *Annual Review of Entomology* 29: 25–46.

Sharpley, A. (2000). Phosphorus availability. In: *Handbook of Soil Science* (ed. M.E. Sumner), D18–D37. Boca Raton: CRC Press.

Siepel, H. (1995). Applications of microarthropod life history tactics in nature management and ecotoxicology. *Biology and Fertility of Soils* 19: 75–83.

Simfukwe P, Griffiths RI, Emmett BA, et al. (2010) Prediction and inter-dependence of soil quality, function and diversity at a national scale. DEFRA Project SP1602, Centre for Ecology and Hydrology and Bangor University.

Six, J., Conant, R.T., Paul, E.A., and Paustian, K. (2002). Stabilization mechanisms of soil organic matter: implications for C-saturation of soils. *Plant and Soil* 241: 155–176.

Smeck, N.E. (1985). Phosphorus dynamics in soils and landscapes. *Geoderma* 36: 185–199.

Smit, E., Lefflang, P., Gommans, S. et al. (2001). Diversity and seasonal fluctuations of the dominant members of the bacterial soil community in a wheat field as determined by cultivation and molecular methods. *Applied and Environmental Microbiology* 67: 2284–2291.

Smith, K.A., Brewer, A.J., Crabb, J., and Dauven, A. (2001). A survey of the production and use of animal manures in England and Wales. III. Cattle manures. *Soil Use and Management* 17: 77–87.

Smith, J., Potts, S.G., Woodcock, B.A., and Eggleton, P. (2008). Can arable field margins be managed to enhance their biodiversity, conservation and functional value for soil macrofauna? *Journal of Applied Ecology* 45: 269–278.

Sparks, D.L. (2000). Bioavailability of potassium. In: *Handbook of Soil Science* (ed. M.E. Sumner), D38–D53. Boca Raton: CRC Press.

Sparks, D.L. and Huang, P.M. (1985). Physical chemistry of soil potassium. In: *Potassium in Agriculture* (ed. R.D. Munson), 201–276. Madison: SSSA.

Spedding, T.A., Hamel, C., Mehuys, G.R., and Madramootoo, C.A. (2004). Soil microbial dynamics in maize-growing soil under different tillage and residue management systems. *Soil Biology and Biochemistry* 36: 499–512.

Spiers, G.A. and McGill, W.B. (1979). Effects of phosphorus addition and energy supply on acid phosphatase production and activity in soils. *Soil Biology and Biochemistry* 11: 3–8.

St John, M.G., Orwin, K.H., and Dickie, I.A. (2011). No 'home' versus 'away' effects of decomposition found in a grassland-forest reciprocal litter transplant study. *Soil Biology and Biochemistry* 43: 1482–1489.

Stevens, R.J., Laughlin, R.J., Burns, L.C. et al. (1997). Measuring the contribution of nitrification and denitrification to the flux of nitrous oxide from soil. *Soil Biology and Biochemistry* 29: 139–151.

Stockdale, E.A. (2011). Organic farming: pros and cons for soil health and climate change. In: *Soil Health and Climate Change*. Soil Biology 29 (ed. B.P. Singh, A.L. Cowie and K.Y. Chan), 317–343. Berlin: Springer-Verlag.

Stockdale, E.A. and Brookes, P.C. (2006). Detection and quantification of the soil microbial biomass – impacts on the management of agricultural soils. *Journal of Agricultural Science* 144: 285–302.

Stockdale, E.A. and Watson, C.A. (2009). Biological indicators of soil quality in organic farming systems. *Renewable Agriculture and Food Systems* 24: 308–318.

Stockdale, E.A., Goulding, K.W.T., George, T.S., and Murphy, D.V. (2013). Soil fertility. In: *Soil Conditions and Plant Growth* (ed. P.J. Gregory and S. Nortcliff), 49–85. Oxford: Blackwell.

Stockdale, E.A., Shepherd, M.A., Fortune, S., and Cuttle, S.P. (2002). Soil fertility in organic farming systems – fundamentally different? *Soil Use and Management* 18: 301–308.

Stockdale EA, Watson CA, Black HIJ, Philipps L (2006) Do farm management practices alter below-ground biodiversity and ecosystem function? Implications for sustainable land management. JNCC Report No. 364. Peterborough: Joint Nature Conservation Committee.

Strickland, M. and Rousk, J. (2010). Considering fungal: bacterial dominance in soils – methods, controls, and ecosystem implications. *Soil Biology and Biochemistry* 42: 1385–1395.

Sun, H.Y., Deng, S.P., and Raun, W.R. (2004). Bacterial community structure and diversity in a century-old manure-treated agroecosystem. *Applied and Environmental Microbiology* 70: 5868–5874.

Svendsen, T.S., Hansen, P.E., Sommer, C. et al. (2005). Life history characteristics of *Lumbricus terrestris* and effects of the veterinary antiparasitic compounds ivermectin and fenbendazole. *Soil Biology and Biochemistry* 37: 927–936.

Syers, J.K. (1998). *Soil and plant potassium in agriculture*, Proceedings No. 411. York: International Fertiliser Society.

Tisdall, J.M. and Oades, J.M. (1982). Organic matter and water stable aggregates in soils. *Journal of Soil Science* 33: 141–163.

Torsvik, V., Øvreås, L., and Thingstad, T.F. (2002). Prokaryotic diversity – magnitude, dynamics, and controlling factors. *Science* 296: 1064–1066.

Tunney, H., Breeuwsma, A., Withers, P.J.A., and Ehlert, P.A.I. (1997). Phosphorus fertilizer strategies: present and future. In: *Phosphorus Loss from Soil to Water* (ed. H. Tunney, O.T. Carton, P.C. Brookes and A.E. Johnston), 177–203. Wallingford: CAB International.

van der Putten, W.H., Bardgett, R.D., de Ruiter, P.C. et al. (2009). Empirical and theoretical challenges in aboveground-belowground ecology. *Oecologia* 161: 1–14.

Van Faassen, H.G. and van Dijk, H. (1987). Manure as a source of nitrogen and phosphorus in soils. In: *Animal Manure on Grassland and Fodder Crops* (ed. H.E. van der Meer, R.J. Unwin, T.A. van Dijk and G.C. Ennik), 27–45. Dordrecht: Martinus Nijhoff.

Van Noordwijk, M., de Ruiter, P.C., Zwart, K.B. et al. (1993). Synlocation of biological activity, roots, cracks and recent organic inputs in a sugar beet field. *Geoderma* 56: 265–276.

Verheijen, F.G.A., Bellamy, P.H., Kibblewhite, M.G., and Gaunt, J.L. (2005). Organic carbon ranges in arable soils of England and Wales. *Soil Use and Management* 21: 2–9.

Wardle, D.A. (1992). A comparative assessment of factors which influence microbial biomass carbon and nitrogen levels in soil. *Biological Reviews of the Cambridge Philosophical Society* 67: 321–358.

Wardle, D.A. (1995). Impacts of disturbance on detritus food webs in agro-ecosystems of contrasting tillage and weed management practices. *Advances in Ecological Research* 26: 105–185.

Wardle, D.A. (2002). *Communities and Ecosystems: Linking the Aboveground and Below Ground Components*. Monographs in Population Biology 34. Princeton: Princeton University Press.

Wardle, D.A. and Giller, K.E. (1996). The quest for a contemporary ecological dimension to soil biology – discussion. *Soil Biology and Biochemistry* 28: 1549–1554.

Watson, C.A., Bengtsson, H., Ebbesvik, M. et al. (2002). A review of farm-scale nutrient budgets for organic farms as a tool for management of soil fertility. *Soil Use and Management* 18: 264–273.

Weinberg, G.M. (1975). *Introduction to General Systems Thinking*. New York: Wiley.

Westergaard, K., Muller, A.K., Christensen, S. et al. (2001). Effects of tylosin as a disturbance on the soil microbial community. *Soil Biology and Biochemistry* 33: 2061–2071.

Wild, A. (1988). Plant nutrients in soil:phosphate. In: *Russell's Soil Conditions and Plant Growth*, 11th edition (ed. A. Wild), 695–742. Harlow: Longman.

Williams, S.E., Marsh, H., and Winter, J. (2002). Spatial scale, species diversity and habitat structure: small mammals in Australian tropical rainforest. *Ecology* 83: 1317–1329.

Winter, J.P., Voroney, R.P., and Ainsworth, D.A. (1990). Soil microarthropods in long-term no-tillage and conventional tillage corn production. *Canadian Journal of Soil Science* 70: 641–653.

Wolters, V. (2000). Invertebrate control of soil organic matter stability. *Biology and Fertility of Soils* 31: 1–19.

Wu, J. and David, J.L. (2002). A spatially explicit hierarchical approach to modeling complex ecological systems: theory and applications. *Ecological Modelling* 153: 7–26.

Young, I.M. and Crawford, J.W. (2004). Interactions and self-organisation in the soil- microbe complex. *Science* 304: 1634–1637.

Young, I.M. and Ritz, K. (1998). Can there be a contemporary ecological dimension to soil biology without a habitat? *Soil Biology and Biochemistry* 30: 1229–1232.

Young, I.M. and Ritz, K. (2000). Tillage, habitat space and function of soil microbes. *Soil & Tillage Research* 53: 201–213.

Young, I.M., Roberts, A., Griffiths, B.S., and Caul, S. (1994). Growth of a ciliate protozoan in model ballotini systems of different particle sizes. *Soil Biology and Biochemistry* 26: 1173–1178.

5

Cropping Systems and Crop Choice

Robin L. Walker

SRUC, Aberdeen, Scotland, UK

5.1 Farming Systems

A clear definition of what is an agricultural or farming production system is both difficult to achieve and technical in nature. When looking at the cropping elements of systems, it is often easier to understand how everything links together by breaking these down into different components, enterprises and activities. There are a number of farming systems which are defined under government rules based around the proportion of different enterprises contributing to the farm's overall income. Some of the classifications suit organic production approaches and others are less appropriate.

The farming type is a way of categorising how a particular piece of land is used to generate income. In Scotland, farm type is determined by the relative contribution of each holding's product mix or activity to its standard gross margin (SGM). If a particular activity accounts for two-thirds of the holding's total SGM then the holding is assigned to that specific farm type. There are 10 farm types under the system defined by RESAS (2011) for Scotland (cereals, general cropping, horticulture, specialist pigs, specialist poultry, dairy, LFA cattle and sheep, lowland cattle and sheep, mixed and other). So to be identified as a cereals holding, for example, cereal crops must account for two-thirds of the holding's total SGM. Other EU countries have similar definitions for farm type or system.

In the UK, the income of many farms is based around a mixture of enterprises, comprising land used for animal production as well as land used for growing crops, which may or may not be used to feed the livestock. The vast majority of organic farms fall into this mixed farm category, although the relative proportion of livestock and cropping enterprises can vary. The characteristics of the main simplified UK farming systems are given in Table 5.1, with an indication of their suitability for organic production.

Mixed farming systems are the most common type of organic production system in the UK, with the livestock and crop components managed in order to balance the system in terms of productivity, profit and risk, and based primarily around the conservation and recycling of nutrients. There are fewer arable or horticulture dominated farms under organic production,

The Science Beneath Organic Production, First Edition. Edited by David Atkinson and Christine A. Watson.
© 2020 John Wiley & Sons Ltd. Published 2020 by John Wiley & Sons Ltd.

Table 5.1 Typical characteristics of the main UK farming systems with an indication of their suitability for organic production

Farming system	Typical characteristics and suitability for organic production
Mainly arable cropping	• Most of the land is cultivated, with few livestock • Mainly lowland/eastern areas of the UK • Relatively low annual rainfall (600–900 mm) • Drier conditions at key times of year (cultivation, fieldwork, crop ripening, crop drying at harvest) • Typical crops include cereals, potatoes, oilseeds, sugar beet, pulses/vining peas • Warmer regions in the south of the UK can grow the widest range of crops • Organic suitability: low
Intensive horticulture/arable	• High-value horticulture and arable crops, with few livestock • Often situated close to areas of higher population • Easily worked, fertile lowland soils • Range of vegetable crops, depending on temperature, some under polytunnels or glass • Irrigation may be required in drier regions • Organic suitability: low to moderate
Mixed farming	• Arable crops, with forage crops and grassland to support livestock enterprises • Areas less suited to intensive cropping due to climate and soil limitations • Typical higher rainfall favours shallow-rooted grassland and fodder crops • Typical crops include permanent grass, temporary grass for grazing and silage, maize (in the south), swedes, fodder beet, cereals, oilseeds and potatoes • Organic suitability: high
Mainly livestock	• Few arable or vegetable crops • Primary products are typically beef, sheep or dairy • Areas often have higher rainfall, or poorer soils not suited to arable production, but suitable for grassland • Typical crops include temporary and permanent grassland as well as some fodder crops • Organic suitability: high
Upland	• Located in upland areas of the UK, e.g. Scotland, Wales and the Pennines • Shallow soils and climate limit crop cultivations to grassland reseeding • Lower temperatures, higher rainfall and exposed fields restrict other cropping, e.g. arable • Majority of cropping is permanent or rough grazing • Organic suitability: high

with the exception of some smaller scale vegetable and fruit producers who manage soil fertility through judicious and skilled use of composting and imported manures, as well as having fertility-building breaks at key points in their rotation. Getting nitrogen into the system remains a key issue for all production systems (Walker and Watson 2011).

Livestock farms, which utilise fodder crops and grassland of various types, are also well adapted to organic production systems. Intensive farming systems that rely on a wide range of agrochemical inputs in order to produce higher outputs are often less diverse in the range of crops that they grow. Organic farming systems are extremely limited in the types and quantities of inputs that they can use to modify the soil and growing crop environment, and rely much more heavily on crop rotations for fertility and crop protection. As such, organic systems sit towards the extensive or low-input end of the scale and rely less on inputs, and accept lower yields as a consequence.

5.2 Land Capability and Cropping System Choice

In the UK, different regions tend to have their own characteristic farming systems (Hay et al. 2000), which have been developed around the typical soil, topography and climate anticipated for most seasons. The most obvious difference is the relative proportion of the farm landscape taken up by grassland or arable crops and, to a lesser extent, vegetables – the cropping system. Climate and topography influence the ability to undertake practical field operations, as well as the growth and development of arable and vegetable crops, grassland and rough grazing. The choice of cropping system thus depends on a range of geographical, economic and social factors which include the following: rainfall, temperature, altitude and topography, soils, markets and government policy. The impact of these key elements on the potential for organic production is detailed below (Table 5.2). The Land Capability Classification for Scotland is given here as an example of a classification system. The ways in which they affect crop production are covered in later chapters in this volume. However, these tend to be factors which are outwith the ability of the producer to change and so provide very real limits to both what can be produced and how much on a particular area of land.

5.2.1 Rainfall

Water is required for seed germination, but excess rain and wet soils can limit mechanised operations such as ground cultivations and sowing, and can also lead to soil erosion issues. Water is also a requirement of photosynthesis, and therefore crop growth, but too much water can encourage disease, particularly in combination with temperature. Dry conditions are required for ripening of the crops and harvesting, especially in the case of combinable crops such as cereals or oilseed rape. Wet soils can also affect the mechanised lifting of root crops such as swedes and potatoes. Use of machinery on wet ground, at sowing, during the season or at harvest time, will result in soil compaction and soil structure damage, which can reduce yields for many years. Rainfall is also one of the factors that dictate land class (Table 5.2). The higher organic matter content, which is a result of organic farming, will interact with rainfall patterns as a factor influencing production.

5.2.2 Temperature

The majority of UK farmed crops do not grow below 5 °C. Physiologically, this baseline temperature is the lower temperature at which phenological development ceases (Morrison et al. 1989). Extremely high temperatures can also affect crop growth but this is usually associated with drought in the UK, rather than being a direct influence on crop development.

Temperature is a critical determinant of the length of the growing season, as sufficient warmth is required for a crop to progress through its different phenological stages up to

Table 5.2 Land capability classification for agriculture in Scotland

Class and description	Remarks
Land capable of supporting arable agriculture (Class 1 to Class 3.1)	Land in these classes, often referred to as prime agricultural land, can be used to produce a wide range of crops. The climate is favourable, slopes are no greater than 7° and the soils are at least 45 cm deep and are imperfectly drained at worst. This land is highly flexible for other uses as well, such as for biofuel crops and woodland, although current management may make other options, such as heathland restoration, difficult in the short term
Land capable of supporting mixed agriculture (Class 3.2 to Class 4.2)	Land in these classes can be used to grow a moderate range of crops, including cereals (primarily barley), forage crops and grass. Grass becomes predominant in the rotation in Class 4.2 whilst other more demanding crops such as potatoes can be grown in Class 3.2. The climate is less favourable than on prime land, slopes up to 15° are included and many soils exhibit drainage limitations
Land capable of supporting improved grassland (Class 5.1 to Class 5.3)	Land in this class has the potential for use as improved grassland. A range of different limitation types, either operating singly or in combination, can restrict the land capability to this class. These limitations include climate, slope, wetness and often a heterogeneous pattern of conditions that render even occasional cultivation unsuitable. Land which has had this potential for improvement exploited is much more productive than land which remains in its unimproved state
Land capable of supporting only rough grazing (Class 6.1 to Class 7)	This land has very severe limitations that prevent sward improvement by mechanical means. This land is steep, very poorly drained, has very acid or shallow soils and occurs in wet, cool or cold climate zones. In many circumstances, these limitations operate together. The existing vegetation is assessed for its grazing quality (Class 6.1 is of high grazing value, for example but Class 7 land is of very limited agriculture value). Nonetheless, this ground often has a high value, for example in terms of storing carbon in its organic soils and supporting rare species and habitats

Source: Abridged from Bibby et al. (1991).
A similar system is used in England and Wales (MAFF 1988).

maturity and subsequently harvest. In cold conditions, crops ripen later in the season, and the crop and/or soils may not dry adequately for machinery to harvest the crop without causing damage to the harvested crop or the soils. Frosty conditions can also have an impact at various stages of crop growth; for example, autumn-sown crops can suffer from 'winter kill' or 'bolting', and late frosts can delay early sowing of spring crops. Temperature in combination with rainfall can create a microclimate within the crop canopy that can encourage a range of crop pathogens (mildew, potato blight, etc.). Temperature is another factor that dictates land class (Table 5.2).

Soil temperature is important to the mobilisation of mineral nutrients and this makes temperature amongst the more important factors affecting the viability of organic production.

5.2.3 Altitude and Topography

Altitude affects a number of factors which influence crop growth. Mean air temperature drops approximately 1 °C for every 160 m increase in altitude above sea level, which will have an impact on crop growth. Rainfall is generally greater on the windward side of a hill or mountain, and this, in combination with the fact that wind tends to be stronger as altitude increases, can affect soil erosion. Upland soils are usually shallower, and so winds are likely to be more damaging to both crop and soil. Upland soils generally have more of a gradient to them and can be rocky. Gradient can also be an issue on land at lower altitude. Altitude and gradient also affect land class (Table 5.2) and are a significant consideration for all forms of agriculture.

5.2.4 Soil

Very shallow or rocky soils are unsuitable for cultivation, and therefore severely limit cropping. Soil texture also influences which crops can be grown where. Very heavy (clays) or light soils (sands) have a major impact on which arable crops will grow successfully, predominantly through effects on water, temperature and nutrient availability. Machinery use on heavy soils can be a problem in the spring, and particularly in the autumn around harvest time for many crops, whereas light soils can be affected by drought during the summer months, a critical time for crop development. Poorly drained soils also restrict land use and the crops that can be grown. These factors have a major impact on the suitability of land for organic production as a consequence of organic production's greater reliance on mechanical cultivation. Soil is the fourth factor that dictates land class (Table 5.2).

5.2.5 Markets

The availability of a market for the crop at harvest, which can command a good price, has an important role to play in crop choice. The distance of the market or processor from the farm will also influence crop choice, as transport costs may be prohibitive. A farming system that includes high-value fresh vegetables will ideally be within easy reach of urban locations where the produce can be sold before quality issues arise. Farmgate sales to local outlets may be a valuable source of revenue, particularly for produce such as milk, meat, eggs and vegetables, and farmers may wish to continue or adopt this approach for social as much as economic reasons. While this is a key issue for all agriculture, it is perhaps of relatively greater importance for organic production because of the need for greater financial returns to cover the increased costs of production. Other issues linked to this are discussed in Chapter 3.

5.2.6 Traditions

This is often neglected as a key factor but is of increasing importance where the age of practitioners is high. Some farmers prefer to continue the systems that they have been brought up with, know and understand. They can be resistant to change. Many farmers like to have at least some livestock on the farm, even though the enterprise might not be especially profitable. Labour and management skills or experience of the crop need to be available on-farm or locally. This restricts which crops are actually grown. This argument also holds true for any livestock enterprises, especially where infrastructure such as buildings and fencing is required. Tradition is a significant element for some in relation to organic production although the need for change in organic methodology is as important. Other issues linked to this are discussed in Chapter 2.

5.2.7 Government Policy

There are a number of drivers emanating from EU and UK governments that influence the choice of crops. This is a particularly important area of concern currently for the UK as it contemplates production outwith EU regulations. Crop choices are affected by legislation, grants and subsidies that may relate to key criteria and to farm management practices. That schemes and their potential impacts change from time to time emphasises the importance of policy, which can be linked to social science issues (Chapter 3), rather than conventional science as the major driver of production. Organic farming is governed by a more complex set of regulations than is the case for 'conventional' farming. This is needed so that at market, organic food can be distinguished from other produce, which may look no different. Trade requires international agreement on many of these rules and definitions. Government policy and stability in the key elements of that policy are thus critical to organic production.

5.3 How Land Capability is Used in Practice

A crop rotation is a series of crops grown in a particular sequence on similar areas of land around a farm, such that each crop, or type of crop, is grown every year, and rotated in the desired sequence each season. The role of crop rotations is discussed in detail in Chapter 6 but many of the decisions which define a rotation are intimately linked to land capability and so it is appropriate to discuss these here as part of examining the decisions which are the responsibility of the producer and those forced by climate and soil.

The key objectives of the system include maintaining soil condition, adding and recycling nutrients and feeding livestock. These require a number of decisions such as what crops are to be grown, how much grass/clover is to be included in the rotation?

Climate and soil play a major role in the ability of land to support a particular range of crops. Some soils are able to produce a wide range of high-value crops while others are more limited (Agricultural Land Classification) (MAFF 1988; Bibby et al. 1991). Crops such as spring wheat, linseed or winter beans that require a long growing season will tend to prosper in drier, warmer areas of the UK. Combinable crops (cereals, oilseeds and pulses) which are normally stored dry (approximately 9–15% moisture), in order to reduce grain quality deterioration through crop pathogens, are best not grown where ripening will be delayed and where harvest is likely to occur in cool or damp autumn weather. However the cooler, longer days of northern UK encourage later ripening of such crops with the potential for very high yields as a consequence of the total radiation intercepted. This potential can be reduced by harvesting later in the autumn when poorer crop drying conditions may make combining difficult. In the same way cash root crops mature in the autumn when ground conditions may impede harvesting.

Soil texture is a significant factor in land capability which will also influence crop choice. Heavier soils, with a higher proportion of silt and/or clay, are generally more fertile than lighter, sandier soils, but are much slower to dry which can make their management difficult, particularly in relation to the timing of critical field operations such as sowing and harvesting. Soil pH will also have a big impact on which crops are likely to grow well as it plays an important role in nutrient availability. The soils in the UK used for arable and vegetable cropping typically have a range of pH from 5.0 up to 7.5, depending on their location.

5.4 Conclusion

The ability to grow a crop on a particular soil and to produce food is one of the key elements of farming practice. It depends on crop selection, on the particular physiological and agronomic characteristics of that crop and on how the crop is managed. This is true for all forms of farming. There are some different issues where production is on the basis of organic standards, discussed elsewhere in this volume, but land capability is a factor which needs to be at the heart of all the above decisions. Success in farming is not uncommonly related to the ability to produce crop through management, which reduces the potential impact of the factors which govern land capability. Understanding the significance of soil quality and of the classification scheme helps to set boundaries for production. Organic production emphasises management over inputs and so being able to beneficially manage factors such as temperature, water supply and the timing of farm operations is of very major significance.

References

Bibby, J.S., Douglas, H.A., Thomasson, A.J., and Robertson, J.S. (1991). *Land Capability Classification for Agriculture.* Aberdeen: Macaulay Land Use Research Institute.

Hay, R.K.M., Russell, G., and Edwards, T.W. (2000). *Crop Production in the East of Scotland*, 1–61. Edinburgh: SASA.

Lampkin, N. (2002). *Organic Farming.* Ipswich: Farming Press.

MAFF (1988). *Agricultural Land Classification of England and Wales.* London: MAFF.

Morrison, M.J., McVetty, P.B.E., and Shaykewich, C.F. (1989). The determination and verification of a baseline temperature for the growth of Westar summer rape. *Canadian Journal of Plant Science* 69: 455–464.

RESAS (2011). *Economic Report on Scottish Agriculture 2011.* Edinburgh: APS Group Scotland.

Walker, R.L. and Watson, C.A. (2011). Are inadequate soil populations of complimentary rhizobia restricting UK self-sufficiency in legume protein? Agricultural ecology research: its role in delivering sustainable far systems. *Aspects of Applied Biology* 109: 71–77.

6

Crop Rotations

The Core of Organic Production

David Atkinson and Robin L. Walker

SRUC, Aberdeen, Scotland, UK

6.1 Introduction

What is a crop rotation? What do we expect it to do? Where is the science? Why is it particularly important to organic production? How does it fit into the wider concept of an organic system? The subject of crop rotations has been reviewed by a number of writers, of which those by Karlen et al. (1994), Hennessy (2004), Knox et al. (2011) and Bruns (2012) are amongst the most recent. Here we aim not to give a review of rotations in general but to focus explicitly on what we can learn about the underpinning science base of rotations within the concept of organic production. Even in this context, the available literature is large, so we have chosen only selected material, which covers important points, rather than trying to be encyclopaedically encompassing.

Rotations have implications for all types of crop production. These include how the crop affects soil organic matter levels and soil structure, how the rotation affects crop protection, how it affects the availability of nutrients and whether it affects soil erosion and other forms of soil degradation (Table 6.1). Choice of crop will affect these factors differently. Whether the rotation needs to deal only with crop production or whether animal production is also a key aim of the farming system will have a major impact on the design of the rotation and what we need from it. The selection of a rotation will be influenced by land capability and soil type, as discussed in Chapter 5.

So what is a crop rotation? It is the practice of growing a sequence of dissimilar crops on the same area of land at different times on a planned basis, with the aim that the different crops are not only able to take advantage of the soil conditions at planting but deliver benefits to subsequent crops. Their role within organic production is highlighted in Chapter 1. It would be usual for each crop in a rotation to be present within the farm every year on approximately the same area, enabling the farm to be managed in 'blocks' of similar-sized areas of land. The sequence of crops and their integration aims to make the whole system more productive than individual elements within it.

The Science Beneath Organic Production, First Edition. Edited by David Atkinson and Christine A. Watson.
© 2020 John Wiley & Sons Ltd. Published 2020 by John Wiley & Sons Ltd.

Table 6.1 Key factors influencing the functioning of a rotation

Feature	Impact and mechanisms
Rotation sequence of crop species	Speciation of nutrients, e.g. P and trace elements, availability of infective propagates of AMF, activity of microbial species, carbon inputs to soil, weed species; which may also influence AMF and microbes
Intercropping	Improved nutrient retention, slower in release of nutrients to soil
Crop and variety selection	Needs of nutrients and pesticides, AMF infection dependent
Cultivations	AMF infection, CO_2 release from soil, N availability, especially in spring
Nutrient management	Losses due to leaching and denitrification. Chemical species in soil, AMF status, soil microbe communities

Source: From Atkinson and Wilkins (2004).
AMF, arbuscular mycorrhizal fungi.

What do we expect the rotation to do? A single crop will deplete the soil of particular nutrients and can lead to the build-up of particular weeds and diseases, or perhaps damage soil structure. This can be mitigated by a rotation because the different needs and impacts of different crops can be harmonised to alleviate issues over time. The sequences of crops can vary as can the duration of the presence of an individual crop; for example, grass-clover leys may be present for several years, whereas a cereal crop will be grown for just one season.

Why do rotations matter to organic production? Crop rotations are commonly practised by all forms of agriculture but are essential to organic production. A primary reason for this is the need to include legumes and other beneficial crops, like deep-rooting forbs, that can bring new nutrients, mainly nitrogen, into the system for exploitation. Non-organic systems can meet this need through the use of mineral fertilisers, but there are other benefits of a rotation beyond crop nutrition. Even where fertilisers are being applied, crops still obtain many nutrients from pre-existing soil reserves. Similarly, even when pesticides are applied to crops, the crop sequences will affect weed and diseases, with management aiming to use this advantageously rather than increase problems. Thus there are elements of a rotation shared by all forms of farming, but others especially linked to nutrient supply, weed, disease and pest control where the science base in respect of organic production differs from that of other approaches.

Some of the principal differences induced by the use of rotations were summarised by Atkinson and Wilkins (2004). These are shown in Table 6.1 as an agenda for those parts of crop functioning which can be delivered by system management.

A consideration of the history of crop rotations helps us to identify why they were seen as important, what they were intended to do and points us to the areas of relevant science.

6.2 The History of Crop Rotations

Why have crop rotations always been important? Crop rotations have been at the core of crop production since settled agriculture began, and a number of ancient historical texts refer to the use of rotations. Advice on crop production is to be found in Hebrew Scripture, for example

Leviticus 25, 2–7, and is the basis of many of the parables in the Christian New Testament, such as *Matthew* 13, 3–9, 24–30. From the Middle Ages onwards, Europe followed a pattern of agriculture based around the rotation of three crops. Land was divided into three parts, with a section planted with cereals such as wheat or rye followed by spring barley or spring oats. A second section was cropped with legumes such as peas or beans while a third was left fallow. Every three years, a single field would be rested and be fallowed, with no crop present during this phase. The basic principles of this were followed in the nineteenth century in USA with the alternation of depleting crops such as cotton with enriching crops, certainly in terms of nitrogen, such as the legumes, peanuts or peas.

So what are rotations expected to do? They build up and maintain soil fertility, provide an opportunity for weed control, institute breaks between crops susceptible to similar disease or pest issues, utilise labour more evenly throughout a calendar year and spread financial and biological risks.

In the UK, rotations became important at the time when the regulated basis of agriculture moved from traditional open field systems to the era of enclosures. One producer using the same area of land for crop production in successive years led to a need to consider crop sequences, and this gave birth to the Norfolk four-course rotation. Different crops were identified as having different effects on the soil. Different crops need the same basic nutritional elements but in different amounts, resulting in distinct shifts in the balance due to nutrient extraction. Contrasting crop types will also remove nutrients at various times during the growing season and with different levels of intensity, both of which affect soil chemistry and the dynamics of soil microbial communities. Some crops, such as cereals, tend to exhaust the soil while others such as grass leys and legumes restore its vitality through nitrogen fixation and improved soil structure.

The history of rotation design and its potential applicability in modern UK agriculture was reviewed by Knox et al. (2011). Rotations had been used in continental Europe from the fourteenth century and permitted the integration of arable crop production, especially wheat, and animal, especially sheep, production (Price 2014). The rotations introduced into the UK by Coke of Holkham and Townsend of Norfolk in the late eighteenth century included turnips, barley, seeds (grass and clover) and wheat (Howard 1945; Hay et al. 2000). The introduction of rotations was responsible for providing sustainable levels of production together with planned care for the soil (Howard 1945). This also made use of the knowledge that crops like legumes had the ability to bring nitrogen into the system, something critical to organic systems. The role of these different crops is highlighted in Table 6.1.

The data in Table 6.2 indicate the distinctive roles of the various crops within a simple rotation. Most commonly, in a rotation 40–60% will be grassland based, including a legume such as clover, and a similar percentage devoted to arable or vegetable cultivation.

The amounts of N fixed or removed will vary due to a wide range of factors. While more recent systems are likely to result in greater amounts being removed, Lidgate (1982) reported around 280 kg N/ha being removed by winter wheat but current recommendations in conventional farming suggest 185–245 kg N/ha in grain plus straw depending on soil type (HGCA 2009). In organic systems, N uptake has been shown to be up to 200 kg N/ha, depending on the approach to fertilisation (Petersen et al. 2013; Tosti et al. 2016). The historic data set of Fried and Broeshart (1967) is used here because of the number of studies included and the range of crops referenced which provide a useful comparison.

Table 6.2 The role and impact of different crops within a traditional crop rotation

Crop	Role in rotation	Nitrogen removal kg/ha	Ratio of N/P/K in crop off-take
Turnip	The major eighteenth century addition. It replaced a fallow period, which could be a third of the length of the rotation, with a crop useable as sheep feed particularly over the winter period. The manure generated by the sheep fertilised the following arable crop. It restored fertility after wheat and gave scope for weed control and the reduction of soil borne diseases. Turnip could be planted immediately after wheat	130	1/0.08/0.6
Barley	Most often grown as spring barley, which could be planted to follow the turnip crop. It gave a second cash-generating crop with a wide range of uses from animal feed, use in brewing and distilling, as well as being an alternative basis for bread making	60	1/0.22/0.87
Grass/clover	The principal means of getting nitrogen into the soil to drive cereal production and to increase the amount of organic matter in soil, to help sustain good soil structure. Essential to the animal production element, especially ruminants, and could be varied in length depending on the amount of restoration needed after cereals	88	1/0.12/0.7
Wheat	The principal cash crop and important for bread making. The crop which probably has the greatest single impact on the duration of other elements of the rotation and cash flows. It was sown in the autumn and harvested towards the end of the following summer during which time soil organic matter levels and available nitrogen tended to have been depleted	76	1/0.18/0.55

Source: Nutrient data from Fried and Broeshart (1967).

Maximum nitrogen removal will vary with variety and fertiliser application, but the relative needs for other elements, such as phosphorus (P) and potassium (K), seem more stable. Relative to nitrogen uptake, the four major crops within the typical rotation described in Table 6.2 removed different amounts of P and K. They also removed nitrogen at different rates and at different times in the growing season. Winter wheat takes up available N from the soil over the period from autumn of one year into May the following year, while spring barley takes up available N from the soil at a much greater rate but only in May and June. Contrastingly, turnips remove N from the soil at an intermediate rate and generally through late summer and into the autumn (Atkinson 1991). Nitrogen absorbed at different times during the year is likely to have been derived from different N pools within the soil (Jenkinson 1982). Different crops thus have the ability to make very different demands on the soil's nutrient reserves both in terms of amounts and chemical forms and to vary the importance of microbial transformations which will affect the crop growth characteristics.

In the USA, there have been several long-term (over 100 years) studies of rotations (Bruns 2012). A long-term study at the University of Alabama, begun in 1896, showed that winter legumes could be as effective as fertiliser N in relation to the yield of the following cotton. Including legumes in the rotation also increased soil organic matter (SOM) from 0.8% to 1.8%

and in more varied systems to 2.3%. In studies in Illinois, begun in 1876, corn yields also benefited from a legume-based rotation. Again increased SOM and provision of N were key issues. This is likely to be particularly important on lighter soils where SOM affects nutrient retention and resistance to erosion.

Historical references such as Howard (1945) focus attention on the influence of rotations on both soil fertility (and the linked issue of nutrient supply to crops) and adaptation to/protection from the impact of pests, diseases and weeds. These remain key issues in agriculture today irrespective of the system employed. In an organic context, they encompass the idea that over time natural processes will build up fertility and that the correct agricultural practice will work in harmony with such processes, enhancing their effects and in doing so aiding production.

The use of the term 'natural processes' identifies that one of our key tasks here is to give a scientific basis to what is regarded by some as an ambiguous expression. The nature of the term suggests that its underpinning basis is most likely to be found in ecological science. The perceived value of rotations also depends on the ideas that the composition of vegetation will affect the performance of other plant species, such as those regarded as being weeds in agricultural production, and that changing the crop annually will reduce the build-up of pests and diseases. In the absence of chemical crop protection, which is very restricted for organic production, these attributes of the rotation are especially valuable.

6.3 Rotations in Organic Production

Organic production can be viewed negatively in terms of inputs, which are very restricted, such as fertilisers and pesticides, rather than in relation to its positive approach to the development of soil fertility through the maintenance of organic matter and its associated microbial populations. These positive attributes are at the heart of the organic science base and were identified by Howard (1945) as critical to sustainable crop protection. This was because they emulated the natural cycles, which he saw as responsible for the development of fertility in relatively unmanaged situations such as forests. At the centre of organic production is a healthy soil with a diverse and sustainable microbial population. Creating the right environment for the functioning of the plant root system and the activities of organisms such as arbuscular mycorrhizal fungi was identified by both Howard (1945) and Balfour (1943) as critical to organic production. In this volume, issues linked to nutrient availability are discussed in Chapter 4, issues linked to the functioning of roots are discussed in Chapter 10 while soil microbial activity is discussed in Chapter 11. They are not therefore considered in detail here.

One of the earliest comprehensive reports on the functioning of a crop rotation in organic production is to be found in Balfour's 1976 account of the results of the Haughley Experiment. The experiment began in 1939 and continued until 1969 although detailed measurements only cover the period from 1947 and are only available in full for the first decade. The aim of the study was a long-term field investigation, which looked not only at agricultural issues but also at the related issue of human health. Delivering improved public health through good crop production was an important issue for the founders of the organic movement (Balfour 1943). The experiment aimed to challenge the prevailing medical model, which focused on the role of the disease-causing organism and its elimination rather than on the contribution made by the health of the host, and so the host's role in resisting the pathogen. These issues are discussed in this volume in Chapter 12, which deals with food quality and health.

The Haughley Experiment ran at a full farming scale, almost 100 ha (216 acres), and did not utilise a random plot design, as the results were not destined for statistical analysis. The view of the originators of the experiment was that attempts to eliminate variables inherent in biological functioning would destroy the concept of the whole, which was seen as being critical to the organic approach. It is important to note that this continues to be an issue in current studies (Watson et al. 2006). The experimental farm was divided into three units, which were run as self-contained small farmlets. Two sections, each of around 75 acres, were stocked while a third section of around 32 acres was stockless. The stocked sections had self-contained herds and a minimum of three species of livestock (Table 6.3). One stocked section received only the return of crop residues and animal manure produced on that section; this is referred to as the 'organic section'. The other stocked section also received mineral fertilisers. This is referred to as the 'mixed section'. The stockless section received mineral fertilisers in addition to crop residues.

The experiment ran under a protocol of strict rules.

1) All three sections were to be run as rotations, which were identical for the two livestock sections. The stocked rotations are detailed in Table 6.3.
2) Importation of feedstuffs was not permitted for either stock section.
3) The organic section aimed to run as a closed system with only food exported and seed and seaweed imported. The closed nature of the system meant that the history of all within it was known and thus minimised the import of viruses and other pathogens. The export of food was seen as being small relative to the nutrients contained within the soil system.
4) The assessment period was the duration of the complete rotation. This was selected so that the impact of a crop not only on the next crop but also on later crops and the consequences of the grazing regime could be assessed. This allowed for the full impact of the rotation and eliminated the impact of the weather in particular years.

In any given year this resulted in 20–25 acres under leys and 40–50 acres under arable crops in each section of the experiment. The stockless rotation was different to the stocked rotations and involved four years of arable cash crops, wheat, sugar beet, barley and undersown barley followed by clover ploughed in. In the first rotation, the yields of wheat, winter oats and spring barley were similar between the two stocked rotations whether or not they also received fertiliser. The yields of winter beans, arable silage and ley hay were higher for the mixed section. Milk production was higher on the organic section. During the second set of rotations, crop production for all but leguminous crops was higher for the mixed section.

Table 6.3 The stocked rotations employed at Haughley in different periods

Year	1	2	3	4	5	6	7 8 9 10
Rotation 1 (1952–1959)	Oats	Barley	Pulses	Arable silage under sown	Ley (4 years)		
Rotation 2 (1960–1969)	Wheat	Roots and forage	Barley	Pulses	Oat arable silage under sown	Ley (4 years)	Wheat

Source: Derived from information in Balfour (1976).

In addition, the key findings were given as

1) Where fertiliser was omitted from any part of the mixed section, the resulting yield was less than that of the organic section while the best organic yields came from the areas which had the longest history of no fertiliser use (some areas of the farm had a very long history of no fertiliser use). Organic fields had an increased workability under all weather conditions.
2) The levels of available minerals, especially phosphorus and potassium and particularly under the arable crops, fluctuated over the season, with peak levels in May to July, and particularly with the organic treatment.
3) Crops in the mixed section seemed to become increasingly reliant on the input of mineral fertiliser.
4) Soil organic matter levels were highest in the organic section.
5) The SOM concentrations were 3.6% in soil from the organic plots and 3.3% in the mixed plots. Early growth of autumn-sown cereals was delayed in the organic section compared to the mixed sections. This is consistent with the effects of mineral fertiliser, which are discussed in Chapter 8.

The study thus confirmed that there seemed to be, at least in some crops, a penalty for using the organic rotation rather than combining the rotation with the use of fertilisers. It identified changes in the soil which gave stability such as the rise in SOM matter but this also had the effect of slowing the release of nitrogen to the developing cash crop. Chapter 8 identifies this as the key element responsible for the variable yields found in different rotations. The organic rotation seemed to have a positive impact on both animal health and productivity.

A more recent rotational experiment conducted at the University of Newcastle's Nafferton farm over the period 2001–2011 has produced results which confirm some of the conclusions of the Haughley study (Cooper et al. 2011). Organically fertilised crops had lower yields than conventional. Soil N was higher and extractable P lower in the organic compared to mineral fertilised plots. In both the Nafferton and Haughley studies, it is hard to assess the extent to which nutrient applications were or were not balanced. The differences in the chemistry of organic and inorganic forms of the same nutrient make strict comparisons difficult. Olesen et al. (2007) illustrate the importance of the sequence design in organic rotations in balancing fertility-building and fertility-exploiting crops but also the need to balance the crop off-takes with manure inputs in organic systems. Managing biological nitrogen fixation (BNF) through legumes for grazing or silage in livestock-based systems is important to providing nitrogen for the complete rotation. In stockless systems, cutting and mulching of green manures or cover crops can be managed to provide enough nitrogen for sustained crop production, as demonstrated by Doltra and Olesen (2013).

Lampkin (2002) identified the following as key requirements for organic rotations.

- Use deeper rooting crops after crops with shallow roots.
- Alternating high and low root biomass crops.
- Alternate N-fixing crops with crops known to have a high N requirement.
- Utilise catch crops, green manures and undersowing at key points.
- Alternate leafy crops with straw crops to reduce weed issues.
- Leave appropriate breaks between crops susceptible to the same disease or soil-borne pests.
- Use mixtures of varieties and crops (intercropping) to maximise resource use efficiency.

- Consider alternating spring-sown and autumn-sown crops.
- Use slow-developing or wide-spaced crops after weed-suppressing crops.
- Use nutrient-demanding crops, with high potential profit, after fertility-building phases of the rotation to maximise nutrient use.
- Use set-aside or other subsidy/grant schemes to improve soil nutrient status if possible.
- Don't over exploit soils (nutrients and structure) and risk weed, pest and disease problems by growing the same type of crop too close together in rotation.

6.4   The Ecological Science Base of Organic Production

The introduction to the results of the Haughley Experiment (Balfour 1976) emphasised the ecological nature of the study and the importance of relationships and the mutuality of organisms. This is a different approach to science from that which underpins contemporary conventional agriculture with its strong chemical base. The results of the Haughley Experiment suggest links to several of the most important key ecological concepts.

Some years ago, ecologists identified the most important concepts in ecology (Cherrett 1989). These are discussed in Chapter 1. Those with key importance to the functioning of the rotation are the ecosystem, the ecological niche, the importance of diversity and stability and the processes governing competition. Carbon and nutrient cycles within ecosystems have been discussed by Waring (1989). He reported on a number of key processes within natural ecosystems which seem equally important to crop rotations.

1) Different ecosystems exhibit large differences in primary production. While much of this is the result of differences in the interception of radiant energy, at least part of it is linked to the relatively large amount of carbon allocated to small-diameter roots. This part of total gross primary production represents the mechanism governing the amount of carbon moved to the soil. This makes it important to SOM, which is at the heart of both sustainable soil structure and fertility.
2) The amount of primary production is related to the accumulation of nitrogen in the system. This emphasises the importance of the management of nitrogen use, which is identified elsewhere in this volume as being critical to high yields (Chapter 7).
3) The ability to support photosynthetic area with a minimum of structural maintenance cost seems to be important to provide sustainable presence and to the relationship between photosynthesis and net primary production. Rotations allow choices of this type to be made through crop and variety selection.
4) The ability to recover following disturbance is a feature of a balanced agricultural system where resilience, the ability to rapidly recover, is important. The buffering of the soil microbial community is important to achieve this in the context of the rotation.

The importance of the niche as an ecological concept was reviewed by Schoener (1989). A niche is a place, sometimes described as a physical part, in a biological community defined by the role of the organisms which occupy it, and commonly as a result of competition (Schoener 1989). The resources it is able to provide and its microhabitat features can also define it. These elements apply to the phases of a rotation. Competitive control affects the opportunities for the organisms within a niche to use resources and to avoid adverse organisms, predators, pathogens or weeds in our context.

Competition is a key ecological concept in attempting to understand the rotation. As an ecological concept it has been reviewed by Law and Watkinson (1989). The impact of increasing competition is easily demonstrated. The removal of unwanted vegetation, such as weeds, improves the growth of the desired vegetation – crops. Altering crop density can have similar effects, for example by increasing seed rates. The review emphasised the importance of competition for resources and its components in reducing the performance of other organisms (effects) and continuing to perform well in the presence of others (responses). This introduces the concept of the controlling factor, a shared need, the same class of resources, commonly soil nitrogen in agricultural production, which restricts growth more severely as population sizes increase. The more similar the needs of the individuals (and many of the worst weeds closely resemble crops), the more intense competition is likely to be. However, this identifies the challenge of identifying the exact resource being competed for and, for example in the case of plant nutrition, identifies the changes in nutrient speciation of a single element, which develop within a rotation and can result in different components occupying different niches, including differences across time.

It is clear that as a consequence of those who practise organic approaches to agriculture, seeking to base it on the management of natural processes, the principal science base of organic production is that of the ecological sciences, with the rotation as a primary means of influencing this effect.

We now explore the major components of the rotation's effects on the soil and the results of these effects on the supply of mineral nutrients and the ability of different crops to co-exist with other organisms such as weeds and pathogens.

6.5   Impact of Rotations on Soil Properties

Basic issues in relation to soils are outlined in Chapter 4. The ways in which a crop rotation can affect the soil are summarised in Table 6.4.

Most of the changes in soil physical condition are a result of changes in soil structure and so these are discussed together. Similarly, changes in soil microbiology drive many of the changes in soil nutrient supply and so these are discussed together.

6.5.1   Impact of Rotations on Soil Condition

Soil erosion (also discussed in Chapter 5) has long been identified as one of the more serious risks facing global cropland productivity. The risk of soil erosion has been shown to be related to both soil type and cropping patterns (Hennessy 2004). The Haughley Experiment (Balfour 1976) pointed to the effects which an organic rotation, relative to either a mineral fertilised stocked or stockless rotation, could have on soil condition. Organically managed soils in the Haughly Experiment, as well as numerous other system comparison experiments, frequently have higher organic matter content. The pattern of change between the two animal-based rotations at Haughley was similar although the organic rotation's soils consistently had higher SOM compared to the mixed rotation, which also received mineral fertilisers. Levels of SOM increased as the rotational sequence (Table 6.2) moved from ley to wheat, from 3.1% to 3.4% in the mixed rotation and 3.3% to 3.8% in the organic rotation. In the organic rotation SOM fell to 3.4% following the wheat crop. Even where soils of relatively similar organic content were

Table 6.4 The impact of crop rotations on soil properties

Soil property	Mechanism leading to change
Structure	Changes in soil bulk density resulting from the different cultivation practices associated with different crops as well as their diverse rooting architecture
	Different amounts and qualities of organic matter input to soil
Physical	Different heat retention and transfer properties resulting from changes in organic matter content and bulk density
	Changes in surface soil aggregation
	Differences in soil water-holding capacity and ability to buffer against flooding and drought
Chemical	Changes in soil pH
	Changes in anion/cation balances in the soil solution
	Changes in the speciation of nutrients which can exist in different forms in soil
	Changes in the amounts of nutrients held in organic complexes
Biological	Changes in ability for arbuscular mycorrhizal infection
	Changes in bacterial and fungal composition
	Changes in macro fauna, e.g. earthworms

compared, those from the organic section of the experiment had better water-holding capacity. Soils from the stockless section were significantly poorer than those from the stocked rotations. This identifies that the impact of organic production is likely to be complex and questions our knowledge of the key elements of soil architecture.

The role and importance of SOM were reviewed by Greenland (2000). He linked organic matter and the supply of nutrients, concluding:

> If soil is to remain productive over many years, nutrients must be replaced and the soil must be maintained in a favourable physical condition. Cultivation of the soil is normally used to create a suitable structural condition for plant growth but the condition can be unstable unless there is sufficient organic matter in the soil.

He divided the main effects of organic matter on soil properties into chemical, physical and biological components. The principal chemical attributes involved in providing a balanced supply of plant nutrients are reviewed in Chapter 4. Biologically providing the source of food for soil animals and micro-organisms was seen as critical and this is reviewed in Chapter 11.

The physical attributes were seen as centring on stabilising transmission pores which allow the easy movement of liquids and gases throughout the soil profile, strengthening soil aggregates against the impact of rain and increasing water retention.

These issues are important in the context of past and current concerns over the condition of UK soils. A UK Government committee which reported in 1970 (Anon 1970) looked at the impact of the three major changes in post-WW2 farming: more powerful cultivation, the use of mineral fertilisers and the increasing use of chemical crop protection materials, and concluded that:

1) the move away from diverse rotations and from significant amounts of ley arable-based production had not adversely reduced inherent soil fertility. There was, however, some concern over trace elements

2) there was no evidence that organic matter was intrinsically a better source of nutrients
3) there was concern over soil structure related to a loss of SOM, especially on inherently unstable soils, and that the reduction in the use of leys had resulted in damage as a consequence of machinery use under wet conditions.

Increased specialisation of agriculture in the conventional sector and, to some extent, also in the organic sector has been a feature of modernisation of agriculture since that report was written in 1970. In some parts of Europe, there is now evidence of declining soil fertility in arable areas, which may be a result of declining use of leys but also reduced reliance on animal manures (Lemaire et al. 2014; Peyraud et al. 2014).

All these concerns relate to the role of SOM and its impact on the physical structure of the soil which is linked to microbial activity, as soil microbes are intrinsically responsible for transformations of carbon which enters the soil as plant residues – both roots and unharvested crop residues such as straw. The range of materials which can be released by roots was reviewed by Manoharachary and Mukerji (2006). Compounds include those which act as carbon sources for rhizosphere organisms and those which have more specific effects such as siderophores, which chelate iron, and flavonoids, which are important routes of communication between plants and associated microbes.

The impact of straw, one of the more common residues returned to the soil, on microbes was discussed by Brookes et al. (1991). They identified that the microbial biomass could account for 1–3% of SOM and that it was influenced by both the input of fresh organic material, such as crop residues, and by soil texture, with microbial populations being generally higher for clay soils than for sandy soils under comparable management. SOM was high under long-term grassland with a value of 2000 kg/ha measured for permanent grassland compared to 400 kg/ha for an unfertilised arable soil. An arable crop receiving farmyard manure (FYM) was intermediate. The addition of straw represents a pulse of substrate being added to the soil and a significant addition to soil C reserves, given that straw incorporation can add around 5 t/ha of dry matter compared to the 1 t/ha estimated to be released by root decomposition. Microbial activity and soil N are closely related, with around 3–5% of total soil N being found in the microbial biomass. This turns over in around 1.5–2 years compared to some of the less labile and more structural SOM, which can be present for thousands of years.

Understanding the contribution of the rotation and its effect on this long-lived SOM matter is important in enabling a better understanding of soil structural stability and the sustainability of production. When a soil is cultivated, microbial decomposition of SOM is increased. It also increases when the soil is undisturbed such as in the ley phase of the rotation. As with plant growth, nitrogen availability influences the production of both microbial biomass and humus (Greenland 2000). It is linked to the concept of soil quality, which has been reviewed by Karlen et al. (1997). They saw soil quality as having three emphases: sustained biological productivity, environmental quality and plant and animal health; all elements identified here as being part of the role of the rotation in organic production. Amongst the indicators of soil quality they identified were SOM, infiltration, aggregation, pH, microbial biomass, bulk density and forms of nitrogen. In a comparison of organic and conventional cropping, they identified increases in soil respiration, faunal populations and infiltration rates as being indicative of the higher biological activity associated with organic production. The types of both organic and conventional management reported in the study affected microbial biomass N, potentially mineralisable N and soil NO_3. Ground cover and increased SOM reduced N leaching in winter but led to a reduction in available N at the beginning of the growing season, which consequently reduced crop production.

Due to the importance of SOM, there have been many studies of the impact of organic production and of different elements within a rotation on it. Armstrong Brown et al. (1993) compared a series of paired farms across the south east of England in relation to levels of SOM and some of its properties. They found that in horticultural, agricultural and pasture phases of production, SOM was generally higher under organic than conventional production. There was some variation depending on how comparisons were made. Under 'pasture', the SOM contents were 4.71% and 5.67% respectively. The lower value for organic production was a consequence of the inclusion of values for the ley element in organic pasture compared with a pure pasture situation with conventional production. Under the arable phase, values were 4.08% and 3.04% respectively. Humic acid, which accounts for around two-thirds of SOM and is its more stable component, was also higher in the organically managed soils, as was the infrared absorbance which characterises the age of the SOM.

6.5.2 Impact of Rotations on Nutrient Availability

As a key element of assessing the impact of rotations on soil nutrient supply, it is important to identify how nutrients enter an organic system. The principal sources are as follows.

- Biological nitrogen fixation (BNF), predominantly from the inclusion of legumes in the rotation and their symbiotic relationship with N-fixing rhizobia in root nodules.
- Bought-in animal feed, which can be an important source of macronutrients as well as trace elements such as copper, zinc, cobalt, manganese, boron, etc.
- Permitted nutrients, which will be limited to a few products, often with low crop availability, and may require derogation from the farm's organic certification body.
- Bought-in manures, which may be from organic farms with surplus livestock wastes or from conventional farms with certain limitations, for example length of storage prior to application on organic land, dictated by the certification body.
- Composts, where again certification bodies will have certain restrictions on the source and type of composts that can be applied to organic land.

The history of the use of rotations emphasises nutrient supply as perhaps the area of its greatest contribution to production. The rotation will influence both the fate of inputs such as the above and the extent to which they may be needed by the crops in any particular phase of the rotation. Effects and mechanisms differ for different nutrients. Here we consider some of the more important elements only. Previous cropping patterns and inputs are known to have continuing effects on subsequent production. In 1948, the Agriculture Department for Scotland set up a committee to review, on an annual basis, the residual value of nutrient inputs into farming. While this was done primarily to inform financial discussions when farms changed hands, it provides information about nutrient carryover between parts of the system. Inorganic N and K fertilisers were not regarded as having residual value but organic materials were identified as having a measurable residual value, as were all forms of phosphate. Organic materials with crop nutritional value included both bought-in feeds and manures produced on-farm (Anon 1998). The use of such materials needs to be seen as an adjunct to what can be achieved through the use of the rotation. The role of such inputs is discussed more fully in Chapter 9.

6.5.3 Nitrogen Supply in Rotations

Nitrogen is key to all crop production and its cycling in soils is discussed in Chapter 4 and its impact on crop growth and development in Chapter 8. Nitrogen in soil exists in a number of chemical forms. Most soil N is present in an organic form, much of which seems to be proteinaceous in nature and which is not directly available to plants. For it to become easily available, it needs to be in an ionic form, first converted to ammonium and then oxidised to nitrate through microbial activity. Crop plants are known to prefer nitrogen uptake in the nitrate form.

Organic nitrogen enters the soil pool in a number of ways. The principal routes are:

- through the decomposition of plant roots and their associated microbes
- the incorporation of above-ground material either through natural processes or through the mechanical incorporation of crop residues or manures which will also contain animal wastes but primarily derived from crop materials
- with leguminous crops through the fixation of atmospheric nitrogen.

The importance of nitrogen means that the quantity of legumes to include is one of the key decisions in the design of the rotation. Thus rotations can influence the availability of nitrogen to a succeeding crop in the following ways.

- Through the amount of organic material made available as a result of root, mycorrhizal and bacterial material produced and released for decomposition.
- Through the composition of this organic material, which can vary in both its carbon to nitrogen ratio (usually around 10:1 but variable) and the nature of the nitrogen components, typically 50–60% protein but with variable potential for decomposition.
- Through the ability of the crop to fix nitrogen through leguminous symbiosis or through other associated micro-organisms with the potential to increase the nitrogen content of soil.
- Through their ability to trap nitrate which might otherwise have been leached to ground water or denitrified to a nitrogen oxide. This is one of the principal roles of a cover crop and a benefit of a winter-sown crop.
- Through a role in conserving soil microbial populations so as to provide the potential for easier establishment of symbiotic relationships with the following crops, such as legumes or those with mycorrhizal associations.

The availability of nitrogen early in the season is critical to crop performance (see Chapter 8). Within organic production, the rotation, and its ley phases, are key to the provision of nitrogen and other elements. The data from the Haughley Experiment indicate that the amount of nitrate N available was affected by the rotation sequence (Balfour 1976). A key feature of the Haughley study was the avoidance of nutrient inputs from outwith the rotation, so finding ways to maximise the efficiency of the internal nutrient pathways was important.

The importance of N input to all farming systems has thus been recognised for much of the history of farming, with this need most commonly met through the use of legumes, until more recent times when N fertilisers have been readily available. Pasture legumes are known to be able to add between 100 and 200 kg N/ha per year (Greenland 2000). While the grass-legume phases consistently increase SOM, other than on soils of poor physical structure, they tend not to outyield those receiving mineral fertilisers. At higher yield levels, over time, biological or physical conditions can start to affect production and it is here that the use of an appropriate rotation can help. Available evidence suggests that leaching losses from grass-clover and N-fertilised grass-only pasture are similar at similar total N inputs (Ledgard et al. 2009). Cameron et al. (2013) provide a useful review of N losses, including leaching, from agricultural systems. This issue was also addressed by Phillips et al. (1993).

Another of the key questions in this area relates to the role of the soil microbial population and the ways in which it is affected by different phases of the rotation and, in an organic context, its ability to release N and other nutrients. Brookes et al. (1991) discussed the contribution of soil microbes to the supply of N for crop production. They identified that N held in the microbial biomass was more labile than N in the rest of the SOM. This pool can contain as much as 30–150 kg N/ha. The turnover time of this biomass is typically adjudged to be around two years, which suggests that the flux available for crop growth from this source could be as large as 100 kg N/ha. Adding straw to soils increased the amount of biomass N within the system. Stockdale et al. (1993) showed that the link between organic manure inputs and crop response is complex. While the mineral N content of manure had little impact on production, N uptake was related to the amount of N recovered from the profile following manure application. Poultry manure was more effective than FYM, perhaps because of its higher microbial activity, possibly linked to its generally higher P content. Cover crops can certainly be important in reducing overwinter leaching. Using the design of the rotation to minimise winter losses of N is especially important to organic production.

A further core issue in the construction of a rotation is the nature of the impact of the preceding crop on the nutrient supply, both in amount and chemical form, to the next crop in the sequence. Williams et al. (1993) assessed the impact of the previous crop or soil treatment, such as different forms of nutrient or straw additions, on the release of N from an incorporated green manure derived from mustard. Nitrogen was released as volatile N compounds as well as nitrate and ammonium. The addition of green manure stimulated the release of all three N forms although this was not affected by past history. Although there was no difference in the total amounts being released, differences in the 15 N/14 N ratio, which indicates the extent to which the heavier isotope is discriminated against during metabolic processes, suggested that there was scope for variation in mechanisms in these processes in different crop situations.

An assessment of the impact of the different crops which preceded a crop of durum wheat (Ercoli et al. 2017) indicated that crop effects interacted with the tillage system being employed and varied between years. Alfalfa as the preceding crop resulted in significantly higher, 30 kg N/ha concentrations in the soil than where wheat was the preceding crop. This difference continued beyond the initial crop into subsequent seasons. Higher N levels in soil were also associated with increased concentrations of N in grain, typically in the form of protein.

6.5.4 Phosphorus Supply in Rotations

Of all the nutrients supplied by the soil and critical to production, phosphorus probably is chemically amongst the most diverse (also discussed in Chapter 4). Phosphorus can exist in soil in a wide range of organic compounds and in three major inorganic forms (PO_4^{3-}, HPO_4^{2-},

$H_2PO_4^-$) which exist in equilibrium in soil solution and which also form a range of complexed inorganic compounds with iron, aluminium and calcium. A small amount of P is present in the soil solutions and is available for root uptake by crop plants. This is in equilibrium with what has been termed the labile phosphate pool, which releases P to the soil solution to replace that absorbed by the crop. Beyond that is the major P pool, which releases P only slowly into the labile pool. Additions of inorganic P to organic rotations tend to be in the form of rock phosphate, which is very slowly available to crops and becomes part of this non-labile or slowly soluble P pool. Organic P will feed into both the labile and slowly labile pools, depending on its chemical form within the organic matter. Nye (1979) showed that while the uptake of nitrogen and potassium could be easily estimated for a typical root system, this was not possible for phosphorus. Problems related to the buffer power of the aggregates surrounding the crop roots, which made P uptake lower than might have been predicted, and at very low concentrations of available P by the activities of the rhizosphere micro-organisms.

For some crops, the release of organic chelating anions, which exchange with surfaces binding phosphate and from slowly soluble forms such as rock phosphate, can have a major impact on P availability to the crop. Kepert et al. (1997) showed that significant amounts of material, perhaps as much as 40% of carbon allocated to the root system, can be deposited in soil, and that such material can aid the dissolution of materials such as rock phosphate. This being the case, the ways in which the rotation can influence the supply of P to succeeding crops would seem to include the following.

1) Through effects on soil pH, which in turn will affect the balance of phosphate speciation and the mobilisation of P from slowly soluble sources such as rock phosphate.
2) Through changes in the balance between cations and anions in the soil solution which affect factors such as soil pH but also the availability of other nutrients.
3) Through changes in the make-up of rhizosphere organisms and, perhaps most importantly, arbuscular mycorrhizal fungi, which have been shown to have a significant impact on the ability of plants to absorb P.
4) Through the length and 3D architecture of roots produced by the crop which both affects the amount of P absorbed by that crop and also contributes to the organic P pool, some of which will be more soluble than prior to crop growth.

Some of these issues were discussed by Dissanayake (1992) who showed that different plant species left different nutritional 'footprints' in the soil (Table 6.5). There was a clear relationship across all species studied between crop growth and the amount of P absorbed which would have the effect of transferring inorganic P into organic P. Some crop species were better able to use rock phosphate as a P source than others, with the brassica species being particularly good at P uptake from this source, with an added uptake of calcium in tandem with the P. The plant species varied in their effect on soil pH which was lowered by the brassica species and pansy, especially when rock phosphate was applied. The balance of the uptake of measured cations (K, Ca, Mg, NH, Na, Al) and anions (NO_3, H_2PO_4, Cl, SO_4) differed between species (Table 6.6). It is clear that different crop species can have an impact on the nutrient regime and its availability to the incoming crop, and of all the effects on soil, pH and the development of the organic matter pool seem likely to be of major importance for both N and P.

Arbuscular mycorrhizal fungi (AMF) have long been known to be important to the composition of multispecies plant communities such as those found in the ley phases of the

Table 6.5 The uptake of phosphate and calcium by different plant species and their impact upon soil pH and nutrient levels when supplied with additional P in the form of Gafsa rock phosphate in a study lasting three months

Plant species	Increase in P uptake, resulting from the addition of rock phosphate, relative to control, mg/pot	Soil pH at conclusion of study	Ca uptake g/pot	Soil P at end of study mg/pot	Soil Ca at end of study g/kg soil
Oil seed rape	18.5	5.05	1.65	10.3	1.49
Cabbage	17.1	5.12	1.52	10.9	1.51
Sunflower	0.1	5.22	0.88	8.1	1.62
Pea	0.6	5.32	0.80	7.9	1.64
Carrot	1.1	5.20	–	5.9	1.79
Pansy	0.7	5.17	0.50	5.1	1.84

Source: Data from Dissanayake (1992).

Table 6.6 The sum of the uptake of measured anions and cations (meqiv/kg plant dry weight) and the estimated cation/anion ratio for a number of plant species

Plant species	Cation control	Uptake Gafsa	Anion control	Uptake Gafsa	Calculated cation/anion ratio
Oil seed rape	4033	3760	725	645	5.75
Cabbage	3691	3299	796	615	5.07
Sunflower	2529	2690	439	457	5.9
Pea	2911	2984	838	851	3.53
Carrot	3487	3248	699	640	5.1
Pansy	3324	3096	848	192	3.53

Source: Data from Dissanayake (1992).

rotation. The species composition, quantities and activity of AMF species can be affected by the crop species of the rotation (Jakobsen et al. 2002). Different AMF species differ in the amount of P they transport to the plant, while different plants affect the amounts of P being transferred by a given fungal species. This is linked to interactions between plant and fungal transporter genes. The composition of the AMF community can affect the floral composition of stages within the rotation where there is scope for species diversity to be present, such as in the pasture phases (van der Heijden 2002). AMF can aid promotion and sustainability of species diversity. The mycorrhizal dependency of a plant species is linked to the amount of P obtained from the fungus and these also affect the quantities of P transported through hyphal networks between individuals and species and the ability to respond to different AMF species.

6.6   Impact of Rotations on Crop Protection

The potential incidence of pests and diseases has long been a driver for the use of crop rotations. More general issues in respect of crop protection are discussed in Chapter 12. The approach taken by organic farming to what in other forms of agriculture would be identified as crop protection is probably better described as co-existence and linked to the concept of a healthy crop, as a result of its soil microbiology, having a greater ability to co-exist with other organisms and without the impact of what other approaches would describe as disease pressure.

The use of crops with different life cycles and growth habits is a fundamental facet of crop rotation design for the control of weeds, pests and diseases (Cook 2003; Garrison et al. 2014). The ways in which crops interact with pathogens and weeds are discussed in Chapter 12, and so here only broad principles and effects which specifically relate to rotational design are briefly identified to give a sense of what such an approach can deliver. This consideration, however, emphasises some of the major differences in total approach to food production between organic and conventional systems. While conventionally, most approaches to crop protection are targeted with specific chemicals or actions aimed at dealing with a single disease or weed, in organic production the effect achieved is the sum of a series of linked actions, of which the design of the crop rotation is one.

Weed management has often been identified as the single greatest problem in organic production. Weeds, as much as crops, are plant species which have been selected by the cultural practices employed (Harper 1977). Weeds are known to commonly reduce yields by around 25–30% although the design and successful implementation of the rotation are known to affect the size of such effects. Yields in a four-year corn, soya, oat and alfalfa rotation were better than in a two-year corn/soya rotation (Bruns 2012), with the improvement attributed in part to a reduction in weed pressure across the rotation. The preceding crop and its management in a rotation have been shown to affect weed development. Ercoli et al. (2017) found that weed biomass was affected by both tillage system and preceding crop, with wheat and alfalfa resulting in increased weed development where minimum tillage was in use.

Bastiaans and Berghuijs (2011a) reported that losses from weeds, pests and diseases in global agriculture were relatively similar. They assessed what was needed to optimise the ability of a rotation to maximise its impact on weed problems. They reported that the length of the rotation, the number and the order of rotational crops all had a significant impact. They suggested that at worst, the potential impact of weeds could be a yield reduction of 34% compared to potential losses of 16% from pathogens and 18% from pests. They estimated that losses resulting from weeds in rotational systems were generally smaller than this, which suggested that currently used rotational systems were having a significant impact on the influence of weeds. They also emphasised the importance of a range of approaches within the rotation which could be employed to reduce the risk of the weed flora adapting to a particular approach and therefore becoming a problem. A key need was preventing the development of connections between the weed flora and the core elements of the rotation's design. This meant that variations in planting dates and crop growth cycles were able to disrupt weed development, resulting in lower weed densities within a particular crop in a rotation compared to the same crop in a monoculture. Thus the selection of crops and their duration within the rotation, especially the duration and the management of the grass-clover ley phase, have a significant effect on the dynamics of the soil seedbank. Some crops, such as potatoes and swedes, allow cultivations during the cropping period, which can also enhance weed reduction over several seasons of the rotation.

However, the impact of a rotation is not always positive in relation to weed control, as volunteer crops surviving from the previous crop in the cycle can become a significant weed problem. Rotations seem likely to have the greatest impact on annual weeds, as opposed to perennial weeds, through a range of strategies associated with different crops that can help deplete the weed seedbank.

Similar issues apply to the control of pests and diseases. Pests and diseases can be very difficult to control in some organically grown crops, as application of control products is extremely restricted. As with weed control, the selection of crop for the various phases of the rotation is important. Potatoes are at risk from late blight but the use of early potatoes, bulked up before the disease usually becomes a major issue, can reduce the problem. Similarly, winter barley is typically sown in September in the UK, and so is exposed to disease for a long period before its harvest the following August. An alternative choice of winter cereal in the rotation, like wheat, triticale or oats, which are sown later in the year, may be beneficial. Crops susceptible to similar pests and diseases that are grown too close together in time within the rotation will encourage pest and disease problems. Cyst nematodes of potatoes require a gap of several years between potato crops to prevent populations becoming too damaging. Likewise, at least four years, if not longer, between brassica crops such as swedes, kale or oilseed rape should be encouraged in order to reduce club root. Breaking the life cycle of the pest or disease in order to confer longer-term control can include managing the weeds that act as hosts to pathogens. Shepherds purse, for example, can host the club root fungus.

Thus where crops are grown as a monoculture, crop pest and disease cycles have more opportunity to develop (Bruns 2012) while rotations have been shown to have significant impacts on diseases of wheat such as take-all (*Gaeumannomyces graminis*) and on nematode problems in maize. Bennett et al. (2012) showed that crops grown in short rotations or monoculture often suffer from yield decline compared to those grown in longer rotations, with greater breaks between susceptible crops. These researchers also showed that crops introduced into a rotation for the first time gave increased yield over repeated production. They suggest this is a combined effect related to soil microbiology as well as other cropping system factors. For many crops, including them too frequently in the rotation has the effect of reducing their yields, with grain legumes being one example (Pfender and Hagedorn 1985). In reviewing the conclusions of a conference assessing how a more integrated approach to crop protection could be made sustainable, Atkinson and McKinlay (1995) stressed the importance of crop management based on ecological principles. This contrasts with control in chemically based systems where little knowledge of weed or pathogen biology is needed relative to a detailed understanding of how and when to apply the pesticide.

Biological control has often seemed to be an attractive alternative to the use of chemicals, but practical experience has suggested that getting an applied organism to establish in the presence of related organisms in the field situation can be difficult. A similar conclusion was reached by Alteri and Nicholls (2012).

6.7 Stockless Rotations

Although the majority of organic farms have a mix of cropping (arable and/or grassland) and livestock rearing enterprises, there are some stockless organic cropping systems. These systems have to utilise their crop rotation in a way that is productive but does not include the phase where grassland-orientated components improve soil fertility in readiness for the nutrient exploitation cropping phase.

In areas with more productive soils, which lend themselves to arable and vegetable production on a large scale, the idea of reverting to a mixed arable and livestock system may deter conventional arable farmers from converting to organically certified production systems. Reasons which might deter a farmer from organic conversion might include the following.

- Cost of purchasing livestock.
- Fencing, water and winter housing requirements, which if not already in place would need to be provided, with costs potentially being prohibitive. Even if fencing exists on the farm, the likelihood of it being stock proof is low assuming the farm had only been growing combinable crops or vegetables for a while.
- There may be limitations in experience of livestock management and animal welfare, which are also very demanding in terms of time.

In circumstances such as these, a stockless organic system might be employed. In terms of crop nutrition, maintenance of soil fertility in a stockless system is still based around BNF (Biological Nitrogen Fixation). Stockless systems utilise a number of approaches to manage soil fertility including:

- nitrogen-fixing green manures such as clover or vetch, sometimes grown as an intercrop/ companion crop with a cash crop
- catch crop such as forage rye or forage rape, to retain nutrients in the system prior to the establishment of another cash crop
- grain legumes such as peas, beans or lupins, with N-fixing capabilities
- using grant schemes, such as set-aside, to maximise potential for use of N-fixing crops where appropriate.

6.8 Conclusion

For those engaged in conventional agriculture, whether or not to use a rotation or to employ some form of monoculture will always be a choice, and one which will involve decisions related to financial returns for particular crops at the time and the amount of management time or equipment involved with a more varied pattern of cropping. For those involved in organic production, rotations are important because they adhere to the basic philosophy and principles behind organic farming and because they are one of the proven methods of influencing both nutrient management and crop health (Hennessy 2004).

In Chapter 10 we discuss the specific potential role of the crop root system in making rotations and organic production function. In concluding our consideration of rotations, it is helpful to set the context for this discussion and of that relating to soil microbes (Chapter 11) and consider the potential root system contribution to the functioning of the rotation.

Table 6.7 The relationship between the design of an organic production system and the importance of potentially variable root system characteristics

Design feature of organic system	Potentially variable root system properties of significance	System need to be met through root selection
Rotational design, especially the selection, sequence and timing of crops	Size of root system Variability in distribution with depth Seasonal periodicity matched to the availability of nutrient release	Input of organic material into the soil Accessing nutrients though the whole soil profile Optimised ability to utilise available nutrients
Intercropping and green manuring use	Rapid initial root development Uptake potential Spatial variation in root distribution and timing	Rapid absorption of available nutrients released by mineralisation Rapid uptake of nutrients within absorption zone Co-existence with crop species
Crop variety selection	Variation in root system properties with a crop species Plasticity of response to varying soil conditions	Options for the design of particular points in the rotation Ability to deal with varying weather within a season
Planned nutrient additions (see also Chapter 9)	Ability to expand the root system to match nutrient availability Ability to modify the dominant speciation of the nutrient addition	Retention of added nutrient within the system Increasing the availability of the complex nutrient addition to the crop
Livestock use	Plasticity of growth and timing	Ability to absorb nutrients returned to the soil through animal manure
Integration of other vegetation presence (weed control strategy) (see also Chapter 12)	Profile of distribution with depth which differs from that of other species Variation in the timing of development	Obtaining nutrients in competition with other species Complementarily in growth periods
Integration with other organisms (pest and disease strategy) (see also Chapter 12)	The ability to develop effective AMF symbioses	The ability to tolerate diseases and pests

AMF, arbuscular mycorrhizal fungi.

Which of the needs of rotational systems can be met either by the selection of the right species of plant or cultivar or by causing the inbuilt plasticity within the root system to be expressed by the selection of the appropriate management? This is summarised in Table 6.7.

Organic producers use a minimal amount of approved crop protection chemicals, and only when deemed necessary, because they see them as incompatible with organic principles. Much of this role is filled in organic systems by rotations. Perhaps the emergence of problems with a range of crop protection practices in conventional production may lead to increased interest by conventional farmers in crop rotations as an important part of their future strategies. In addition, and as part of discussions of global climate change and its mitigation, rotations have again become of greater general interest because of their potential to increase carbon storage in the soil. Rotations, which have been vital to agricultural production for millennia, seem likely to continue in importance, so necessitating a better understanding of how they function and what they can deliver.

References

Alteri, M.A. and Nicholls, C.I. (2012). *Acta Horticulturae* 933: 35–41.

Anon (1970). *Modern Farming and the Soil*. London: MAFF.

Anon 1998, Annual Report of the Scottish Standing Committee for the Calculation of the Residual Values of Fertilizers and Feeding Stuffs, Scottish Government, Edinburgh

Armstrong Brown, S., Cook, H.F., and McRae, S.G. (1993). Investigations into soil organic matter as affected by organic farming in south East England. In: *Soil Management in Sustainable Agriculture* (ed. H.F. Cook and H.C. Lee), 189–200. Kent: Wye College Press.

Atkinson, D. (1991). Influence of root system morphology and development on the need for fertilisers and the efficiency of use. In: *Crops as Enhancers of Nutrient Use* (ed. V.C. Baligar and R.R. Duncan), 411–452. San Diego: Academic Press.

Atkinson, D. and McKinlay, R.G. (1995). Crop protection in sustainable farming systems. *BCPC Symposium Proceedings* 63: 483–488.

Atkinson, D. and Wilkins, R.J. (2004). *Eco-efficiency in the Future Pattern of British Agriculture*. BCPC Position Paper. Farnham: BCPC.

Balfour, E.B. (1943). *The Living Soil*. London: Faber and Faber.

Balfour, E.B. (1976). *The Living Soil and the Haughley Experiment*. New York: Universe Books.

Bastiaans, L. and Berghuijs, H.N.C. (2011b). Delivering the barebones for designing more weed suppressive crop rotations. *Aspects of Applied Biology* 113: 45–52.

Bennett, A.J., Bending, G.D., Chandler, D. et al. (2012). Meeting the demand for crop production: the challenge of yield decline in crops grown in short rotations. *Biological Reviews* 87: 52–71.

Brookes, P.C., WU, J., and Ocio, J.A. (1991). Soil microbial biomass dynamics following the addition of cereal straw and other substrates to soil. In: *The Ecology of Temperate Cereal Fields* (ed. L.G. Firbank, N. Carter, J.F. Darbyshire and G.R. Potts), 95–137. Oxford: Blackwell.

Bruns, H.A. (2012). Concepts in crop rotations. In: *Agricultural Science* (ed. G. Aflakpui), 25–48. London: Intech Books.

Cameron, K.C., Di, H.J., and Moir, J.L. (2013). Nitrogen losses from the soil/plant system: a review. *Annals of Applied Biology* 162: 145–173.

Cherrett, J.M. (1989). *Ecological Concepts*. Oxford: Blackwell.

Cook, R.J. (2003). Take-all of wheat. *Physiological and Molecular Plant Pathology* 62: 73–86.

Cooper, J.M., Baranski, M., Carmichael, A. et al. (2011). Effects of diverse and cereal intensive crop rotations on crop yields and soil fertility- reflections after the first cycle of the Nafferton long term trials. *Aspects of Applied Biology* 113: 37–43.

Dissanayake, DMAP 1992 Plant and soil factors influencing the availability of phosphorus from natural phosphate sources. PhD Thesis, University of Aberdeen.

Doltra, J. and Olesen, J.E. (2013). The role of catch crops in the ecological intensification of spring cereals in organic farming under Nordic climate. *European Journal of Agronomy* 44: 98–108.

Ercoli, L., Masoni, A., Mariotti, M. et al. (2017). Effect of preceding crop on the agronomic and economic performance of durum wheat in the transition from conventional to reduced tillage. *European Journal of Agronomy* 82: 125–133.

Fried, M. and Broeshart, H. (1967). *The Soil Plant System*. New York: Academic Press.

Garrison, A.J., Miller, A.D., Ryan, M.R. et al. (2014). Stacked crop rotations exploit weed-weed competition for sustainable Weed Management. *Weed Science* 62: 166–176.

Greenland, D. (2000). Effects on soils and plant nutrition. In: *Shades of Green* (ed. P.B. Tinker), 6–20. Stoneleigh: RASE.

Harper, J.L. (1977). *Population Biology of Plants*. London: Academic Press.

Hay, R.K.M., Russell, G., and Edwards, T.W. (2000). *Crop Production in the East of Scotland*, 1–61. Edinburgh: SASA.

Hennessy, DA, 2004, On monoculture and the structure of rotations. Working Paper 04-wp 369. Iowa State University, Ames.

HGCA 2009 Nitrogen for winter wheat– management guidelines. HGCA, Stoneleigh.

Howard, A. (1945). *Farming and Gardening for Health or Disease*. London: Faber and Faber.

Jakobsen, I., Smith, S.E., and Smith, F.A. (2002). Function and diversity of arbuscular mycorrhizae in carbon and mineral nutrition. In: *Mycorrhizal Ecology* (ed. M.G.A. van der Heijden and I.R. Sanders), 75–92. Berlin: Springer.

Jenkinson, D.S. (1982). The supply of nitrogen from the soil. In: *The Nitrogen Requirements of Cereals*, 79–94. London: MAFF.

Karlen, D.L., Mausbach, M.J., Doran, J.W. et al. (1997). Soil quality: a concept, definition and framework for evaluation. *Soil Science Society of America Journal* 61: 4–10.

Karlen, D.L., Varvel, G.E., Bullock, D.G., and Cruse, R.M. (1994). Crop rotations for the 21st century. *Advances in Agronomy* 53: 1–45.

Kepert, D.G., Robson, A.D., and Posner, A.M. (1997). The effect of organic root products on the availability of phosphorus to plants. In: *The Soil Root Interface* (ed. J.L. Harley and R.S. Russell), 115–123. London: Academic Press.

Knox, O.G.G., Leake, A.R., Walker, R.L. et al. (2011). Revisiting the multiple benefits of historical crop rotations within contemporary UK agricultural systems. *Journal of Sustainable Agriculture* 35: 163–179.

Lampkin, N. (2002). *Organic Farming*. Ipswich: Farming Press Books.

Law, R. and Watkinson, A.R. (1989). Competition. In: *Ecological Concepts* (ed. J.M. Cherrett), 243–284. Oxford: Blackwell.

Ledgard, S.F., Schils, R., Eriksen, J., and Luo, J. (2009). Environmental impacts of grazed clover/grass pastures. *Irish Journal of Agricultural Research* 48: 209–226.

Lemaire, G., Franzluebbers, A., de Faccio Carvalho, P.C., and Dedieu, B. (2014). Integrated crop–livestock systems: strategies to achieve synergy between agricultural production and environmental quality. *Agriculture, Ecosystems & Environment* 190: 4–8.

Lidgate, H.J. (1982). Nitrogen uptake of winter wheat. In: *The Nitrogen Requirements of Cereals*, 177–182. London: MAFF.

Manoharachary, C. and Mukerji, K. (2006). Rhizosphere biology – an overview. In: *Microbial Activity in the Rhizosphere* (ed. K.G. Mukerji, C. Manoharachary and J. Singh), 1–15. Berlin: Springer.

Nye, P.H. (1979). Soil properties controlling the supply of nutrients to the soil surface. In: *The Soil Root Interface* (ed. J.L. Harley and R.S. Russell), 39–49. London: Academic Press.

Olesen, J.E., Hansen, E.M., Askegaard, M., and Rasmussen, I.A. (2007). The value of catch crops and organic manures for spring barley in organic arable farming. *Field Crops Research* 100: 168–178.

Petersen, S.O., Schjonning, P., Olesen, J.E. et al. (2013). Sources of nitrogen for winter wheat in organic cropping systems. *Soil Science Society of America Journal* 77: 155–165.

Peyraud, J.-L., Taboada, M., and Delaby, L. (2014). Integrated crop and livestock systems in Western Europe and South America: a review. *European Journal of Agronomy* 57: 31–42.

Pfender, W.F. and Hagedorn, D.J. (1985). Aphanomyces as a component of the bean root rot complex in Wisconsin. *Annual Report of the Bean Improvement Cooperative* 24: 125.

Philipps, L., Stopes, C., and Woodward, L. (1993). The impact of cultivation practice on nitrate leaching from organic farming systems. In: *Soil Management in Sustainable Agriculture* (ed. H.F. Cook and H.C. Lee), 488–496. Kent: Wye College Press.

Price, W. (2014). *Fifty Foods that Changed the Course of History*. London: Apple Press.

Schoener, T.W. (1989). The ecological niche. In: *Ecological Concepts* (ed. J.M. Cherrett), 79–114. Oxford: Blackwell.

Stockdale, E.A., Rees, R.M., and Davies, M.G. (1993). Nitrogen supply for organic cereal production in Scotland. In: *Soil Management in Sustainable Agriculture* (ed. H.F. Cook and H.C. Lee), 254–264. Kent: Wye College Press.

Tosti, G., Farneselli, M., Benincasa, P., and Guiducci, M. (2016). Nitrogen fertilization strategies for organic wheat production: crop yield and nitrate leaching. *Agronomy Journal* 108: 770–781.

Van der Heijden, M.G.A. (2002). Arbuscular mycorrhizal fungi as a determinant of plant diversity: in search for underlying mechanisms and general principles. In: *Mycorrhizal Ecology* (ed. M.G.A. van der Heijden and I.R. Sanders), 243–266. Berlin: Springer.

Waring, R.H. (1989). Ecosystems: fluxes of matter and energy. In: *Ecological Concepts* (ed. J.M. Cherrett), 17–41. Oxford: Blackwell.

Watson, C.A., Kristensen, E.S., and Alroe, H.F. (2006). Research to support the development of organic food and farming. In: *Organic Agriculture: A Global Perspective* (ed. P. Kristiansen, A. Taji and J. Reganold), 361–383. Clayton: CSIRO Publishing.

Williams, S., Atkinson, D., and Sinclair, A.H. (1993). The effect of past soil management on the release of nitrogen from green manure. In: *Soil Management in Sustainable Agriculture* (ed. H.F. Cook and H.C. Lee), 497–502. Kent: Wye College Press.

7

What Can Organic Farming Contribute to Biodiversity Restoration?

Ruth E. Feber, Paul J. Johnson and David W. Macdonald

Wildlife Conservation Research Unit, The Recanati-Kaplan Centre, Department of Zoology, University of Oxford, Oxford, UK

> ... solitude in the presence of natural beauty and grandeur, is the cradle of thoughts and aspirations which are not only good for the individual, but which society could ill do without. Nor is there much satisfaction in contemplating the world with nothing left to the spontaneous activity of nature; with every rood of land brought into cultivation, which is capable of growing food for human beings; every flowery waste or natural pasture ploughed up, all quadrupeds or birds which are not domesticated for man's use exterminated as his rivals for food, every hedgerow or superfluous tree rooted out, and scarcely a place left where a wild shrub or flower could grow without being eradicated as a weed in the name of improved agriculture. If the earth must lose that great portion of its pleasantness which it owes to things that the unlimited increase of wealth and population would extirpate from it, for the mere purpose of enabling it to support a larger, but not a better or a happier population, I sincerely hope, for the sake of posterity, that they will be content to be stationary, long before necessity compels them to it. (John Stuart Mill. *Principles of Political Economy with some of their Applications to Social Philosophy, 1848*)

7.1 Why Conserve Farmland Biodiversity?

Mill was one of the great philosophers of the nineteenth century, and his influence today pervades much political, ethical and economic thinking. In this extended quote from one of his best known works, we see several strands of thinking with great relevance for the crisis facing modern agriculture. Mill's thoughts here acknowledge the conflict between food production and nature, and at a point in history where the conflict was a new problem. (He was writing at a time in the nineteenth century when changes to the UK countryside were becoming conspicuous.) He asks what is nature *for*? In his final words in the passage above, Mill also clearly considers how future generations would balance the needs of increasing numbers of humans with the destruction of nature necessitated by feeding more mouths. He ponders the

The Science Beneath Organic Production, First Edition. Edited by David Atkinson and Christine A. Watson.
© 2020 John Wiley & Sons Ltd. Published 2020 by John Wiley & Sons Ltd.

quality of life that such people would experience in the presence of the environmental degradation he foresees (with startling prescience). We can detect precursors of the idea of 'sustainability' in this passage. It is exactly these issues that confront modern farming, but rather more urgently.

The issues Mill raised are relevant for the current crisis in the sustainability of modern farming. In this chapter we look at those issues, and particularly closely at what the science of modern organic farming has to offer for contributing to a sustainable future, through the conservation of biodiversity. We focus here on the UK, but the central issues are global.

Mill's musings raise the fundamental question concerning the purpose of nature and biodiversity[1] (a word which did not, of course, exist until recently), a question to which we do not yet have a settled answer – a deficiency which continues to complicate policy making. The dominant western tradition holds that the world exists for the benefit of man. As Peter Singer discusses (Singer 1993), this worldview has religious roots: God gave man dominion over nature and the destruction of plants and animals cannot be sinful unless harm to humans results. Preservation of nature is, under this philosophy, encouraged only insofar as it benefits human well-being. Modern conservation ethics, a rather neglected field of inquiry, retains some echoes of the traditional worldview. One prominent school of thought, in the relatively new discipline of conservation science, is 'anthropocentrism'. This holds that all values are ultimately dependent on human interests and perceptions (Sarkar and David 2012). Some ethicists maintain that nature has an intrinsic value, and that this leads to a human obligation for its conservation (Vucetich et al. 2015). These ideas are clearly not exclusive.

Mill also alludes to posterity. The extent to which we have a duty to preserve biodiversity for future generations has been much debated. The idea has wide appeal: May (2007) finds this ethical basis for conservation 'more compelling' than purely utilitarian considerations[2] (although the basis of it remains anthropocentric in being founded on benefit to people, albeit those not currently existing). The 'stewardship ethic' has been enshrined in the environmental polices of both the UK and the USA (Hambler and Canney 2013).

A narrowly utilitarian basis for biodiversity conservation is based on the value of some services provided by the biosphere: those that can be costed in market terms (such as food, clean air and water). Farming is clearly a vital provider of food, and an influence on the others. One often quoted service provided by farmland biodiversity is that of pollinating insects for food production; this is of the order of billions of dollars globally each year (Gallai et al. 2009) and hundreds of millions of pounds annually in the UK (Garratt et al. 2014). A broader, but still utilitarian, perspective encompasses values that cannot be traded in the marketplace, such as spiritual needs (Sarkar and David 2012). Mill is strongly attracted to these in the passage we quote – he muses on the poverty of the experience provided by a countryside

1 We adopt the meaning derived from the Convention on Biological Diversity, as given on p. 15 of UK National Ecosystem Assessment, World Conservation Monitoring Centre (2011). UK National Ecosystem Assessment: technical report. United Nations Environment Programme World Conservation Monitoring Centre, Cambridge. 'Variability among living organisms from all sources including, inter alia, terrestrial, marine and other aquatic ecosystems and the ecological complexes of which they are part. This includes diversity within species, between species and of ecosystems'.

2 Mill is often characterised as a utilitarian philosopher but he rejected the cruder forms of utilitarianism originating with Jeremy Bentham (King 2004).

lacking in birds, shrubs and hedgerows. The term *biophilia* was coined by E.O. Wilson in the modern era to encapsulate the dependence of humans on engagement with nature; the science exploring the links between psychology, stress, and nature is an active area in current research (Hughes et al. 2013).

All of this also prompts the philosophical question of exactly what a conservationist, in this case one concerned with farmland, is aiming at. Is the aim to restore degraded habitats to their condition at a point before the degradation started (and, if so, is climate change influential?)? British farmland is a highly perturbed system and has been so for millennia. This has led some to question conservation efforts on the grounds that the landscape was rendered by enclosure, or is of necessity always changing – the 'enclosure' and 'kaleidoscope' myths respectively of Rackham (1997). Utilitarianism provides one answer, at least in theory, in that we can state what exactly we want nature to deliver and then estimate the habitats and species which will do the job. The reality of human affairs, and the distinction between asking what we need and what we want, leads to many trade-offs and inevitably to a political dimension to conservation policy (Macdonald and Willis 2013). It is clear that farmland, certainly in the UK, is expected to provide multiple functions, and that providing food is only one of them. In deciding what we mean by also providing a 'healthy' environment, the choice of baseline is important (Bull et al. 2014). But, as we shall see, the scale and timing of biodiversity loss on UK farmland constitute a continuing crisis such that this is not in practice a serious policy issue.

What are the roots of the current crisis, and by implication a serious failure in the stewardship ethic for mainstream farming? We need to understand this, and the manifest failures and unsustainability of conventional farming, to consider whether organic farming, or attributes of it, could be used to reverse current trends in biodiversity loss. Farming, by necessity, often results in the simplification of an ecosystem – some, usually most, of the natural flora and fauna are removed (Conway 2007) (what constitutes 'natural' is also debatable).

In the UK, the history of farming can be simplified into rather unequal phases (a longer account can be found in Macdonald et al. (2015)). In the first, which extended through much of the historical period up to the later part of the eighteenth century, farm structure had much in common with land that was cleared from woodland in the Iron Age. This began to change with the Enclosure Acts of the eighteenth century, when much, previously common, land was taken into private hands (well over half a million acres between 1845 and 1864 according to Shoard (1987)). By the nineteenth century, wild scenery was cherished because it provided a refuge from the cities (Thomas 1983). Under the influence of poets such as Wordsworth, common land near towns had acquired importance for the growing urban populations for fresh air and exercise, a turning away from the previously utilitarian outlook (Shoard 1987). At this time there was some habitat destruction under pressure from farming (the Corn Laws exacerbated pressure by limiting food imports (Macdonald et al. 2015)); Oliver Rackham points to 1851, for example, as a black year, as significant habitats were lost for farming (Rackham 1997). Mill had plainly seen some similar losses.

But a hundred years later, destruction began on a scale which Mill would have found unimaginable. The intervening years had been relatively uneventful; very few hedges, for example, were lost between 1870 and 1945 (Rackham 1997). Mabey (1996) notes that, in 1926, Sir Edward Salisbury recalled of a cornfield near Oxford when the Prince of Wales was visiting that it '...looked as if it had been sown for the occasion, since it was red, white and blue with Poppies, Mayweed and *Centaurea cyanus* [cornflower]'. Indeed, the cornflower was so common in Britain until the mid-nineteenth century that it was considered 'a pernicious weed injurious to the corn and blunting the reapers' sickles'. But between 1950 and 1975, the destruction of wildlife on farmland occurred on such a scale that Rackham refers to this period as the 'locust' years. By the end of the 1970s, the once pestilential cornflower had become a nationally scarce species in a transformed landscape. The political stimulus for this was provided by the deprivation of the Second World War: the Scott Report of 1942 into UK land use urged farmers to 'make every acre count'. Pressure for higher yields provoked spiralling use of pesticides and fertilisers, semi-natural grassland was drained and tilled, and woods, hedges and fens were destroyed. After three millennia of co-existence with agriculture, UK wildlife had been decimated in less than 30 years (Mabey and Nature Conservancy Council 1980).

The size of the effect of farming on biodiversity in the second half of the twentieth century is now known to be enormous (see Macdonald et al. 2015). During the twentieth century, approximately one species of plant was lost every two years (average from 23 English counties), with greatest losses in the 1960s (NE 2010). Over the past 50 years, 60% of those species for which trends can be quantified[3] have declined, and 31% have declined strongly (Burns et al. 2013). Where long-term data exist, for birds, moths and butterflies for example, severe downward trends are recorded. Fox et al. (2011) reported that 72% of butterfly species have decreased in abundance over 10 years and 54% decreased in distribution at the UK level. The numbers of butterflies fell by 24% over 10 years and the rate of decline for some butterfly species appears to be increasing. Farmland birds are one of the best monitored and most important biological indicators of environmental health. Of 19 UK indicator bird species, 12 declined in numbers between 1970 and 2010 (Eaton et al. 2013). Similarly, many of Britain's larger moths, another well-monitored group, though still widespread (and in some cases still relatively common), have been in decline since recording began in 1968. A recent summary of the Rothamsted national moth light trap network data indicated that approximately two-thirds of 337 species had declined over a 35-year period (Fox et al. 2013). Some of these have shown declines of over 70% over this period (Figure 7.1).

In Europe, the agricultural regime responsible for this scale of biodiversity loss was supported by the Common Agricultural Policy (CAP). This subsidised production was grossly inefficient, eventually leading, at great expense, to lakes and mountains of unwanted produce. By the 1980s, it was clear this was both financially and environmentally catastrophic. Several fundamental changes in policy followed (Macdonald et al. 2015). In 1988, a budget ceiling was introduced, based on a maximum produce quantity for which support payments were guaranteed (e.g. quotas for milk). A compulsory 'set-aside' policy was also introduced, which paid farmers for taking land out of production, with the aim of reducing the grain mountains (as it happened, the policy also delivered some environmental benefits) (Firbank et al. 2003).[4] Production subsidies were phased out and, at the same time, opportunities were introduced for farmers to be paid for conservation measures on their land. By the late 1980s a framework of agri-environment schemes (AES) was being developed (European Commission 2013). The detail of the schemes and how they are administered has undergone a number of changes, but

---

3 This is 3148 species, 5% of the 59 000 UK terrestrial and aquatic species.
4 'Set-aside' became compulsory in 1992, but was abolished in 2008 to increase EU cereal supply to the market and hence reduce prices, which had been high after low EU harvests.

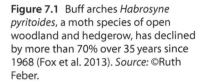

Figure 7.1 Buff arches *Habrosyne pyritoides*, a moth species of open woodland and hedgerow, has declined by more than 70% over 35 years since 1968 (Fox et al. 2013). *Source:* ©Ruth Feber.

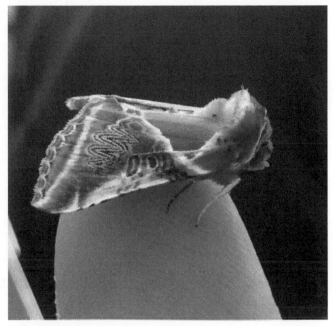

essentially they provide financial support for the delivery of environmental, particularly biodiversity, benefits. Options include, for example, the enhanced management of boundary strips, sowing seed mixes for birds and pollinators, and restoring features such as hedgerows and ponds. At the time of writing, how schemes like these will be funded in the UK is in some doubt; following a referendum in June 2016, the UK plans to leave the EU in March 2019.

Agricultural intensification, widely believed to be responsible for many species declines, encompasses an array of changes in farming practice. Increased use of pesticides and fertilisers, changes in sowing dates (autumn sowing has greatly increased over the past 60 years while spring sowing has decreased), fewer overwinter stubbles and the loss of traditional crop rotations have accompanied land drainage, pasture improvement and hedgerow and pond removal (Fuller 1987; Vickery et al. 2001). All of these have affected biodiversity. There are many mechanisms for the effects (Chamberlain et al. 2000). Benton et al. (2003) reviewed the evidence for the impact of removal of non-cropped habitats from farmland on birds, concluding that loss of heterogeneity at multiple spatial scales was one of the causes. Of 13 components of intensification, Geiger et al. (2010) found that insecticides and fungicides have had the most consistent negative effects. Recently, the widely used neonicotinoid insecticides have been the focus of particular attention due to their effects on non-target species, especially pollinators (Goulson 2013).

Agri-environment schemes aimed at countering the effects of intensification, and promoted both by economic and environmental motives, have brought some biodiversity benefits. There have been local successes for some species (e.g. the cirl bunting, *Emberiza cirlus*). A recent review concluded that, properly deployed, these schemes are useful (Batáry et al. 2015). However, they do not tackle the crisis: there is little evidence that widespread declines for many species are being reversed. The recent revolution in policy concerning farming and conservation (and the expansion of AES which came with it) is clearly not enough. Goulson's (2015) compelling explanation has, in our view, interesting implications for the potential for organic farming philosophy to tackle the problems. He argues that much of the existing AES land is contributing little to biodiversity, and that there is little policing of what farmers are doing. The second and more serious strand to his thesis concerns the lack of any incentives for conventional farmers to abandon reliance on pesticides. Government funding for research into crop rotations, biological control and so on has drastically reduced. The dominance of the private sector in agronomic funding means the profit motive holds sway – the goal of farming remains then to maximise profit, and not to protect wildlife or to ensure sustainability. Goulson also argues that the current UK regulatory system is inadequate. Neither chronic long-term exposure nor interactions between different applications are considered. The number of pesticide applications per field more than doubled between 1990 and 2015 (Goulson et al. 2018). It is clear that the market alone will not deliver what is needed – many of the ecosystem services provided by the landscape are still not valued in monetary terms (Godfrey and Garnett 2014).

7.2 What Can Organic Farming Contribute to Biodiversity Conservation?

The 'business as usual' model is not working. Organic farming, with its strong emphasis on sustainability, may have lessons for policy makers seeking to meet the current challenges. We do not purport to be advocates of organic farming. As currently defined, organic farming has rigorous restrictions on permissible and prohibited practices – its proponents defend it as a 'holistic' system. We have treated it here as an observational experiment on the effects of low-intensity farming. We ask whether organic farms are more wildlife friendly and, if so, what is it that makes them so? This may simply reflect the absence of the changes that intensification effected on much of the agricultural landscape. But how big are the differences? Perhaps most obviously, organic farming differs from conventional farming with regard to pesticide and fertiliser inputs; some types of fertilisers and pesticides are used but they are strictly limited compared to conventional farming (see Chapters 8, 10, 12). Organic farming takes a systems approach to management and relies on the use of crop rotations, green manure, compost, and biological pest control. Cropping patterns differ: there is a greater emphasis on spring sowing of cereals, whereas autumn sowing in conventional farming has greatly increased over the past 60 years. Organic crop rotations typically include a fertility-building ley.

Norton et al. (2009), in a study of 89 pairs of organic and conventional farms across England, quantified many of the differences that might influence biodiversity. Fields were smaller on organic farms and boundary density (km/ha) of all boundaries (including hedgerows) was consequently higher. Organic hedgerows had fewer gaps and were higher and wider than those surrounding non-organic fields. Organic farmers cut their hedges less often and were more likely to use a traditional hedge management method (laying). Organic farms also had a higher proportion of grassland compared to non-organic farms. Organic farmers' crop rotations differed, with organic systems always including a grass ley as part of a cereal/vegetable rotation. Approximately a fifth of non-organic farms cropped continuously, but no organic farmers did this. Organic farms were more likely to have livestock (and a wider variety of types of livestock) and were more likely to use them on arable land. More organic farms had agri-environment agreements (in addition to the Organic Farming Scheme) than non-organic (Norton et al. 2009).

Many of these differences are likely to affect biodiversity. Some of the links between farm management practices and wildlife are well established. For example, a number of bird species may be favoured by bigger hedges (Hinsley and Bellamy 2000; Macdonald and Johnson 1995). Small mammal biomass has been shown to be higher when there is more hedgerow habitat available (as measured by hedgerow width, height and length) (Gelling et al. 2007). Heterogeneity resulting from mixed cropping and the presence of livestock is also important for biodiversity (Benton et al. 2003; Henderson et al. 2009). We also know that communities of plants and invertebrates can be more diverse where pesticide inputs are reduced or removed (Pelosi et al. 2013; Petit et al. 2015). So we expect organic farms to be biodiversity friendly for a variety of reasons, and many studies suggest this to be the case.

Plants (Batáry et al. 2013; Gabriel et al. 2006; Hardman et al. 2016; Hyvonen et al. 2003; Rundlöf et al. 2010) and invertebrates (Happe et al. 2018; Holzschuh et al. 2008; Kragten et al. 2011; Lichtenberg et al. 2017) have generally been found to benefit most from organic farming. Positive effects of organic management have also been demonstrated for small mammals (Coda et al. 2015; Fischer et al. 2011), birds (Chamberlain et al. 1999; Kragten and de Snoo 2008; Smith et al. 2010), and bats (Fuller et al. 2005; Wickramasinghe et al. 2003). A meta-analysis of published studies by Bengtsson et al. (2005) suggested that organic farming was associated with increased species richness and abundance of plants, predatory invertebrates and birds. Hole et al.'s (2005) meta-analysis also provides clear evidence for a beneficial effect of organic farming on biodiversity. Tuck et al. (2014) found that, on average, organic farming increased species richness by about 30%.

Butterflies have complex life cycles, with egg, larval, pupal and adult stages, and so require larval foodplants (often species specific), nectar sources for the adult stages during the spring and summer, and habitats in which the egg, larva, pupa or adult can overwinter. They are particularly sensitive to habitat change. We monitored butterflies on 12 pairs of organic and conventional farms across southern England over three years (Feber et al. 2007) and found butterfly abundance to be higher on organic compared to conventional farms in all three years of the study. There were also more butterfly species on organic farms in each year, significantly so in one of the years.[5] Most species were more abundant on organic farms in most years, particularly three of the less mobile species: large skipper *Ochlodes sylvanus*,[6] common blue *Polyommatus icarus*[7] and meadow brown *Maniola jurtina*.[8] Several features of organic farms are likely to have contributed to these differences. Loss of larval foodplants and adult nectar sources are significant threats for butterflies. The aspects of intensification most relevant for affecting these are grassland improvement (Vickery et al. 2001), herbicide use and fertiliser

---

5 $t = 7.48$, d.f. = 7, $P<0.001$: Feber et al. (2007).

6 1994: mean abundance per km (S.E. in parentheses): 3.2 (0.98) on organic, 1.5 (0.66) on non-organic. $F_{(1,7)} = 6.17$, $P = 0.042$.

7 1994: mean abundance per km (S.E. in parentheses): 7.4 (2.33) on organic, 2.5 (0.98) on non-organic, $F_{(1,7)} = 6.95$, $P = 0.034$.

8 1995: mean abundance per km (S.E. in parentheses: 15.5 (1.54) on organic, 9.1 (1.07) on non-organic, $F_{(1,7)} = 21.51$, $P = 0.001$.

application (Smart et al. 2000), and reduction in uncropped habitats (Feber and Smith 1995). Concerning the last of these, our organic farms did have larger hedgerows and more perennial field-edge plant communities, providing shelter as well as food. The greater prevalence of grassland on organic farms was also likely to have had benefits for butterflies. Larvae of the meadow brown and gatekeeper, for example, feed on grass and overwinter in the larval stage. Stable grass leys, in place for two or more years on organic farms, and the presence of more perennial field-edge plant communities are likely to favour these species.

7.3 Effects of Organic Farming Vary with Taxa

Many studies of organic farming have been small scale and geographically restricted, a short-coming highlighted by Hole et al. (2005). They are also often taxonomically restricted. Recognising this problem, we sampled 89 pairs of English farms (Fuller et al. 2005) to find out if the organic effect was consistent across taxa and landscape types. Higher plants, spiders, ground (carabid) beetles, wintering birds and bats were studied to represent a spectrum of trophic levels, niches and ecological requirements. We focused on farms that had cereal fields, and we sampled both autumn-sown, referred to as 'winter', cereals (commonly grown on non-organic farms), and spring-sown, referred to as 'spring', cereals (more often grown on organic farms). The organic farms in the study were (i) at least 30 ha in area, (ii) not highly fragmented holdings (i.e. where organic fields were interspersed with non-organic fields), and (iii) not predominantly agro-forestry or horticultural. Virtually all suitable organic farms in England growing winter-sown wheat and spring cereals at the time of the study were examined. The organic farms were paired with non-organic farms using a procedure that was purely geographical and not based on any attributes of either system. The farms were widely distributed, but there was a cluster to the east and south of Bristol.

Plants and invertebrates were sampled at the 'field' scale; cereal fields were sampled over three years. Both the spring cereal and the winter cereal fields were approximately equally divided between recently converted (<5 years) and old organic (>5 years). Plants were recorded in plots in the field boundary, the crop edge and within the field. Spiders and carabid beetles were sampled using pitfall trapping, before and after harvest. Eighteen traps per field sampled the crop and uncropped boundary habitats, with nine traps in each habitat. Spiders and carabid beetles were identified to species. Bats and birds were studied at the 'farm' scale. Summer bat surveys were completed pre-harvest on 65 farm pairs between June and August in 2002 and 2003. Winter surveys of birds were carried out on 61 farm pairs on the target field and up to five adjacent fields. Birds were mapped on large-scale maps and individual records were subsequently allocated to habitat categories. The focus was on wintering, rather than breeding, birds because it was thought likely that, in winter, flocking seed-eating birds in particular might be attracted to organic farms if these farms provided concentrations of seeds in the wider landscape.

Organic farming was mainly associated with positive effects on biodiversity, but the effect varied substantially amongst the different taxonomic groups (Fuller et al. 2005). Species density and abundance were typically higher on organic farms, but patterns of diversity were less clear. The direction of the effects was consistent, with all but one of the significant differences relating to higher diversity or higher abundance on organic farms compared to non-organic farms (Figure 7.2). The largest and most consistent effects were for plants. Organic fields were estimated to hold 68–105% more plant species and 74–153% greater abundance of weeds (measured as cover) than non-organic fields, and cover of weeds was consistently higher at all distances into the crop (Fuller et al. 2005). The least consistent effects were observed for carabid (ground) beetles. Schneider et al. (2014) also observed consistent effects for plants and bees, with less consistent effects for earthworms and spiders. A review concerning central and northern Europe found that predatory groups were more likely to be 'losers' than 'winners' of organic farming where diversity was concerned, and carabid beetles were one of the groups affected negatively (Birkhofer et al. 2014). Batáry et al. (2012) found that functional groups may be affected in different ways by organic farming.

The effect of farming system on birds (see Figure 7.2) was studied in more detail by Chamberlain et al. (2010). Of 16 species considered, none was more abundant on non-organic farms, and six showed statistically significant effects. Variation in habitat abundance of the type detailed by Norton et al. (2009) was thought to be a plausible explanation although, for some species, there were no habitat effects. Hedgerow density, the proportion of arable area at the farm scale (for stock dove *Columba oenas* and jackdaw *Corvus monedula*), and the grass/arable ratio at the landscape scale (for woodpigeon *Columba palumbus* and jackdaw) were influential.

Non-cropped habitat features were also important for the other, more mobile, groups that we studied. Our analyses suggested that presence of ponds, hedgerows and livestock promoted bat activity. Bat activity was higher on farms with livestock, and the effect was more marked on

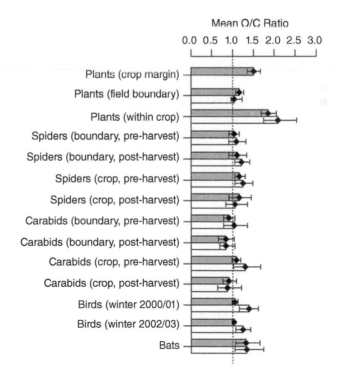

Figure 7.2 Effect of farming system on numbers of species (*grey bars*) and abundance (*open bars*) with confidence intervals. Dotted reference line at ratio = 1.0 indicates no system effect. No data for plant (crop margin) abundance. *Source:* Feber et al. (2015a).

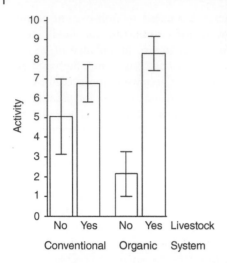

Figure 7.3 The association between the presence of stock on a farm and mean bat activity (± SE). Bat activity is number of passes expressed per 3 km. *Source:* Feber et al. (2015a).

organic farms (Figure 7.3). Overall, however, bat activity was low on all farms, and the differences in bat activity seen between conventional and organic farms were modest (Feber et al. 2015a). Our analysis suggested that much of the British landscape, and particularly that used for cereal production, offers rather unfavourable habitat for bats regardless of farming system. Attributes of farms that were favourable to bats were also, to some extent, confounded with farming system, as all were more common on organic than non-organic farms. Nevertheless, our data suggest that the quality of these features was greater on organic than conventional systems. Organic farming may not offer a simple panacea to the decline in bat populations; better hedgerow and waterway management on all farms and an overall increase in the number of mixed rather than purely arable enterprises would be likely to yield benefits.

7.4 How Rapid is the Effect of Conversion to Organic on Biodiversity?

Studying how plant and animal communities change as a farm converts to an organic regime can be informative. Such studies are scarce, perhaps because they need to be long term. Jonason et al. (2011) examined the effects of time since transition to organic farming on plants and butterflies in Sweden. The organic farms had transferred from conventional management between one and 25 years before the survey took place. They found that the effect of organic farming on diversity was rapid for plant and butterfly species richness, whereas butterfly abundance increased gradually with time since transition. A shift towards increased biodiversity might be expected to occur very soon after conversion and the removal of synthetic inputs. Damgaard et al. (2014) found that the species richness of hedgerow ground flora increased since conversion to organic of the adjacent fields, until a maximum was reached. Preconversion farm management may be influential. For example, farmers sympathetic to organic philosophy may be converting farms which already have low inputs or are managed in wildlife-friendly ways. On such farms, conversion may have little effect, particularly in less favoured areas such as uplands (Fraser et al. 2014).

The effect of herbicide removal on botanical diversity is a conspicuous change associated with conversion, and with clear implications for the habitats and foods of many other organisms. Jonason et al. (2013) found that moths were more abundant on recently converted organic farms. Withdrawal of herbicide led to high densities of *Cirsium arvense*, a pernicious weed, but one which can be an important nectar source. Pollinator abundance and community composition within 2–4 years since conversion to organic have been evidenced by significant improvements in the pollination success of strawberry crops (Andersson et al. 2012).

Fuller et al. (2005) found no effect of time since conversion, but the metrics they report were restricted to abundance and species richness. These may not be sensitive for detecting community changes following conversion. Schroter and Irmler (2013) showed that carabid communities changed following conversion. They speculated that, under conventional farming, arable fields are barriers for most species of dispersing ground beetles, while conversion to organic farming supports their invasion into the arable fields by offering more food resources for predators or seed feeders. We have also recently observed a tendency for low-mobility hunting spiders to increase in prevalence with time since conversion (unpublished data). The changes were not abrupt.

7.5 Landscape Context and Species Traits

In our study, we found that organic farms were not distributed in the landscape in the same way as were conventional farms, and that this could affect biodiversity. For example, when we explored the links between bat abundance and the complexity of the landscape, measured at three spatial scales, 1, 9 and 25 km^2, we found a clear link between landscape complexity and bat activity; some of the benefits of organic farming could therefore be due to off-farm influences. Other studies of organic farming and biodiversity have reported similar effects. Schmidt et al. (2005) found more spider species where the surrounding landscape had a higher proportionate area of non-cropped habitats, and this was true regardless of farming system. They attributed this to a higher availability of refuge and overwintering habitats. Spider density was affected more by local management practices, including organic farming. Carabid assemblages are strongly influenced by the quantity and arrangement of habitat elements at the scale of the landscape (Burel 1989; Millán de la Peña et al. 2003); Jonason et al. (2013) showed that, for carabid beetles, species richness (and weed seed predation by carabids) was influenced more by wider landscape context than by local factors. Plant species richness in arable fields is greatly influenced by processes operating at the landscape scale (Gabriel et al. 2006) and similar observations have been made for arable weeds (Roschewitz et al. 2005). Henckel et al. (2015) observed that the presence of organic farming was associated with higher weed densities on conventional farms. Butterfly abundance and species richness were significantly greater only on organic farms in homogeneous landscapes in Sweden (Rundlof and Smith 2006; Rundlof et al. 2008), and Holzschuh et al. (2008) demonstrated that an increase in organic farming in the surrounding landscape increased bee species richness and bumblebee density at the local level.

Landscape context was influential for invertebrates in our study. We asked whether we could detect responses of spiders to organic farming, and whether these differed at local and landscape scales (Feber et al. 2015b). We also considered whether responses to organic farming differed according to spider dispersal strategies. Spider species were assigned to one of two broad guilds. Hunting spiders, which include the wolf spiders (Lycosidae), generally disperse on the ground by walking. Web builders, which include the linyphiids, frequently disperse long distances by a method known as ballooning (Bell et al. 2005). This classification (also used in other studies (e.g. Batáry et al. 2008)) allowed us to explore patterns in the data which might suggest different local and landscape impacts of organic farming on groups with potential differences in dispersal characteristics.

Our results showed that both farming system and landscape affected spiders, but spider dispersal ability was clearly influential. For the hunting spiders, which have more limited dispersal, farming system and landscape (extent of arable land) were both statistically significant predictors. More hunting spiders were captured, and more species of hunting spiders captured, on organic compared to conventional farms; the farming system effect was particularly marked in the cropped area before harvest – an average of 77% more individuals and 36% more hunting spider species were captured on organic farms (Feber et al. 2015b). No effects of farming system or landscape type were detectable for web-building spiders, many of which have high dispersal abilities, suggesting that local impacts of pesticides, for example, on conventional farms are buffered by rapid recolonisation. A benefit of organic farming to spiders, restricted to hunting spiders, has also been observed in Germany (Batáry et al. 2012), and Schmidt et al. (2005) reported similar system effects for two taxa of spiders which walked into fields from overwintering sites, and which were not observed for ballooning spiders.

The farming system difference (more hunting spiders captured on organic compared to conventional farms) was much more marked in less arable (more complex) landscapes. In the cropped area, it resulted from the presence of an upward trend in activity density and species density of hunting spiders in more arable landscapes on conventional farms but not organic farms. This greater effect of organic farming on hunting spiders in more complex landscapes was unexpected, and counter to the effects observed by Schmidt et al. (2005), who reported more spiders on conventional farms in complex landscapes but no such effect on organic farms, and Batáry et al. (2012), who found hunting spider abundance to be lower in more intensively managed landscapes, the effect more marked on conventional farms.

One possible explanation for the trend we observed (consistent over two years) could be an increasing reliance on the cropped area for resources in arable-dominated, conventionally farmed landscapes, a trend not observed on organic crops where abundance and diversity were independent of landscape complexity. Species that dominate arable landscapes appear adapted to disturbed habitats (Samu and Szinetár 2002). Female spiders of one such genus, *Pardosa* (Lycosidae), which are hunting spiders, have been found to be in better condition in more arable landscapes (Oberg 2009). Schmidt et al. (2008) found that, while most spider species in arable fields benefited from non-crop habitats in the surrounding landscape, some arable species declined when landscapes became dominated by non-crop habitats. The interactions between landscape and farming system highlight the importance of developing strategies for managing farmland at the landscape scale for effective conservation.

In the UK, organic farms are concentrated in the more heterogeneous landscapes of southwestern England where our results, for invertebrates at least, indicate the benefit of organic farming to biodiversity are likely to be greatest. Our observations support the conclusions of Gabriel et al. (2009), who showed that environmental factors associated with lower agricultural potential predispose conversion of farms to organic production, and that this results in regions with a high prevalence of organic farms. They argue that the most efficient conservation strategy for English farmland biodiversity would be to encourage the conversion of non-organic farms to organic production in regions where organic farming is already prevalent, and to maximise the intensity of production in areas where it is not.

7.6 Wider Considerations

The extent of the challenges facing agriculture and its impact on the environment are now well understood (Poppy et al. (2014) list recent reports on this). For example, food production currently relies on unsustainable soil management and often involves excessive water extraction. The concept of sustainable intensification (SI) is intended to encapsulate a range of methods for tackling these challenges, acknowledging that increased food production must be part of the solution (Godfrey and Garnett 2014). Increasing production without environmental harm is the aspiration. While we know that current world calorific production exceeds that needed to feed the world's population, social and economic factors prevent its distribution (Ponisio et al. 2014). Sustainable intensification (SI) does encompass the need to change human behaviour, to reduce waste and the amount of meat consumption (Godfrey and Garnett 2014). But food production will still need to increase before these issues are solved.

The range of options to achieve SI presents a complex of decisions concerning technology, and balancing the functions of agricultural land other than providing food. The development of artificial meat, for example, may reduce pressure on farmed land (Tuomisto and Teixeira de Mattos 2011). The current philosophy of organic farming (see Chapters 1, 2) as defined in the UK by the Soil Association (see Chapter 1) precludes the use of some technology options, novel genetically modified (GM) crops for example. This has been said to be paradoxical (Krebs 2013), there being no rational reason for organic farmers to oppose the use of pesticide-resistant crops. It is of course possible to envisage low-intensity farming using GM crops. Other elements of SI are more aligned with organic farming philosophy, in addition to the emphasis on wildlife. Garnett et al. (2013) argue for SI to embrace a wide ethical framework, barring some options, those with unacceptable livestock welfare standards, for example.

Sustainable intensification considers, with obvious relevance for organic farming, the appropriate balance between intensification (increasing yields on existing land) and extensification (bringing more land into production). Related to this is 'land sharing' versus 'land sparing' – under what conditions is it optimal to adopt low-intensity farming across a wide area, compared with the option of farming intensively in some areas while sparing land elsewhere from agriculture altogether? A continuum of options ranges from extensification of the entire farm area through, for example, low-intensity grazing and various low-input systems, to fine-scale land sparing. Organic farming can be thought of as a version of the 'land sharing' option. As Godfrey and Garnett (2014) point out, in Europe the main policy question is not one of simply sparing versus sharing *per se* but whether to focus investment in the areas most benefiting or to spread it more thinly.

Organic yields are lower compared to non-organic; for example, Gabriel et al. (2013) found organic crop yields to be around 55% of those of non-organic farms in two areas of England. The yield differences differ between sites, as we would expect (Seufert et al. 2012). The size of the effect may have been overestimated. Ponisio et al. (2014) put it about 20% lower than conventional; they also found that multicropping and crop rotations reduced the gap, and suggested that organic management schemes could be refined to reduce or even eliminate the yield gap for some crops or regions.

Whether organic farming can contribute to biodiversity without unacceptable yield loss depends on a variety of factors. For example, Phalan et al. (2014) found that for tropical birds, land sparing was best. Hodgson et al. (2010) explored the relative merits of balancing wildlife-friendly land versus productive land under different crop yield scenarios. Their analysis suggested organic farming could be optimal for much lower yield gaps for butterflies in UK farmland; the critical yield ratio for winter cereals, for example, was 45%. For organic yield higher than that, organic farming was optimal. Gabriel et al. (2013) assessed the trade-off between yield and biodiversity in organic and conventional farms in lowland England. They concluded that organic farming did not have an effect other than via reducing yields and therefore increasing biodiversity. Only plants benefited substantially from organic farming.

In Europe, multifunctionality (that is, that farmland must deliver a whole range of ecosystem services, of which biodiversity is just one) has largely motivated the development of agri-environment schemes. This is not so everywhere. In the US, for example, farmed land is generally viewed as being for food production, with specific non-farmed wild areas (national parks/reserves) set aside for wildlife. In Africa, the agricultural land of a smallholder may be expected to produce only enough food for a single family.

What of wildlife in the wake of intensification? From developed agricultural systems in Europe and from subsistence farming in developing countries, the answer is always the same: agricultural intensification coincides with biodiversity loss (Mattison and Norris 2005). Farming methods and the affected species may differ but there is always competition between species for nutrients, space, light and water. So, too, similar problems recur: eutrophication and pollution of waters by agrochemicals, fertilisers, pest and weed control in Europe and, by crop encroachment into forests and wetlands, nutrient depletion and soil erosion in Africa. The difference between farming systems lies in how the different components of farming are valued. Wildlife on farmland can be either desirable (e.g. pollinators) or undesirable (pernicious weeds, some invertebrates (slugs) and some birds (pigeons) because they reduce production). How do we achieve the balance between controlling those aspects we do not want, but with minimal damage to wildlife that we view as desirable or essential to ecosystem function? Food security and environmental protection are overlapping spheres, but not inevitably harmonious.

Much recent thinking on the economics of conservation has been heavily influenced by the Millennium Ecosystem Assessment (MEA 2005). This analyses the full spectrum of links between ecosystem health and human well-being, including providing recreational places. In the UK, it led to the National Ecosystem Assessment (UK National Ecosystem Assessment, World Conservation Monitoring Centre 2011). This emphasises the value of natural capital and all the ecosystem services it provides. One basic example is that agricultural production is supported by complex communities of soil organisms and natural enemies of crop pests. One influential view is that the underpricing of nature leads to its overexploitation (Dasgupta 2010). There are some fears that this leads to an over-reliance on the commoditisation of wildlife and to unforeseen outcomes. Corporate finance might, for example, seek to influence habitat destruction if it could be shown that its value could be offset by creating new habitat elsewhere. The policy and practice of biodiversity offsetting are fraught with uncertainties (Curran et al. 2014; Evans et al. 2015). Some argue that instrumental arguments for protecting nature are inadequate and that the ethical basis for conservation has underplayed the intrinsic values of nature (Vucetich et al. 2015).

The understanding that everything is linked to everything else has led to life cycle analysis (LCA). LCA assesses the overall environmental impact of products, processes or services. It follows a product 'from cradle to grave'. The impact of resource extraction (e.g. land use), material production, manufacture, consumption/use and the end of life of the product (for example, whether it is recycled, how it is collected and how the waste is disposed of) are all taken into account. The aim is to estimate all the environmental impacts, including climate change, eutrophication, acidification, effects on human health and ecosystems, depletion of resources, water and land use, and noise (Rebitzer et al. 2004). LCA has been used to compare organic versus conventional farming in terms of greenhouse gas emissions from milk production (Flysjö et al. 2012), and between integrated, organic, extensive and intensive farming production (Nemecek et al. 2011a, 2011b; Tuomisto et al. 2012). The latter authors concluded that, for the range of scenarios they considered, integrated farming systems that incorporated some elements of organic regimes were, overall, preferable to either conventional or organic farming.

The world's human population continues to grow inexorably greater and with it, the pressure to produce food. As Mill saw, there is a conflict between this and a healthy environment. The century after Mill's comments was dominated by habitat destruction and loss of biodiversity – the conflict was one-sided. Despite the recent realisation that this is unsustainable, and a revolution in the economics of balancing food and biodiversity, the trends in biodiversity are still, with few exceptions, downward. The picture is complicated by man-made climate change which, in the relative short term, will appear to create some 'winners' as well as 'losers' as species ranges change. But there is little doubt that substantial challenges remain to be met.

Optimists may believe that changes in human behaviour and technological advances will at least partly prevent the realisation of Mill's vision of an ever more hungry populace inhabiting an increasingly sterile planet. Changes in diet (particularly switching to lower protein diets) may help, as would reduction of waste (Bajzelj et al. 2014). The continued development of GM crops is another promising area, as is that of artificial meat. But it looks unlikely that these will be enough by themselves to establish the much hoped for sustainable intensification.

Our synopsis of organic farming as a laboratory for the merits of low-intensity farming to contribute to a solution leads to a number of connected ideas. The first of these is the basic one that biodiversity is higher on organic farms, and that the effect varies substantially depending on taxonomic group. The second is that spatial and taxonomic variability in effects points to pesticides and herbicides as the dominant threat. Insecticides and fungicides have had the most consistent negative effects on biodiversity (Geiger et al. 2010). The biggest differences on organic farms are consequently in the cropped area, particularly for plants, as seen in our study of UK organic farms. Further, our observations on spiders suggest that trophic knock-on effects may be greatest for low-mobility groups that cannot rapidly recolonize from the surrounding countryside. Many of the invertebrates sampled in the cropped area may have very recently dispersed from the surrounding countryside. Goulson et al.'s (2018) observation on the frequency of applications to arable crops in the UK is striking. This could also account for the most marked effect of organic farming on low-mobility spiders occurring in landscapes that are not dominated by arable crops, and which therefore have local populations of potential colonists. In the short term, there are implications for the targeting of low-intensity farming.

While organic farming as a certified system may contribute to a solution, the bigger picture leads to the conclusion that sustainability may be achieved only if conventional farming is transformed into something more like organic farming, and that there are ways to do this without unacceptable loss of yield. More sensitive management of non-cropped habitats in European agriculture is likely to be part of this, but the use of pesticides may be a wider and more urgent concern. A recent conference in Bonn saw the third plenary of an international initiative for biodiversity protection – the Intergovernmental Platform on Biodiversity and Ecosystem Services (IPBES). This initiative aims to enforce international standards for protection of nature. One of its first reports will tackle the state of pollinating insects (Massood 2015). The initiative faces powerful enemies, as Massood points out. More locally, in the UK, a 2018 agriculture bill[9] has described plans to make payments to farmers conditional on specific environmental protection targets. There are grounds for optimism. But if these and other similar schemes fail to deliver, and business as usual prevails, the future looks bleak.

Acknowledgements

We are grateful to all our colleagues, in particular those from the Wildlife Conservation Research Unit at Oxford University, the British Trust for Ornithology, the Centre for Ecology and Hydrology, and the Royal Agricultural University, for their contributions to the studies described in this chapter. We thank David Atkinson and Christine Watson for inviting us to contribute to their book.

References

Andersson, G.K.S., Rundlöf, M., and Smith, H.G. (2012). Organic farming improves pollination success in strawberries. *PLoS One* 7 (2): e31599.

Bajzelj, B., Richards, K.S., Allwood, J.M. et al. (2014). Importance of food-demand management for climate mitigation. *Nature Climate Change* 4: 924–929.

Batáry, P., Baldi, A., Samu, F. et al. (2008). Are spiders reacting to local or landscape scale effects in Hungarian pastures? *Biological Conservation* 141: 2062–2070.

Batáry, P., Holzschuh, A., Orci, K.M. et al. (2012). Responses of plant, insect and spider biodiversity to local and landscape scale management intensity in cereal crops and grasslands. *Agriculture, Ecosystems and Environment* 146: 130–136.

Batáry, P., Sutcliffe, L., Dormann, C.F., and Tscharntke, T. (2013). Organic farming favours insect-pollinated over non-insect pollinated forbs in meadows and wheat fields. *PLoS One* 8 (1): e54818.

Batáry, P., Dicks, L.V., Kleijn, D., and Sutherland, W.J. (2015). The role of agri-environment schemes in conservation and environmental management. *Conservation Biology* 29: 1006–1016.

Bell, J.R., Bohan, D.A., Shaw, E.M., and Weyman, G.S. (2005). Ballooning dispersal using silk: world fauna, phylogenies, genetics and models. *Bulletin of Entomological Research* 46: 69–114.

---

9 www.theguardian.com/environment/2018/sep/12/
gove-hails-plans-to-reward-uk-farmers-for-adopting-green-policies.

Bengtsson, J., Ahnstrom, J., and Weibull, A.C. (2005). The effects of organic agriculture on biodiversity and abundance: a meta-analysis. *Journal of Applied Ecology* 42: 261–269.

Benton, T.G., Vickery, J.A., and Wilson, J.D. (2003). Farmland biodiversity: is habitat heterogeneity the key? *Trends in Ecology & Evolution* 18: 182–188.

Birkhofer, K., Ekroos, J., Corlett, B., and Smith, H.G. (2014). Winners and losers of organic cereal farming in animal communities across central and northern Europe. *Biological Conservation* 175: 25–33.

Bull, J.W., Gordon, A., Law, E.A. et al. (2014). Importance of baseline specification in evaluating conservation interventions and achieving no net loss of biodiversity. *Conservation Biology* 28: 799–809.

Burel, F. (1989). Landscape structure effects on carabid beetles spatial patterns in Western France. *Landscape Ecology* 2: 215–226.

Burns, F.; Eaton, M.A.; Gregory, R.D.; et al. (2013) State of Nature. State of Nature Partnership, CEH Project No. C04535. Available at: http://nora.nerc.ac.uk/id/eprint/502092

Chamberlain, D.E., Wilson, J.D., and Fuller, R.J. (1999). A comparison of bird populations on organic and conventional farm systems in southern Britain. *Biological Conservation* 88: 307–320.

Chamberlain, D.E., Fuller, R.J., Bunce, R.G.H. et al. (2000). Changes in the abundance of farmland birds in relation to the timing of agricultural intensification in England and Wales. *Journal of Applied Ecology* 37: 771–788.

Chamberlain, D.E., Joys, A., Johnson, P.J. et al. (2010). Does organic farming benefit farmland birds in winter? *Biology Letters* 6: 82–84.

Coda, J., Gomez, D., Steinmann, A.R., and Priotto, J. (2015). Small mammals in farmlands of Argentina: responses to organic and conventional farming. *Agriculture Ecosystems & Environment* 211: 17–23.

Conway, G. (2007). A doubly green revolution: ecology and food production. In: *Theoretical Ecology: Principles and Applications* (ed. R.M. May and A. McLean), 158–171. Oxford: Oxford University Press.

Curran, M., Hellweg, S., and Beck, J. (2014). Is there any empirical support for biodiversity offset policy? *Ecological Applications* 24 (4): 617–632.

Damgaard, C., Strandberg, B., Strandberg, M.T. et al. (2014). Selection on plant traits in hedgerow ground vegetation: the effect of time since conversion from conventional to organic farming. *Basic and Applied Ecology* 15: 250–259.

Dasgupta, P. (2010). Nature's role in sustaining economic development. *Philosophical Transactions of the Royal Society, B: Biological Sciences* 365: 5–11.

Eaton MA, Balmer DE, Bright J, et al. (2013) The state of the UK's birds 2013. RSPB, BTO, WWT, NRW, JNCC, NE, NIEA and SNH. Sandy, Bedfordshire, UK.

European Commission (2013) Agri-environment measures. Available at: http://ec.europa.eu.

Evans, D.M., Altwegg, R., Garner, T.W.J. et al. (2015). Biodiversity offsetting: what are the challenges, opportunities and research priorities for animal conservation? *Animal Conservation* 18: 1–3.

Feber, R.E. and Smith, H. (1995). Butterfly conservation on arable farmland. In: *Ecology and Conservation of Butterflies* (ed. A.S. Pullen), 84–97. London: Chapman and Hall.

Feber, R.E., Johnson, P.J., Firbank, L.G. et al. (2007). A comparison of butterfly populations on organically and conventionally managed farmland. *Journal of Zoology* 273: 30–39.

Feber, R.E., Johnson, P.J., Firbank, L.G. et al. (2015a). Does organic farming affect biodiversity? In: *Wildlife Conservation on Farmland* (ed. D.W. Macdonald and R.E. Feber), 108–132. Oxford: Oxford University Press.

Feber, R.E., Johnson, P.J., Bell, J.R. et al. (2015b). Organic farming: biodiversity impacts can depend on dispersal characteristics and landscape context. *PLoS One* 10 (8): e0135921.

Firbank, L.G., Smart, S.M., Crabb, J. et al. (2003). Agronomic and ecological costs and benefits of set-aside in England. *Agriculture, Ecosystems & Environment* 95: 73–85.

Fischer, C., Thies, C., and Tscharntke, T. (2011). Small mammals in agricultural landscapes: opposing responses to farming practices and landscape complexity. *Biological Conservation* 144: 1130–1136.

Flysjö, A., Cederberg, C., Henriksson, M., and Ledgard, S. (2012). The interaction between milk and beef production and emissions from land use change – critical considerations in life cycle assessment and carbon footprint studies of milk. *Journal of Cleaner Production* 28: 134–142.

Fox, R., Brereton, T.M., Asher, J. et al. (2011). *The State of the UK's Butterflies 2011*. Wareham: Butterfly Conservation and the Centre for Ecology and Hydrology.

Fox, R., Parsons, M.S., Chapman, J.W. et al. (2013). *The State of Britain's Larger Moths 2013*. Wareham: Butterfly Conservation and Rothamsted Research.

Fraser, M.D., Vale, J.E., and Firbank, L.G. (2014). Effect on habitat diversity of organic conversion within the less favored areas of England and Wales. *Agroecology and Sustainable Food Systems* 38: 243–261.

Fuller, R.J., Norton, L.R., Feber, R.E. et al. (2005). Benefits of organic farming to biodiversity vary among taxa. *Biology Letters* 1: 431–434.

Fuller, R.M. (1987). The changing extent and conservation interest of lowland grasslands in England and Wales: a review of grassland surveys 1930–84. *Biological Conservation* 40: 281–300.

Gabriel, D., Roschewitz, I., Tscharntke, T., and Thies, C. (2006). Beta diversity at different spatial scales: plant communities in organic and conventional agriculture. *Ecological Applications* 16: 2011–2021.

Gabriel, D., Carver, S.J., Durham, H. et al. (2009). The spatial aggregation of organic farming in England and its underlying environmental correlates. *Journal of Applied Ecology* 46: 323–333.

Gabriel, D., Sait, S.M., Kunin, W.E., and Benton, T.G. (2013). Food production vs. biodiversity: comparing organic and conventional agriculture. *Journal of Applied Ecology* 50: 355–364.

Gallai, N., Salles, J.M., Settele, J., and Vaissière, B.E. (2009). Economic valuation of the vulnerability of world agriculture confronted with pollinator decline. *Ecological Economics* 68: 810–821.

Garnett, T., Appleby, M.C., Balmford, A. et al. (2013). Sustainable intensification in agriculture: premises and policies. *Science* 341: 33–34.

Garratt, M.P.D., Breeze, T.D., Jenner, N. et al. (2014). Avoiding a bad apple: insect pollination enhances fruit quality and economic value. *Agriculture, Ecosystems & Environment* 184: 34–40.

Geiger, F., Bengtsson, J., Berendse, F. et al. (2010). Persistent negative effects of pesticides on biodiversity and biological control potential on European farmland. *Basic and Applied Ecology* 11: 97–105.

Gelling, M., Macdonald, D.W., and Mathews, F. (2007). Are hedgerows the route to increased farmland small mammal density? Use of hedgerows in British pastoral habitats. *Landscape Ecology* 22: 1019–1032.

Godfrey, H.G. and Garnett, T. (2014). Food security and sustainable intensification. *Philosophical Transactions of the Royal Society, B: Biological Sciences* 369.

Goulson, D. (2013). An overview of the environmental risks posed by neonicotinoid insecticides. *Journal of Applied Ecology* 50: 977–987.

Goulson D (2015) Biodiversity v Intensive farming: has farming lost its way? Animal Ecology in Focus. https://journalofanimalecology.wordpress.com/2015/01/16/biodiversity-v-intensive-farming-has-farming-lost-its-way

Goulson, D., Thompson, J., and Croombs, A. (2018). Rapid rise in toxic load for bees revealed by analysis of pesticide use in Great Britain. *Peer Journal* 6: e26856v1.

Hambler, C. and Canney, S.M. (2013). *Conservation*. Cambridge: Cambridge University Press.

Happe, A.K., Riesch, F., Rosch, V. et al. (2018). Small-scale agricultural landscapes and organic management support wild bee communities of cereal field boundaries. *Agriculture Ecosystems & Environment* 254: 92–98.

Hardman, C.J., Harrison, D.P.G., Shaw, P.J. et al. (2016). Supporting local diversity of habitats and species on farmland: a comparison of three wildlife-friendly schemes. *Journal of Applied Ecology* 53: 171–180.

Henckel, L., Borger, L., Meiss, H. et al. (2015). Organic fields sustain weed metacommunity dynamics in farmland landscapes. *Proceedings of the Royal Society B* 282: 20150002.

Henderson, I.G., Ravenscroft, N., Smith, G., and Holloway, S. (2009). Effects of crop diversification and low pesticide inputs on bird populations on arable land. *Agriculture Ecosystems and Environment* 129: 149–156.

Hinsley, S.A. and Bellamy, P.E. (2000). The influence of hedge structure, management and landscape context on the value of hedgerows to birds: a review. *Journal of Environmental Management* 60: 33–49.

Hodgson, J.A., Kunin, W.E., Thomas, C.D. et al. (2010). Comparing organic farming and land sparing: optimizing yield and butterfly populations at a landscape scale. *Ecology Letters* 13: 1358–1367.

Hole, D.G., Perkins, A.J., Wilson, J.D. et al. (2005). Does organic farming benefit biodiversity? *Biological Conservation* 122: 113–130.

Holzschuh, A., Steffan-Dewenter, I., and Tscharntke, T. (2008). Agricultural landscapes with organic crops support higher pollinator diversity. *Oikos* 117: 354–361.

Hughes, J., Pretty, J.N., and Macdonald, D.W. (2013). Nature as a source of health and well-being: is this an ecosystem service that could pay for conserving biodiversity? In: *Key Topics in Conservation Biology 2* (ed. D.W. Macdonald and K.J. Willis), 143–160. Oxford: Wiley-Blackwell.

Hyvonen, T., Ketoja, E., Salonen, J. et al. (2003). Weed species diversity and community composition in organic and conventional cropping of spring cereals. *Agriculture, Ecosystems & Environment* 97: 131–149.

Jonason, D., Andersson, G.K.S., Öckinger, E. et al. (2011). Assessing the effect of the time since transition to organic farming on plants and butterflies. *Journal of Applied Ecology* 48: 543–550.

Jonason, D., Smith, H.G., Bengtsson, J., and Birkhofer, K. (2013). Landscape simplification promotes weed seed predation by carabid beetles (Coleoptera: Carabidae). *Landscape Ecology* 28: 487–494.

King, P.J. (2004). *One Hundred Philosophers: A Guide TO THE World's Greatest Thinkers*. Hove: Apple.

Kragten, S. and de Snoo, G.R. (2008). Field-breeding birds on organic and conventional arable farms in the Netherlands. *Agriculture, Ecosystems and Environment* 126: 270–274.

Kragten, S., Tamis, W.L.M., Gertenaar, E. et al. (2011). Abundance of invertebrate prey for birds on organic and conventional arable farms in the Netherlands. *Bird Conservation International* 21: 1–11.

Krebs, J.R. (2013). *Food: A Very Short Introduction*. Oxford: Oxford University Press.

Lichtenberg, E.M., Kennedy, C.M., Kremen, C. et al. (2017). A global synthesis of the effects of diversified farming systems on arthropod diversity within fields and across agricultural landscapes. *Global Change Biology* 23: 4946–4957.

Mabey, R. (1996). *Flora Britannica*. London: Sinclair-Stevenson.

Mabey, R. and Nature Conservancy Council (1980). *The common ground: a place for nature in Britain's future?* London: Hutchinson in association with the Nature Conservancy Council.

Macdonald, D.W. and Johnson, P.J. (1995). The relationship between bird distribution and the botanical and structural characteristics of hedges. *Journal of Applied Ecology* 32: 492–505.

Macdonald, D.W. and Willis, K.J. (2013). *Key Topics in Conservation Biology*. Chichester: Wiley-Blackwell.

Macdonald, D.W., Raebel, E.M., and Feber, R.E. (2015). Farming and wildlife: a perspective on a shared future. In: *Wildlife Conservation on Farmland* (ed. D.W. Macdonald and R.E. Feber), 1–19. Oxford: Oxford University Press.

Massood, E. (2015). Major biodiversity initiative needs support. *Nature* 518: 7.

Mattison, E.H.A. and Norris, K. (2005). Bridging the gaps between agricultural policy, land-use and biodiversity. *Trends in Ecology & Evolution* 20: 610–616.

May, R.M. (2007). Unanswered questions and why they matter. In: *Theoretical Ecology: Principles and Applications* (ed. R.M. May and A. McLean), 205–215. Oxford: Oxford University Press.

MEA (2005). *Ecosystems and Human Well-Being: Synthesis. A Report for the Millennium Ecosystem Assessment*. Washington, D.C.: Island.

Millán de la Peña, N., Butt, A., Delettre, Y. et al. (2003). Landscape context and carabid beetle (Coleoptera: Carabidae) communities of hedgerows in western France. *Agriculture, Ecosystems and Environment* 94: 59–72.

NE (2010) Lost life: England's lost and threatened species (NE233). Available at: http://publications.naturalengland.org.uk/publication/32023?category=10002

Nemecek, T., Hugenin-Elie, O., Dubois, D., and Gaillard, G. (2011a). Life cycle assessment of Swiss farming systems: I. Integrated and organic farming. *Agricultural Systems* 104: 217–232.

Nemecek, T., Hugenin-Elie, O., Dubois, D. et al. (2011b). Life cycle assessment of Swiss farming systems: II. Extensive and intensive production. *Agricultural Systems* 104: 233–245.

Norton, L., Johnson, P.J., Joys, A. et al. (2009). Consequences of organic and non-organic farming practices for field, farm and landscape complexity. *Agriculture, Ecosystems and Environment* 129: 221–227.

Oberg, S. (2009). Influence of landscape structure and farming practice on body condition and fecundity of wolf spiders. *Basic and Applied Ecology* 10: 614–621.

Pelosi, C., Toutous, L., Chiron, F. et al. (2013). Reduction of pesticide use can increase earthworm populations in wheat crops in a European temperate region. *Agriculture Ecosystems & Environment* 181: 223–230.

Petit, S., Munier-Jolain, N., Bretagnolle, V. et al. (2015). Ecological intensification through pesticide reduction: weed control, weed biodiversity and sustainability in arable farming. *Environmental Management* 56: 1078–1090.

Phalan, B., Green, R., and Balmford, A. (2014). Closing yield gaps: perils and possibilities for biodiversity conservation. *Philosophical Transactions of the Royal Society, B: Biological Sciences* 369: 20120285.

Ponisio, L.C., M'Gonigle, L.K., Mace, K.C. et al. (2014). Diversification practices reduce organic to conventional yield gap. *Proceedings of the Royal Society B* 282: 20141396.

Poppy, G.M., Jepson, P.C., Pickett, J.A., and Birkett, M.A. (2014). Achieving food and environmental security: new approaches to close the gap. *Philosophical Transactions of the Royal Society, B: Biological Sciences* 369: 20120272.

Rackham, O. (1997). *The History of the Countryside*. London: Phoenix Giant.

Rebitzer, G., Ekvall, T., Frischknecht, R. et al. (2004). Life cycle assessment part 1: framework, goal and scope definition, inventory analysis, and applications. *Environmental International* 30: 701–720.

Roschewitz, I., Gabriel, D., Tscharntke, T., and Thies, C. (2005). The effects of landscape complexity on arable weed species diversity in organic and non-organic farming. *Journal of Applied Ecology* 42: 873–882.

Rundlof, M. and Smith, H.G. (2006). The effect of organic farming on butterfly diversity depends on landscape context. *Journal of Applied Ecology* 43: 1121–1127.

Rundlof, M., Bengtsson, J., and Smith, H.G. (2008). Local and landscape effects of organic farming on butterfly species richness and abundance. *Journal of Applied Ecology* 45: 813–820.

Rundlöf, M., Edlund, M., and Smith, H.G. (2010). Organic farming at local and landscape scales benefits plant diversity. *Ecography* 33: 514–522.

Samu, F. and Szinetár, C. (2002). On the nature of agrobiont spiders. *Journal of Arachnology* 30: 389–402.

Sarkar, S. and David, F.M. (2012). Conservation biology: ethical foundations. *Nature Education Knowledge* 3: 1–6.

Schmidt, M.H., Roschewitz, I., Thies, C., and Tscharntke, T. (2005). Differential effects of landscape and management on diversity and density of ground-dwelling farmland spiders. *Journal of Applied Ecology* 42: 281–287.

Schmidt, M.H., Thies, C., Nentwig, W., and Tscharntke, T. (2008). Contrasting responses of arable spiders to the landscape matrix at different spatial scales. *Journal of Biogeography* 35: 157–166.

Schneider, M.K., Luscher, G., Jeanneret, P. et al. (2014). Gains to species diversity in organically farmed fields are not propagated to the farm level. *Nature Communications* 5: 4151.

Schroter, L. and Irmler, U. (2013). Organic cultivation reduces barrier effect of arable fields on species diversity. *Agriculture Ecosystems & Environment* 164: 176–180.

Seufert, V., Ramankutty, N., and Foley, J.A. (2012). Comparing the yields of organic and conventional agriculture. *Nature* 485: 229–232.

Shoard, M. (1987). *This Land is our Land*. London: Paladin.

Singer, P. (1993). *Practical Ethics*. Cambridge: Cambridge University Press.

Smart, S.M., Firbank, L.G., Bunce, R.G.H., and Watkins, J.W. (2000). Quantifying changes in abundance of food plants for butterfly larvae and farmland birds. *Journal of Applied Ecology* 37: 398–414.

Smith, H.G., Dänhardt, J., Lindström, Å., and Rundlöf, M. (2010). Consequences of organic farming and landscape heterogeneity for species richness and abundance of farmland birds. *Oecologia* 162: 1071–1079.

Thomas, K. (1983). *Man and the Natural World, Changing Attitudes in England, 1500–1800*. London: Allen Lane.

Tuck, S.L., Winqvist, C., Mota, F. et al. (2014). Land-use intensity and the effects of organic farming on biodiversity: a hierarchical meta-analysis. *Journal of Applied Ecology* 51: 746–755.

Tuomisto, H.L. and Teixeira de Mattos, M.J. (2011). Environmental impacts of cultured meat production. *Environmental Science & Technology* 45: 6117–6123.

Tuomisto, H.L., Hodge, I.D., Riordan, P., and Macdonald, D.W. (2012). Comparing energy balances, greenhouse gas balances and biodiversity impacts of contrasting farming systems with alternative land uses. *Agricultural Systems* 108: 42–49.

UK National Ecosystem Assessment, World Conservation Monitoring Centre (2011). *UK National Ecosystem Assessment: Technical Report*. Cambridge: United Nations Environment Programme World Conservation Monitoring Centre.

Vickery, J.A., Tallowin, J.T., Feber, R.E. et al. (2001). Effects of grassland management on birds and their food resources, with special reference to recent changes in fertiliser, mowing and grazing practices on lowland neutral grasslands in Britain. *Journal of Applied Ecology* 38: 647–664.

Vucetich, J.A., Bruskotter, J.T., and Nelson, M.P. (2015). Evaluating whether nature's intrinsic value is an axiom or anathema to conservation. *Conservation Biology* 29: 321–332.

Wickramasinghe, L.P., Harris, S., Jones, G., and Vaughan, N. (2003). Bat activity and species richness on organic and non-organic farms: impact of agricultural intensification. *Journal of Applied Ecology* 40: 984–993.

8

Optimising Crop Production in Organic Systems

David Atkinson and Robin L. Walker

SRUC, Aberdeen, Scotland, UK

8.1 Introduction

The contribution of organic crops to total global crop production and the extent of the varia-tion in harvested yields between different approaches to crop production vary for different crops. However, the production of all crops, regardless of the approach to farming, depends on the interception of radiant energy and the influence of four major factors on energy capture: selection of crop species and variety, management of the soil, supply of nutrients and protection of the crop from pests, weeds and disease. Generic issues related to these areas are discussed here. More specific issues related to nutrient additions are covered in Chapter 9. Differences in how nutrient additions are made to crops has long been one of the major practical areas separating organic agriculture and horticulture from other systems of production. The other major approach dividing crop production is crop protection, which is discussed in Chapter 12. Light interception is fundamental to all crop production but how it is achieved varies in ways which reflect the wider aims of the producer and the extent to which those aims go beyond a simple maximisation of production (see Chapters 1, 3).

As discussed in Chapter 1, different systems of production emphasise different aspects of the science base. 'Conventional' systems of production have depended upon advances in organic and inorganic synthetic chemistry. Developments in fertilisers, giving more predict-able patterns of nutrient release, and the use of organic pesticides to protect crops from path-ogens and to minimise competition from weeds have been key to the use of higher yielding genotypes of crops such as cereals. Organic production, in contrast, has made greater use of inputs from ecology (see Chapter 6) and from the social sciences (see Chapter 3). As a result, the scope for an increased contribution to total production by organic systems is high for crops grown on a limited scale, such as horticultural crops, where the maintenance of soil condition is important (see Chapter 4) or where human inputs remain important to the inher-ent character of the food. Here there is potential to match the needs of production with the organic approach (Willer et al. 2012). Worldwide, horticultural crops are amongst the most important organic products and their market share is greater than the total organic share of land area. Recently, the potential for organic production to enhance output and profitability in tropical developing countries has been identified and related to improved crop nutrition and greater resilience to pests (Burke 2016).

Assessing the science which underpins organic crop production requires consideration of both agricultural and horticultural crops. In this chapter, we focus on the science which underpins all crop production while drawing attention to those aspects which become critical when crops are produced without chemical inputs. We also stress approaches which have the deliberate intentions of minimising environmental impact and of involving people. A number of other recent texts, such as Barker (2010), have focused on the practical issues associated with organic production. As such issues have recently been reviewed, we have elected to focus on the science base so as to join a basic understanding of science to recent work dealing specifically with organic crop production.

To grow and produce food, all crops need four major resources: solar energy, CO_2, water and mineral nutrients. Energy from the sun powers photosynthesis and heats the atmosphere, the crop and the soil. CO_2 is rarely absolutely limited but its supply to the sites of photosynthesis may be reduced by stomatal closure. Stomatal closure is most commonly the result of restrictions to water supply and a consequence of leaf area and its interception of incoming radiation. Mineral nutrients are held in the soil in a variety of complex forms. In conventional systems of production, the ability of the soil to supply nutrients at an appropriate rate is augmented by the addition of fertilisers. These supply large amounts of nutrients and in chemical forms which do not require the transformations needed to release nutrients from organic compounds or less soluble inorganic sources. Chemical transformations require energy, usually in the form of heat, a product of radiation, and appropriate soil characteristics, some of which, such as porosity, are a product of cultivation. The chemistry of nutrient supply and the materials used to affect soil nutrient status are discussed in Chapter 9.

There is great scope for affecting basic relationships through the management of the system. Management is of importance in all systems but critical to organic production. In addition, many basic processes which are normally described in simple terms, such as the impact of temperature, are in practice complex, with a range of effects and with the scope to be optimised by management. Ultimately, the basis of all production in all systems is the interception of solar radiation by leaves and its conversion into food.

Although basic biology is at the heart of organic production, its delivery differs, for example in relation to the involvement of people. People are seen as integral, not costs to be minimised. This difference in emphasis leads to different priorities. In addition, the rotation (see Chapter 6) is at the heart of organic production, meaning that crops are important both for the food they produce and for their conditioning of the system for the next crop. Key elements affecting production within an organic system are summarised in Table 8.1.

8.2 Basic Issues

The growth of all crops is affected by a number of major environmental factors, which are summarised in Table 8.2. These factors vary between different places and with time on a day-to-day and year-to-year basis. As a result, no crop produces the same yield each time it is grown and the more complex the life cycle, as with fruit crops, the greater the fluctuations.

Table 8.1 Key elements of an organic crop production system

Production decision	Purpose
Crop selection	Selection of species most likely to be fitted to particular site growing characteristics and likely to be able to make use of light and other environmental attributes
Rotation design	The ability to use one crop as a means of effecting the appropriate nutrient speciation and availability for later crops and as a means of reducing the carryover effects of pathogens
Nutrient inputs	Making resources of nutrients from the soil, inputs from other farm activities and waste recycling fit the demand profile of particular crops, especially the early development of leaf area
Cultivations	Important to the control of weeds but also important to moving nutrients within the soil profile
Crop varieties	Key to controlling the impact and spread of pests and diseases and to the delivery of crop quality
Spatial arrangement of crops	Important to light interception, controlling the spread of diseases and allowing the use of cultivation during a cropping period

Table 8.2 Factors in the plant environment affecting growth and productivity

Above ground	Below ground
Radiant energy: day length, spectral composition, intensity	Nutrients: concentration, balance and availability of major and trace elements
Temperature: seasonal and daily fluctuations, incidence of frost	Temperature: absolute and gradients between depths
Humidity (vapour pressure deficit)	Soil moisture: amount and availability (soil moisture potential)
Wind: average speed, gustiness, direction	Soil reaction status (pH)
Cloud, mist and fog	Soil characteristics: depth, physical characteristics (sand/silt/clay) in as far as they affect air water and root growth
Precipitation: dew, rain, sleet, snow, hail	
Atmospheric composition: CO_2, pollutants, ozone	Composition of the soil atmosphere: CO_2/O_2, C_2H_4

Source: Modified from Hudson (1977).

National average crop yields show significant year-to-year variations. Farmers try to minimise such variation by:

1) attempting to match crop to site
2) selecting within a crop species the variety with the highest potential productivity for a particular site
3) carrying out appropriate cultural operations at the optimum time
4) attempting to minimise the adverse effects of weather, for example by the use of irrigation, and recognising that the impact of weather on the plant will differ at different stages in its development; seedlings and mature plants are likely to respond differently (Hudson 1977).

8.3 Light Interception: The Basis of All Production

At its most basic, the principal impact of all the above factors is on the ability of leaves to grow, intercept radiation and absorb CO_2. These basic processes are affected by both crop-specific and environmental factors. The restricted inputs which characterise organic production mean that these factors have greater potential impact than in other approaches to production. The following are amongst the key influences which are particularly likely to affect the yield of organic crops.

8.3.1 Energy Capture

In most leaves, the rate of photosynthesis is related to the incident radiation or photon flux density received. The relation between these is in the form of a two-parameter rectangular hyperbola with photosynthesis initially increasing as photon flux density increases but later being less or unaffected (Incoll 1977). In practice, the relationship between actual photosynthesis and photosynthetically active radiation (Figure 8.1) can be much noisier than is commonly portrayed in such idealised relationships (Dunsiger 1999).

Different crops commonly behave in a similar way in the early part of their growth cycle (Biscoe and Gallagher 1977), with a linear relationship between crop growth rate and intercepted radiation for crops as different as wheat and sugar beet, although the slope of the line describing the relationship varies. In wheat, an interception of $100 \, Mj/m^2/week$ gave growth of $>200 \, g/m^2/week$ while in sugar beet and barley it was significantly lower. Final yield is also

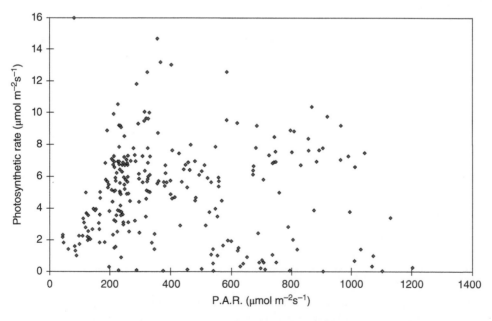

Figure 8.1 The relationship between photosynthetically active radiation and photosynthetic rate in poplar. *Source:* After Dunsiger (1999).

commonly a function of cumulative radiation interception. Radiation interception is thus fundamental to crop yield in all crop production systems. The task of those producing crops is to maximise this within the constraints of the factors which define their production system.

The fraction of radiation intercepted depends on leaf area index (LAI) and the ability of the leaf area to cover the ground surface area. It is also influenced by the chlorophyll content of the leaves, especially at low light intensities (Heath 1969). For most crops, a LAI of between 4 and 5 will give 80% light interception and so anything which restricts leaf area development or maintenance, particularly in the early part of the growth cycle, will result in a reduction in yield. Temperature and water stress both have a significant effect on the development of LAI. In many crops, leaf growth increases exponentially with an increase in temperature (Biscoe and Gallagher 1977). Petcu et al. (2011) found that organic systems showed reduced leaf area development in a number of wheat cultivars. In an organic system, the slower rate of nutrient supply resulted in LAI values of 1.87–3.02. This was associated with the production of a reduced biomass; biomass and LAI were correlated in the organic system but not in conventional plots. Biomass and yield were correlated in both systems of production.

Most agricultural crops rapidly create a continuous canopy so that it is relatively simple to relate leaf area to incoming radiation. In perennial horticultural crops, assessing light interception is more difficult because of the discontinuous canopy. Despite the added complexity due to the discontinuous canopy in an orchard, light interception remains a function of LAI but is modified by the spatial arrangement of the leaves. At low leaf area indices, tree size and spacing have little effect but when LAI is significant then size and arrangement have a significant effect although the more an orchard approximates to a continuous canopy, the less important size and arrangement become (Palmer 1981). A study of the impact of cropping system on the interception of photosynthetically active radiation (PAR) showed that across a range of orchard systems, both total dry matter accumulation and dry matter in fruit were linearly related to PAR interception (Jackson 2003). The slope of the relationship for total dry matter was, however, steeper than that for fruit dry matter, suggesting that fruiting is influenced to a greater extent by factors other than the interception of radiation

8.3.2 Canopy Duration

While the production of leaves is important, so is the persistence of their functioning during the growing season. In most crops, the production of new leaves ends just before flowering and thereafter photosynthesis depends on the activity of leaves produced up to that time. A shortened period for the production of leaves, for example as a result of a low temperature-induced restriction on nitrogen mineralisation, will result in a reduced leaf area with consequences for yield. The need for the nitrogen in organic manures to be mineralised potentially restricts leaf area development to a greater extent than is the case for crops fed with more soluble sources of nitrogen. During this postflowering period, much of the assimilate produced is needed to fill seeds or fruits. The persistence of a functioning leaf area is thus important to yield and is influenced by prevailing weather conditions, especially by high temperatures and soil moisture deficits. It is also affected by diseases and pests, hence the impact of crop protection.

During this period, however, the leaf surface is progressively less receptive to intercepted radiation. In addition, high temperatures reduce dry matter production. In some crops, such as perennial fruits, where there is not a continuous leaf canopy, the relationships between light interception and leaf area (Jackson 1981) and between all the components of yield are more complex as a result of the larger number of components involved in growth and production, that is, fruit, leaves, perennial woody tissues. This has complex effects on the number of fruits although the final weight of individual fruits remains a function of the rate of growth and its duration. The former can be influenced by the potential supply of photosynthates to the fruit (Hansen 1977). In addition, in crops such as apple different types of leaves may supply assimilate to different components within the tree. Spur leaves commonly supply the fruit while much of the assimilate produced by the leaves on growing shoots is retained within the shoot. In addition, because these leaves develop at different times within the season (Faust 1989), there is a more complex pattern than that found in cereals.

Much of the impact of orchard design on crop production is a result of the effects on light interception both in total and in different parts of the canopy (Jackson 2003). Damage to the leaves of a tree due to pests or diseases can reduce leaf area and as a result reduce crop yield if this reduction occurs at a critical stage in the fruit growth cycle (Faust 1989). This represents a key interaction of organic practice with the basic growth cycle.

8.3.3 Stomatal Functioning

Carbon dioxide assimilation is also affected by the resistance of the stomata in the leaves and internal resistances within the leaf. The internal structure of a strawberry leaf is shown in Figure 8.2.

The effect and the degree of control of assimilation due to the stomata vary with factors such as wind speed (Heath 1969). Through its control of water loss, the plant's stomatal opening influences the partition of energy between latent and sensible heat. Latent heat causes a change in the state of a material, such as the change of water from liquid to gas, as in evaporative fluxes without there being a change in temperature. In contrast, sensible heat causes an increase in temperature. The relationship between temperature and the extent of stomatal opening in an idealised form is an 'n'-shaped curve. With increasing temperature, it first rises, then levels and finally decreases. In practice, the relationship is much more variable, indicating both a need for greater understanding and the potential to be changed through optimised management (Figure 8.3). In addition, even under normal conditions not all stomata can be equally open at any point in time (Figure 8.4). The design of the cropping system and the ways in which it manages temperature and air movement can affect stomatal functioning and so water loss and the need for optimised soil management or irrigation.

8.3.4 Crop Species

The relationship between energy and photosynthesis can and does vary between different species, as does the relationship between net photosynthesis and incident radiation. In addition, the relationship between CO_2 absorption and intercepted radiation varies between different plant species (Eckhardt 1977), and across a range of species net photosynthesis can vary with leaf nitrogen content which is often related to chlorophyll content (Ryugo 1988).

The extent to which partition between latent and sensible heat occurs varies between species as a consequence of the LAI, the relationship between leaf area and the ground area they cover and the architecture of the canopy. The erect leaves of a grassland intercept around 30% of light

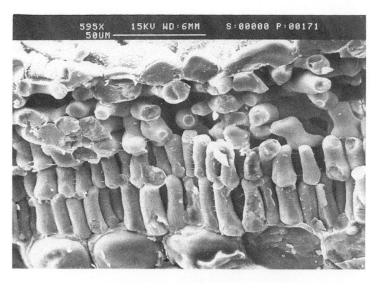

Figure 8.2 A freeze fracture scanning electron micrograph of the interior of a strawberry leaf showing the pathway from the stomata at the top of the picture to the palisade cells at the base of the picture. *Source:* Micrograph by DS Skene, East Malling Research.

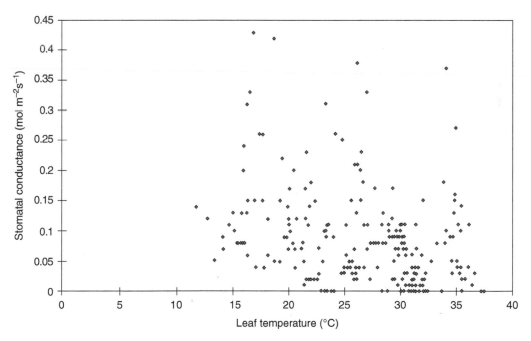

Figure 8.3 The relationship between stomatal conductance and air temperature in poplar. *Source:* After Dunsiger (1999, p. 159).

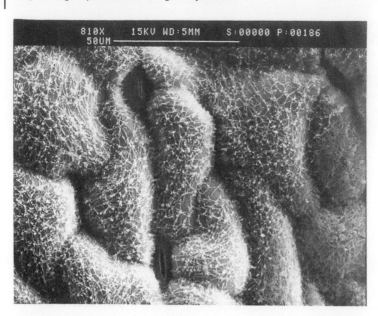

Figure 8.4 A scanning electron micrograph of the lower surface of a strawberry leaf showing epidermal cells and two stomata open to very different extents. *Source:* Micrograph by DS Skene, East Malling Research.

while the more horizontal leaves of a crop like sunflower intercept 70% (Saugier 1977). Different species also change the impact of crop on environmental features such as the vertical profile of radiation, sensible and latent heat fluxes and CO_2 flux. In a tall crop such as sunflower, the crop affects the environment up to a height of 1.5 m while in grassland, impact is only to a height of around 30 cm. As a result, in grassland radiation and CO_2 fall sharply over a small distance while in sunflower similar changes occur over a much greater distance. In the two crops, the profiles of sensible and latent heat fluxes are different. In sunflower, latent heat flux varies greatly with height above the ground while sensible heat is relatively unchanged. In grassland, the opposite is the case. Such differences have an impact on the energy reaching the soil surface, which is important because of the effect of temperature on soil processes, such as the mineralisation of organic matter which controls the release of nitrogen.

Basic processes are similar for all crops but variation has different consequences for organic and conventional systems so that the choice of crop and its effects on the physical environment will have a greater impact on organic systems.

8.3.5 Crop Growth and Resource Partitioning

Dry matter production depends upon a balance between photosynthesis and respiration. Net photosynthesis is driven by the related factors radiation and temperature, which means that early in the crop growth period, net photosynthesis is linearly related to radiation interception. Thus, for crops grown under conditions of good husbandry and in the right place, intercepted radiation determines productivity (Biscoe and Gallagher 1977). However, later in the season, the linear relationship only holds at lower light intensities, with the mechanism becoming light saturated at around half the previous rate.

The yield of saleable produce from a crop is also affected by the partition of assimilate into grain or fruit. In cereals, yield is a function of the number of ears per unit ground area, the number of grains per ear and the weight of individual grains. The number and weight of grains per unit area thus determine final yield. Grain number is usually established by anthesis while much of grain weight is determined after anthesis, that is, later in the season. There is probably more scope to vary grain number than grain weight. The period leading up to anthesis is important to grain numbers. Fast rates of dry matter production in this period are important although there can be significant competition for reserves between stem growth and grain number establishment. The growth of an individual grain can be described in terms of both rate, which is commonly linear during the grain growth period, and duration. Grain growth occurs during the period when crop growth rate is decreasing and so is not determined by current photosynthesis but by temperature (Biscoe and Gallagher 1977). As with total yield, the speed with which N is made available to the crop will have a specific impact on grain number. Managing this effect is an important component in organic cereal production.

In perennial crops, because fruit of high quality is only produced in well-illuminated parts of the canopy, the effect of canopy design on light penetration within the canopy is critical to the yield of quality/marketable fruit. Light interception thus only gives a guide to potential yield because actual yield can be higher or lower depending upon within tree shading (Palmer 1981). Some effects of this are discussed later in relation to fruit quality.

8.3.6 Soil-Related Factors

The supply of nitrogen from the soil has a major impact on the leaf area which can be produced, while the ability of the soil to supply water affects both leaf growth and, as a result of its effect on stomatal opening, the ability to fix atmospheric CO_2. Net radiation decreases as it passes through a plant canopy. In addition to being used for photosynthesis, this energy is dissipated as sensible heat, which warms the air, and at the soil surface, by conduction, and as latent heat, which is used to power transpiration (Saugier 1977). In organic production systems, in addition to its direct effect on leaf growth, temperature will influence the mineralisation of N in the soil, which will also have a major impact on a crop's ability to grow leaves.

In cereal crops, leaf growth is greatly influenced by nitrogen supply, which is usually predictable where the principal source of N is fertilisers. The uptake of N runs in advance of the crop dry matter gain. In a winter barley crop, 50% of the maximum N content had been absorbed by mid-April while 50% of maximum dry weight was only achieved in early May. Both were closely related to radiation interception (Figure 8.5).

Organic systems are particularly dependent on the effects of soil temperature, and so the start of leaf growth is commonly delayed, as soil will usually warm later in the season than air. This is responsible for a significant proportion of the difference in yield between conventional and organic systems. A significant challenge in organic systems is to optimise N release from organic sources and to gear the requirement for nutrients needed for leaf growth to factors such as anticipated weather. This can be helped by the use of simulation models which link leaf area development and canopy N measurements made in the field to expected values produced by the model (Jongschaap 2006).

In addition to effects on leaf area development and so on light interception, increasing the planting density will increase both its need for soil water and the time in the season when water

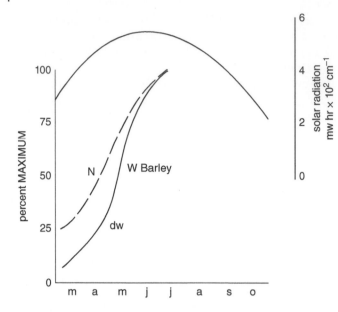

Figure 8.5 The increase with time during a season of N content and dry matter. *Source:* Data from Atkinson (1990).

stress will begin to have a significant impact on fruit production and quality (Atkinson 1981). The amount of growth will be influenced by the supply of nitrogen from the soil which itself will be influenced by soil temperature and the impact of the size of the canopy on amounts of sensible heat reaching the soil. Leaf N is a major factor determining the photosynthetic rate per unit leaf area; the higher the leaf N, the higher the photosynthesis (Faust 1989). A consequence of this complexity is that there is more scope in these systems for organic production to equal that in other systems of production.

8.3.7 Consequences

Leaf area is thus important for crop production and many of the most common farm practices in all agricultural systems are related to optimising leaf area development. The supply of nitrogen, and its availability to the crop at the time when demand is highest, is needed to develop leaves and so maximise light interception. Crop protection strategies, both those related to the removal of competing vegetation (weed control) and to the reduction of other organisms (pest and disease control), have at the heart of their purpose the maximisation of leaf area development and functioning. The principal ways in which these basic factors seem likely to differ between different approaches to crop production are summarised in Table 8.3.

These and related issues are explored in the remaining sections of this chapter.

8.4 What Current Issues Affect Choice of Crop Production System?

There is a general recognition that global climate change (GCC) is one of the greatest challenges facing civilisation (Sicher and Kim 2011). Agriculture, because of its global scale, is inevitably a major source of the greenhouse gases (GHG) which are powering these changes (Table 8.4).

Table 8.3 Environmental and other factors which have different impacts on crops produced in organic farming systems relative to those produced in more intensive agriculture

Factor	Differential impact in organic farming system
Interception of radiant energy	Later developing leaf area may intercept less total energy over the duration of the season, with potential effects on grain size and fruit development
Development and maintenance of leaf area	A limited supply of N reduces leaf area while non-use of pesticides results in a greater loss of leaf area as a result of fungal diseases and insect damage
Soil and air temperatures	The higher organic matter levels found in soils in organic systems take longer to warm and reduce the speed of soil warming at depth. This adversely affects the mineralisation of N
Nutrient availability at key growth stages	Slower soil warming results in reduced and delayed N mineralisation, which reduces the extent and speed of canopy development. Competition from other vegetation (weeds) is likely to further reduce N availability to the crop
Soil chemical and physical condition	A rotational system is likely to result in a more optimal soil pH, improved soil conditions and more effective soil microbiology, especially arbuscular mycorrhizas.
Soil water potential	Potentially improved as a consequence of increased soil carbon but could also result in wetter soil conditions early in the season with implications for soil-borne diseases and slower soil warming in the spring

Table 8.4 The impact of agriculture on factors contributing to or affecting global climate change

Factor	Mechanism of impact	Estimated importance
Soil C & N storage and release	The soil is one of the major reserves of carbon within the system. Crop growth results in the transfer of significant quantities of C to the soil much of which can be retained. Significant amounts of N can also be retained but can be released as the damaging N_2O	The dynamics of soil carbon storage differentiates agriculture from all other industries, plus soil C is one of the major components of the global carbon picture. Importance – High
Cropping system including cultivation and machinery	Perennial crops, such as trees, will hold significant amounts of C. Perennial crops will reduce C release from the soil by reducing soil disturbance	Soil disturbance releases carbon from storage Reduced cultivations decrease this Importance – Medium
Nutrient inputs	The production of N fertilisers for use in agriculture is a major element in the global carbon equation	The production of N fertilisers uses large quantities of fossil fuels. Importance – High
Crop protection	Synthetic pesticides use significant amounts of energy in their production	The use of fossil fuels is less than for fertilisers Importance – Medium
Consequences of biodiversity impact	Systems which plan for high biodiversity tend to retain more C in biomass	Biodiversity is affected by both diversity in the crop and the amount of land used for food production Importance – High
Destination of food products	Food designed for local consumption will minimise transport and shipping costs. Transport interacts with the amount of food waste produced	Transport by air and road are major elements in the global C equation. Food waste is also a significant factor Importance – High

This means that agriculture, as a sector, must assess whether it can minimise its contribution by reducing GHG production and increasing C fixation through changes in how it does things. In this respect, agriculture is almost unique as a business sector. Most areas of economic activity only release CO_2 and other GHGs while agriculture both releases gases and stores them in plant production and soil. Where agriculture is appropriately managed, the storage component can match or exceed the release.

The most commonly examined potential effect of GCC is impact on temperature. Current estimates of changes in mean global temperature in this century are for rises of 2.3–4.5 °C. As temperature is a limiting factor in many temperate crop production areas and as a significant development in crop production has been the introduction of what once were tropical crops into more northerly climates, this is an area with importance for all crop production systems. In most crops, temperature rather than radiation interception drives development although the two are related (Lorgeou and Prioul 2011). Maize was once a tropical crop but developments, primarily through crop breeding, have greatly extended its geographical range. For many crops, adverse effects of suboptimal temperatures on maturity and growth are critical limiting factors. A major area of concern in agriculture is its need for water; altered patterns of rainfall are a consequence of GCC.

The basic philosophy underpinning organic crop production is such that it ought both to offer an approach with lower impact and be willing to seek to reduce that impact to the least possible. In practice, this is likely to require responses from individual producers and through modifications to crop production. Schafer and Blanke (2012) assessed the carbon and water footprints for the production of Hokkaido pumpkin and demonstrated the complexity of assessing the impact of farming systems on GHG production. On-farm, the carbon footprint was better for organic production (240 kg CO_2/ha) than a non-organic farm (448 kg CO_2/ha). The difference was primarily due to nitrous oxide emissions resulting from nitrogen fertiliser use. In the non-organic system, fertiliser and plant protection materials were responsible for 10% and 1% respectively of the total C footprint of the system as a whole. Assessed on the basis of CO_2 produced per unit weight of pumpkin, at the point of sale, a specialised farm was the most effective (139 g CO_2/kg). Differences were greatly influenced by relative yield levels; 18 compared to 5.8 t/ha which were influenced by the use of potassium fertilisers. Imported pumpkins from Argentina, which had been transported by sea, also had a low carbon foot print (243 g CO_2/kg) for this phase of the food production system, suggesting that for some crops importing can reduce the carbon footprint. Ultimately, consumer shopping distances were responsible for 89% of the C footprint. Therefore, it is important to put the various elements of GHG production into context when comparing systems.

8.5 What Options Exist for Regulating Yields?

Progress in increasing maize yields over the past century has been the result of almost equal contributions from improved crop management and the breeding of crop hybrids better adapted to stressful conditions (Lorgeou and Prioul 2011). Crop management in conventional systems has been dominated by the effective use of imported mineral nutrients and crop protection materials. This emphasis has led to less new information becoming available in relation to areas of science which are more appropriate to organic systems of production. Data on the relationship between nutrient availability and crop response and on the impact of different levels of disease are of general value to all systems of production. Crop breeding has sought to increase yields principally through increasing the quantity of fixed carbon directed to edible product. The impact and potential impact of crop breeding are discussed in Chapter 13 but it is important to note here the impact which breeding can have on a crop's need for N and other nutrients, the ability of the crop to resist pests and diseases, the ability to develop symbiotic relationships with organisms such as arbuscular mycorrhizal fungi (AMF) and the need for a modification of both rotations and production systems. These issues are explored in the following sections of this chapter.

Brock et al. (2011) showed that soil organic matter (SOM) levels were important to organically produced crops. A series of studies showed a correlation between yields of non-leguminous crops and SOM in organic but not in conventional systems. Over time, SOM development and yields became negatively related because of the need for SOM to be mineralised so as to provide a nutrient supply.

Currently, most organic agriculture uses the same varieties as are used in conventional production. Vlachostergios et al. (2010) concluded, from a study using lentil as a test crop, that the demands of organic agriculture for increasing yields could only be satisfied partly by varieties developed through conventional breeding programmes. They found that significant interactions between genotype and environment occurred in four of the five environments they studied. Correlation indices between organic and conventional systems for the 20 varieties assessed varied between 0.27 and 0.93. In the top five varieties, the highest yielding under conventional production were not the highest under organic production. This leads to the key questions of whether selection for varieties to be used in organic production must be done under organic conditions or whether, with modified focus and a better understanding of stages of growth, selection could occur within conventional programmes.

8.6 How Different are Conventional and Organic Yields?

The response of a crop to being produced in an organic system compared to conventional production with optimised inputs varies between types of crop and with the complexity of the developmental process. For crops grown for their vegetative parts such as leaves, stems, etc., accommodation seems not to occur, perhaps because of the short growing season for such crops. In grain and related crops, rapid early growth determines the resources which can be made available for the reproductive growth, the food product, and so under temperate growing conditions organic yields are likely always to be limited. For perennial crops, development early in the life of a plant may have an impact which is measurable for some years.

Crop yield is important to feeding the world and to the viability of agricultural businesses. However, as discussed in Chapter 1, it is not the only factor governing the management of any system of production. It has, however, been the basis of most studies which have compared organic and other systems of production. Seufert et al. (2012) compared yields in organic and conventional agriculture. They found that although organic yields were typically lower than conventional yields, the differences were highly contextual and depended on system and site characteristics. Yield differences thus ranged from 5% lower in rain-fed legumes and perennials on weak acidic to alkaline soils, through 13% when all best organic practices were used, to 34%

when practices were most comparable. Thus, in certain conditions, for example with good management practices and for particular crops and growing conditions, organic systems can come close to conventional systems. These authors suggested the importance of those factors which limit organic yields and which represent the basis of improving yields in the future.

A similar analysis of published literature was undertaken by de Ponti et al. (2012). They showed that organic yields of individual crops were on average 80% of conventional yields but with substantial variation; the standard deviation was 21%. The conventional–organic yield gap differed between crops and regions and tended to increase when conventional yields were high. For this to occur, conventional yields must be close to either a potential or a water-limited value with nutrient stress low and pests and diseases well controlled; a situation much more difficult to achieve in an organic system. In a comparison of yield levels obtained under conventional and organic production systems, Klima and Labza (2010) found that organic production was associated with a 12% reduction in yield. However, in their systems reduced inputs to the organic system led to higher financial returns.

A relatively similar assessment of the effect of organic practices was reported by Thorup-Kristensen et al. (2012). For a range of cereal and vegetable crops, organic yields were on average 82% of conventional yields although the differences varied between crops (Table 8.5).

For vegetable crops, absolute yield is not the only characteristic of importance. The value of the crop is affected by the size of the vegetables and the proportion classed as saleable. The system of growing influences this (Table 8.6).

Maggio et al. (2008) assessed the performance of potato under conventional and organic systems. They focused on difference between the systems in product quality stability and its relationship to interactions between agronomic and environmental factors. Yields were reduced

Table 8.5 The impact of crop production system on the yield (Mg/ha) of a range of crops

Crop	Cropping	system	
	Conventional from 88 (oats) to 310 (cabbage) kg N/ha as fertiliser	**Organic with brought-in manure from 26 (oats) to 234 (cabbage) kg N/ha as manure**	**Organic with green manure and cover cropping up to 140 kg N/ha as manure**
Oats after white cabbage	4.5	4.6	4.6
Carrot	103.6	93.5	95.1
Winter rye after carrot	*7.8*	*4.9*	*4.5*
Lettuce	43.4	36.2	32.4
Oats after lettuce	4.5	4.3	4.4
Onion	*73.8*	*51.8*	*54.0*
Winter rye after onion	8.4	6.6	6.0
White cabbage	98.1	80.7	82.0

Source: Data from Thorup-Kristensen et al. (2012).
Differences between treatments in italicised rows significant at $P < 0.05$.

Table 8.6 The effect of growing system on the percentage of the crop which had to be discarded as unfit for sale

Crop	Cropping	system	
	Conventional	Organic with imported manure	Organic with green manure
Carrot	33.3	29.7	38.5
Lettuce	9.0	9.3	10.2
Onion	15.1	19.1	17.0
White cabbage	2.2	4.1	9.2

Source: Data from Thorup-Kristensen et al. (2012).

under conditions of organic production principally through an effect on tuber size. Mourao et al. (2012) found that an increase in the yield of cabbage grown with mineral fertiliser (organic yield around 30% lower) compared to a number of different organic fertilisers was due to increased N accumulation early in the season although this was accentuated by a major period of uptake in the final 10 days of growth before harvest.

8.7 The Environmental Impact of Organic Systems

The overall value of organic production to total global sustainability is commonly questioned on the basis of the area of land required, using an organic approach, to meet the global need for food. It is argued that maximising production on a unit area of land minimises the area which needs to be used for any form of agriculture, so maximising the land area which can then be available for other uses such as by wildlife. This is discussed in Chapter 3. The impact on environmental attributes, such as improvement in soil quality, has also been questioned. On the basis of a reduced substrate use by heterotrophs, Leifeld (2012) suggested it could lead to reduced soil organic matter storage. The energy requirements and different systems of cultivation also affect environmental impact. Soil organic carbon, microbial activity and soil structure are often improved in the surface soil layers by reduced tillage compared to ploughed soils although such systems can result in reduced yields as a result of increased weed pressures (Mader and Berner 2012).

The energy balances of a series of different production systems were assessed by Moreno et al. (2011). They compared conventional, a no-till system and organic production and four different barley crop rotations, including a barley monoculture. Energy inputs were 3–3.5 times higher in conventional (11.7 Gj/ha/year) and no-till (10.4 Gj/ha/year) than in an organic system (3.41 Gj/ha/year). In relation to rotations, energy inputs were lowest for a barley fallow approach (6.19 Gj/ha/year) and highest for continuous barley (11.7 Gj/ha/year). The lowest energy use was by organic barley fallow (2.56 Gj/ha/year) and the highest by continuous barley grown conventionally (16.3 Gj/ha/year) or with no-till (14.9 Gj/ha/year). The reduced barley and hay yields resulted in energy outputs being lowest in organic systems (17.9 Gj/ha/year). Rotations with more than one crop had higher energy outputs than continuous barley. Energy efficiency indicators (net energy 14.5 Gj/ha/year, input/output ratio 5.36, energy productivity 400 kg/GJ) were highest for the organic systems.

Agriculture has always involved a series of trade-offs. Optimising environmental benefits will lead to a reduction in yields both because maximum yield is no longer the primary focus of the production system and because limits to environmental impact will have a negative result for crop performance. Sapkota et al. (2012) found that the use of chicory, radish or rye grass as a catch crop reduced spring barley yields but resulted in an increased storage of both C, by 170–498 kg/ha/year, and N, by 16–46 kg/ha/year, compared to a system without catch crops. Spargo et al. (2011) showed in a comparison of five different soil management systems that potentially mineralisable N and particulate N were higher in organic systems (315 kg/ha) than in either cultivated or no-till conventional production (235 kg/ha). N availability increased with the length of the rotation. Smuckler et al. (2012) assessed the impact of a number of variants of good environmental practice on an irrigated organic farm in California. The use of cover crops compared to fallow reduced winter run-off of suspended solids, ammonium and dissolved organic carbon but lead to increased leaching losses. Approaches to catch run-off and leachate reduced nutrient and solid losses and the combined approach reduced losses of NO_3 and N_2O to 25 and 0.5 kg/ha/year for the entire farm system.

8.8 Conclusion

Under temperate conditions, it seems clear that crops produced using an organic approach are likely to give reduced yields. In this chapter, we have looked at the basic science which underpins all crop production as a guide to both why this is and what might be done to reduce the current yield gap. At the beginning of the growth cycle of an annual crop, the supply of mineral nutrients from the soil is critical. For elements such as nitrogen, this is greatly influenced by soil temperature. Where nitrogen is supplied in a soluble form, as where nitrogen fertilisers are the principal source, the effect of temperature on early season root growth and on absorption will be the key limiting factors for uptake. Both of these are also important for organic crops but because organic crops are dependent on the release of nitrogen from soil organic forms, the impact of temperature on the microbial decomposition of the organic material is also of importance. The lag in the release of N from these microbial processes limits early leaf growth, so reducing the foundation for the supply of assimilate to later growth and grain. Many of the current crop varieties have been selected for performance under conditions where fertilisers are used as routine practice. Evidence suggests that such varieties will always be limited under organic conditions. Improving the production of crop under organic conditions may need varieties for which the limitations of early growth on subsequent performance are less dramatic. This may mean varieties in which the onset of grain establishment is delayed, so allowing a longer time for leaf area to develop.

References

Atkinson, D. (1981). Water use and the control of water stress in high-density plantings. *Acta Horticulturae* 114: 45–56.

Atkinson, D. (1990). Influence of root system morphology and development on the need for fertilisers and the efficiency of use. In: *Crops as Enhancers of Nutrient Use* (ed. V.C. Baligar and R.R. Duncan), 411–452. San Diego: Academic Press.

Barker, A.V. (2010). *Science and Technology of Organic Farming.* Boca Raton: CRC Press.

Biscoe, P.V. and Gallagher, J.N. (1977). Weather dry matter production and yield. In: *Environmental Effects on Crop Physiology* (ed. J.J. Landsberg and C.V. Cutting), 75–100. London: Academic Press.

Brock, C., Fleissbach, A., Oberholzer, H.R. et al. (2011). Relation between soil organic matter and yield levels of non legume crops in organic and conventional farming systems. *Journal of Plant Nutrition and Soil Science* 174: 568–575.

Burke, M. (2016). Tropical organic farming. *Chemistry and Industry Issue* 7: 11.

De Ponti, T., Rijk, B., and van Ittersum, M.K. (2012). The crop yield gap between organic and conventional agriculture. *Agricultural Systems* 108: 1–9.

Dunsiger Z 1999 The influence of arbuscular mycorrhiza on the water relations of trees. PhD Thesis, University of Edinburgh

Eckhardt, F.E. (1977). Physiological behaviour in relation to the environment. A comparison between a crop and various types of natural vegetation. In: *Environmental Effects on Crop Physiology* (ed. J.J. Landsberg and C.V. Cutting), 157–171. London: Academic Press.

Faust, M. (1989). *Physiology of Temperate Zone Fruit Trees.* New York: Wiley.

Hansen, P. (1977). Carbohydrate allocation. In: *Environmental Effects on Crop Physiology* (ed. J.J. Landsberg and C.V. Cutting), 75–100. London: Academic Press.

Heath, O.V.S. (1969). *The Physiological Aspects of Photosynthesis.* London: Heinemann.

Hudson, J.P. (1977). Plants and the weather. In: *Environmental Effects on Crop Physiology* (ed. J.J. Landsberg and C.V. Cutting), 1–22. London: Academic Press.

Incoll, L.D. (1977). Field studies of photosynthesis: monitoring with $14CO_2$. In: *Environmental Effects on Crop Physiology* (ed. J.J. Landsberg and C.V. Cutting), 137–155. London: Academic Press.

Jackson, J.E. (1981). Theory of light interception by orchards and a modelling approach to optimise orchard design. *Acta Horticulturae* 114: 69–79.

Jackson, J.E. (2003). *Biology of Apples and Pears.* Cambridge: Cambridge University Press.

Jongschaap, R.E.E. (2006). Run-time calibration of simulation models by integrating remote sensing estimates of leaf area index and canopy nitrogen. *European Journal of Agronomy* 24: 316–324.

Klima, K. and Labza, T. (2010). Yielding and economic efficiency of oat crops cultivated using pure and mixed sowing stands in organic and conventional farming systems. *Zywnosc-Nauka Technologia Jakosc* 17: 141–147.

Leifeld, J. (2012). How sustainable is organic farming. *Agriculture, Ecosystems and Environment* 150: 121–122.

Lorgeou, J. and Prioul, J.-L. (2011). Agrophysiology: light interception, crop photosynthesis, mineral nutrition and crop managing. In: *Advances in Maize* (ed. J.-L. Prioul, C. Thevenot and T. Molnar), 393–410. London: SEB.

Mader, P. and Berner, A. (2012). Development of reduced tillage systems in organic farming in Europe. *Renewable Agriculture and Food Systems* 27: 7–11.

Maggio, A., Carillo, P., Bulmetti, G.S. et al. (2008). Potato yield and metabolic profiling under conventional and organic farming. *European Journal of Agronomy* 28: 343–350.

Moreno, M.M., Lacastra, C., Meco, R., and Moreno, C. (2011). Rainfed crop energy balance of different farming systems and crop rotations in a semi arid environment: results of a long term trial. *Soils and Tillage Research* 114: 18–27.

Mourao, I., Pinto, R., Brito, L.M., and Rodrigues, R. (2012). Response of protected organic pea crop to increased application rate of manure compost. *Acta Horticulturae* 933: 181–186.

Palmer, J.W. (1981). Computed effects of spacing on light interception and distribution within hedgerow trees in relation to productivity. *Acta Horticulturae* 114: 80–88.

Petcu, E., Toncea, I., Mustatea, P., and Petcu, V. (2011). Effect of organic and conventional farming systems on some physiological indicators of winter wheat. *Romanian Agricultural Research* 28: 131–135.

Ryugo, K. (1988). *Fruit Culture: Its Science and Art*. New York: Wiley.

Sapkota, T.B., Askegaard, M., Laegdsmand, M., and Olesen, J.E. (2012). Effects of catch crop type and root depth on nitrogen leaching and yield of spring barley. *Field Crops Research* 125: 129–138.

Saugier, B. (1977). Micrometeorology on crops and grassland. In: *Environmental Effects on Crop Physiology* (ed. J.J. Landsberg and C.V. Cutting), 39–55. London: Academic Press.

Schafer, F. and Blanke, M. (2012). Farming and marketing system affects carbon and water foot print-a case study using Hokaido pumpkin. *Journal of Cleaner Production* 28: 113–119.

Seufert, V., Ramankutty, N., and Foley, J.A. (2012). Comparing the yields of organic and conventional agriculture. *Nature* 485: 229–232.

Sicher, C. and Kim, S.-H. (2011). Photosynthesis, growth and maize yields in the context of global change. In: *Advances in Maize* (ed. J.-L. Prioul, C. Thevenot and T. Molnar), 273–292. London: SEB.

Smuckler, S.M., O'Geen, A.T., and Jackson, L.E. (2012). Assessment of best management practices for nutrient cycling: a case study on an organic farm in a Mediterranean-type climate. *Journal of Soil and Water Conservation* 67: 16–31.

Spargo, J.T., Cavigelli, M.A., Mirsky, S.B. et al. (2011). Mineralizable soil nitrogen and labile soil organic matter in diverse long-term cropping systems. *Nutrient Cycling and Agroecosystems* 90: 253–266.

Thorup-Kristensen, K., Dresboll, D.B., and Kristensen, H.L. (2012). Crop yield, root growth and nutrient dynamics in a conventional and three organic cropping systems with different levels of external inputs and N recycling through fertility building crops. *European Journal of Agronomy* 37: 66–82.

Vlachostergios, D.N., Lithourgidis, A.S., and Roupakias, D.G. (2010). Adaptability to organic farming of lentil (*Lens culinaris* Medik.) varieties developed from conventional breeding programmes. *Journal of Agricultural Science* 149: 85–93.

Willer, H., Granatstein, D., and Kirby, E. (2012). The global extent and expansion of organic horticultural production. *Acta Horticulturae* 993: 23–33.

9

Crop Production

Meeting the Nutrient Needs

David Atkinson and Robin L. Walker

SRUC, Aberdeen, Scotland, UK

9.1   Introduction

A significant supply of mineral nutrients is vital to all crop production and where the amount and rate of nutrient supply fall behind what growth demands, then production is restricted. Basic issues related to the ability of the soil to supply mineral nutrients are discussed in Chapter 4. The relationship between nutrient supply and growth is explored in Chapter 8. The awareness of the limits to growth resulting from restrictions to nutrient supply from traditional manures resulted in the development and use of mineral fertilisers. Their introduction, initially in the nineteenth century and substantially in the mid-twentieth century, revolutionised crop production by increasing yields. This introduction was not without cost. There were impacts on soil quality, a requirement for major changes in crop protection, new interactions with global industrial concerns and changes in the functioning of rural communities.

All nutrient additions to the soil, in whatever form, affect the ability of a growing crop to satisfy its requirements. This ability can and has been used as a definition of enhanced soil fertility. However, we are also aware that only some forms of addition will have an impact beyond a single season (Anon 1998). Nutrient additions with limited duration have come to be known as fertilisers while those with longer impact tend to be called manures. The move from manures as the major source of inputs to arable production to the use of fertilisers to meet this need changed agriculture and contributed to the establishment of 'organic' as a word, which represented a different way of meeting the nutrient needs of crops. The difference between how fertilisers and manures work is at the heart of the differences in mindsets which direct these two approaches to crop production.

As illustrated in Chapter 8, crop growth is a function of the interception of radiant energy. However, for growth to occur at the rate which energy interception would allow, the crop needs mineral nutrients. Most mineral nutrients are absorbed from the soil by the crop's root system as soluble inorganic anions (see Chapter 10). Fertilisers provide these nutrients largely in such a form while manures require microbial decomposition to release inorganic nutrients. The chemical processes which are part of the microbiology of nutrient transformations are discussed in Chapter 11. The processes which allow crops to absorb nutrients through their roots and associated microbes are discussed in Chapter 10.

Here we seek to look at the science which underpins the provision of mineral nutrients to organic crops. We also look at the range of materials which are used as nutrient sources, and at the processes which influence their ability to release nutrients for crop growth and production.

9.2   Getting Nutrients into Organic Crops

In all systems, crops receive nutrient inputs from both the soil reserves and from external sources. In conventional systems, fertilisers predominate. The provision of nutrients to an organic crop is carried out on a different basis (Table 9.1).

In organic systems run under favourable conditions, all nutrient needs are met from the soil as a result of the design of the rotation. Rotational designs capture nitrogen from the air; this maintains the reserves of N and in addition holds nutrients within the system and changes the chemical form of the nutrient so as to optimise supply to subsequent crops. However, sometimes nutrient resources are inadequate and inputs are required; there is need for 'organic fertilisers'. Organic fertilisers are always based on materials from which the release of materials is slow. Microbial activity is commonly needed to make it useable by the crop. In this sense, such materials feed the soil, not the crop. A consequence of this is that the nutrients have residual value (Anon 1998). Many also have the potential to affect soil structure. Their complex origins mean that they are usually sources of more than one nutrient (Table 9.2). Cattle manure is rich in N and K while pig manure has a relatively higher P content. The composition of poultry manure varies with the type of production system from which it originated.

In addition to the above objectives, organic manures provide a means of moving nutrients from one area of activity, for example housed cattle, to another and a means of recycling

Table 9.1 Available mechanisms for supplying the nutrient needs of crops in organic systems

Nutrient source	Mechanism of effect
Organic 'fertiliser'	Increases the quantity of nutrients within the system although much of the added nutrient will not be immediately available
Crop rotation	Uses the chemical and biological actions of crops within the rotation to retain nutrients within the system, to affect chemical speciation and for leguminous crops to increase nitrogen content
Legumes	To increase the amount of N within the system
Cover crops (green manures)	To retain nutrients which might otherwise be lost as a result of leaching, especially during periods when no main crop is present. To increase the quantity of nutrients available to a following crop in an easily available form
Microbial inoculation	To enhance the soil microbial processes which are responsible for the mineralisation of N, increasing the availability of P and of trace elements

Table 9.2 The quantities of major nutrients available from different sources of farmyard and other manures (kg/t)

Type of manure	N	P	K
Cattle	6.5	1.5	6.6
Pig	7.0	3.1	5.0
Poultry: broiler	24.0	9.6	11.6
Poultry: battery	41.0	14.8	14.1

Source: Recalculated from Anon (1998).

Table 9.3 The quantities of major nutrients (kg/ha) contained in the harvested components of major farm crops

Crop/(yield)	N	P	K
Winter wheat (grain, 10 t/ha, straw, 5.2 t/ha)	203	32	88
Winter barley (grain, 9 t/ha, straw, 5 t/ha)	189	36	102
Spring barley (grain, 7.5 t/ha, straw, 3.9 t/ha)	156	30	82
Winter oats (grain, 9 t/ha, straw, 5.7 t/ha)	159	31	125
Hay (8.5 t/ha)	127	19	153
Potatoes (55 t/ha)	165	21	265
Maize (12 t/ha)	170	31	37

Source: Recalculated from Anon (2012b) and Anon (1998).

materials between activities. Using organic sources of nutrients has a downside: the low nutrient content of organic manures means that they must be applied in bulk. The presence of N in organic forms means that microbial activity is required before it becomes available. This is strongly influenced by soil temperature so the ability of organic manures to meet crop requirements is less effective than with mineral fertilisers.

The addition of P as rock phosphate (apatite) is limited in a similar way, requiring either weathering or microbial activity to make the phosphate available. Organic manures (see Table 9.2) may have a different balance of nutrients from those required by a growing crop. This is complicated by the varied needs for nutrients of different crops (Table 9.3) and the time period over which they are required.

Although the differences in the demand for major nutrients by cereal crops are small, there are major differences in the relative demands of hay and potatoes for N and K (see Table 9.3).

Table 9.4 The response of a range of crops to different forms of nutrient supply

Crop	Nutrient treatment and impact on yield	Impact on nutrient supply	Reference
Cabbage	Composted animal manure (30 Mg/ha) compared to180 kg/ha fertiliser N. Yield: 51.8 Mg/ha with best organic treatment compared to 79.6 Mg/ha for mineral N; unfertilised 28.7	N recovery from mineral fertiliser 55%. Compost efficiency 9–14% depending on maturation. N uptake 156 kg/ha for mineral N and 80–103 for organic N	Mourao et al. (2012)
Cabbage	Composted animal manures (90–40 Mg/ha) compared to 0–180 kg/ha mineral N. Yield on dry weight basis around 10% higher with mineral N	Total N uptake 350 kg/ha for mineral N and 300 for organic	Brito et al. (2012b)
Orange	Various organic fertilisers and equivalent amounts of mineral N. There were no effects on yield	There were no effects on leaf N	Roccuzzo et al. (2012)
Hazelnut	Pig manure (80 kg/ha N) compared to same rate as mineral N. No effect on yield		Cerutti and Beccaro (2012)

These issues, and the differences between organic and mineral fertilisers, have been discussed by Barker (2010).

The impact of source of nutrient input is important to organic–conventional comparisons (Table 9.4). In practice, some forms of organic nutrient input or conservation are better geared to meet the immediate needs of the crop. Crop needs for N are usually in the range of 120–200 kg/ha (see Table 9.3) which can be met by composted animal manures which provide 103–300 kg/ha N (see Table 9.4). In some perennial crops, mineral sources of N gave no advantage (see Table 9.4).

The impact of the various ways in which the availability of nutrients may be influenced is discussed in more detail in the following sections after a consideration of the impact of modifying the supply of nutrients.

9.3 What is the Impact of Differences in Soil Nutrient Supply?

All crops require some available mineral nutrients. An excessive application increases costs and environmental impact while an inadequate supply reduces crop performance. This is true of all systems. For systems where the supply of nutrients from the soil is dependent on microbial processing, soil temperature is important. It varies both over a season and over the course of a single day (Buchan 2001). At a site in north Scotland, temperature at 30 cm depth was a minimum of 3 °C in February, rising to a maximum of 14 °C in August. Over a single day, although temperature was constant below 30 cm depth, it changed significantly above this depth. Temperature at the surface was higher than at other depths (>20 °C) at mid-day but lowest at midnight (<10 °C). While absolute temperatures are important, the differences between soil and air temperature are also important. Air temperatures influence the development of the above-ground part of the crop. Soil temperature influences the ability of

Table 9.5 The mean air temperature and soil temperature °C at 10 and 20 cm depth for a number of months in 1976 at East Malling, Kent, UK

Month	Mean maximum air temperature	Mean air temperature	Soil temperature at 10 cm depth	Soil temperature at 20 cm depth
March	8.9	5.2	3.7	4.1
April	12.9	8.3	8.5	8.4
May	18.5	12.9	14.2	13.3
June	24.3	17.7	20.6	19.3
July	25.2	19.4	21.5	20.9
August	23.8	17.8	19.1	18.8
September	18.4	14.2	13.9	14.1

Source: Data from Wickenden (1977).
1976 was an especially hot year.

the root system to supply nutrients and the ability of soil microbes to transform nutrients. Data for variation in mean soil and air temperature for the main months in which crop growth occurs are shown in Table 9.5.

Srek et al. (2010) assessed the relationship between the size of nutrient additions and their uptake by the potato crop in an experiment which ran for >50 years. P concentrations of <30 mg/kg soil (Mehlich III) and K concentrations <150 mg/kg soil resulted in reduced yields. To obtain a potato yield of >30 Mg/ha, there was need for applications of 140 kg N/ha, 63 kg P/ha and 186 kg K/ha. The highest acceptable application of N on the basis of environmental criteria was 120 kg N/ha.

One of the key aims of organic production is to minimise the import of resources. However, many organic production systems, especially those producing high-value vegetable crops, depend on the import of manures. The need for such imports is influenced by the effectiveness of the rotation, especially the use of green manures and catch crops (Thorup-Kristensen et al. 2012). In a vegetable production system imports of N were 149, 85, 25 and 25 kg/ha in conventional, manure-based organic and two green manure-based systems. Organic yields were on average 82% of conventional yields. The green manure systems increased the exploration of the soil by the crop and the use of subsoil N and had significant effects on the quantities of N absorbed by the crop (Table 9.6).

The system of production also impacted on the difference between total additions and removal (Table 9.7).

9.4 Organic Manures: Recycling of Nutrient Sources

The nutrient inputs required by an organic system can be met in a variety of ways. The approach closest to the basic principles of the organic movement is through the use of the crop rotation, including the use of cover crops. This is discussed in detail in Chapter 6. However, there are

Table 9.6 The impact of system of production on the total N uptake of the crop and the percentage N in the harvested produce

	Production	system				
	Conventional		Organic with	Imported manure	Organic with	Green manures
Crop	N uptake kg/ha	%N	N uptake kg/ha	%N	N uptake kg/ha	%N
Oats after white cabbage	94.3	2.00	77.3	1.63	75.1	1.61
Carrot	*183*	*0.97*	*138.4*	*0.82*	*152.4*	*0.89*
Winter rye after carrot	*144.2*	*1.77*	*76.0*	*1.47*	*70.2*	*1.52*
Lettuce	*114.2*	3.09	93.9	2.91	*86.0*	2.84
Oats after lettuce	87.8	1.89	69.0	1.59	67.0	1.47
Onion	*173.2*	*1.60*	*106.7*	*1.35*	*112.3*	*1.35*
Winter rye after onion	*157.5*	1.78	*111.1*	1.53	*101.7*	1.61
White cabbage	*379.3*	2.15	*309.8*	2.12	*305.3*	2.09

Source: Data from Thorup-Kristensen et al. (2012): values in italics in the same row significantly different at P < 0.005.

Table 9.7 The effect of growing system on the quantity of nutrients brought into the system and the nutrient balance, that is, the difference between addition and off-take in the crop kg/ha/year

	Conventional	Organic with manure	Organic with green manure
N application	149	85	25
N balance	32	1	−58
P application	28	6	2
P balance	8	−10	−15
K application	70	31	9
K balance	−27	−50	−70

Source: Data from Thorup-Kristensen et al. (2012).

times when either the cropping system has needs beyond what can be provided in this way or there is an opportunity to use the production system to recycle materials from outside the production base. Muramoto et al. (2011) found that in a system where a legume cereal cover crop was used to supply and retain N, the addition of an organic fertiliser derived from feather and blood meals and applied at rates between 84 and 252 kg/ha gave added yield. However, in a soil without a long organic history, significant amounts of N were mineralised, so reducing the impact. Correspondingly, only a proportion of nutrient additions of this type become immediately available. With farmyard manure (FYM), commonly only a half is available in the year of application with smaller proportions becoming available over succeeding years (Anon

Table 9.8 Organic manures derived from waste products

Material	Characteristics	Nutrient release and other characteristics	Reference
Feather meal and liquid fish emulsion	Feather meal applied to soil twice a year, fish emulsion applied as drench seven times a year	Supplying 29 and 57 kg/ha N increased blueberry yields	Strik et al. (2012)
Horse manure	Manure composted for eight months and applied with lime	Supplied 7.2 g/kg N. 13–20% of N recovered by crop. Yield increased by 26% without impact on N content of lettuce	Brito et al. (2012a)
Composted animal manures	Manure from both cattle and horses	Supplied 21.3 g/kg N. N accumulation 33 kg/ha compared to 82 with mineral fertiliser. Yield 4.2 Mg/ha compared to 9.3 with mineral fertiliser	Mourao et al. (2012a)
Meat industry wastes	Composted slaughter house waste combined with phosphogypsum and vinasse	1.6–3.7% total N	Ramirez-Guerrero et al. (2012)
Poultry manure	Composted poultry manure	0.95 N content with increasing applications to 20 Mg/ha increasing tissue N. Yield increase in pineapple	Aiyelaagbe et al. (2012)
Sewage sludge	Anaerobically digested sewage sludge co-composted with other materials	5.7% N content. Increased tissue N in tomato from 1.4% to 2.2%	Llorens et al. (2012)
Sheep wool	Pelleted sheep wool with additives of cellulose, starch or casein	10–11% N. Able to substitute for mineral fertilisers giving yield advantages for lettuce, tomato and increased tissue N	Bohme et al. (2012)

1998). Successive applications of such forms, however, create a significant reserve of N. In addition, because of its importance to production, fertiliser applications are commonly based on the crop's need for nitrogen. While this will normally supply all of the K needed by the system (see Tables 9.2, 9.3), it may only meet around a quarter of the P requirement (Barker 2010).

As all organic systems result in the removal of significant quantities of nutrients from the system (see Table 9.7), there comes a time when nutrients need to be replaced. If this can be done in a way which uses nutrients which might otherwise be regarded as waste materials, then this represents a positive link between organic production and total environmental management. Materials which have been used in this way as a source of nutrients are summarised in Table 9.8.

The source of the addition affects the speed of release and soil microbial activity; liquid fertilisers were better at meeting the needs of sweet pepper transplants than were solid materials (Gravel et al. 2012).

9.5 Crop Rotations

The impact of crop rotations is discussed in detail in Chapter 6 and so impacts on nutrient supply are mentioned here to put them into a nutrient-specific context.

The rotation is at the heart of organic farming. The principles of rotations were described by Sanders (1944) who wrote during the Second World War, a time when it was critical to maximise food production. Soil, climate, altitude, accessibility, availability of labour, markets, finance and the inclination of the farmer all affect the rotation selected. Where a rotation includes stock, which is the norm, then the number and type of animals will determine the amount of land needed to provide an annual supply of feed and grazing with what remains as the cropped area. Here we focus on those aspects of rotations which directly influence nutrient supply. Stockdale and Watson (2011) emphasised that there was no simple recipe for a rotation which would best fit all farms.

The design of the rotation has a significant effect on yield and nutrient cycling. Kayser et al. (2010) assessed the impact of crop rotation and nutrient source: FYM compared to arable legumes for the production of triticale. They found yields to be highest with ploughed-in grass or grass-clover which had previously received FYM. The previous crop in the rotation affected both yield and residual soil N. In a conventional production system, the quantities of N and P supplied in fertilisers more than replaced the quantity of nutrient removed in the crop, giving potential for >20% of the added N to be lost through leaching. The application of K was insufficient to counter crop off-take. In the organic system which received imports of manure, the quantity added balanced off-take for N. Where the quantity of manure added was small and the crop depended upon its ability to extract N from that being mineralised, the imported quantity was smaller than the removal by the crop. Here the root system, averaged over the season, occupied 38% of soil depth compared to 21% for a conventional crop. In both organic systems, off-takes of P and K exceeded inputs by a significant margin.

The ways in which nutrients are supplied affect losses. A maize crop absorbed 276 kg N/ha/year. Where N was supplied within a rotation as manure, the concentration of N varied between 627 mg/kg at 10 cm depth and 2.5 mg/kg at 140 cm, both in late summer. Where N was applied as slurry, comparable values were 345 and 8.9 mg/kg (at 60 cm depth). Total profile amounts of N were 1.27 and 1.32 and estimated quantities leached over a 50-year period were 9.3 and 11.1 Mg/ha for manure and slurry applications.

9.6 Cover Crops

Cover crops, which include both green manures and companion crops, significantly influence nutrient supply either by increasing the quantities of N available as a result of N fixation or by removing N and other nutrients from the soil solution and so preventing their loss by leaching (Corre-Hellou et al. 2011). Cover crops have value in a range of crops (Table 9.9).

The impact of nutrients available from the soil differs between different nutrients (Anon 1998).

9.7 Legumes

Legumes have long been used as a means of increasing the total amount of N in soil. The microbial association between leguminous crops such as peas, beans or clover and the *Rhizobium* bacteria allows the crop to fix mineral nitrogen and increase the N in the system. The amount of N which can be added in this way varies between crops, being as high as 200 kg/

Table 9.9 The impact of cover crops on yield and nitrogen supply

Crop	Cover crop/green manure	Effect on yield	Effect on N	Reference
Tomato	Six legumes used as no-till or living mulches. Controls of plastic mulch or no mulch	Increased from 6 to 9.4 kg/m^2 for best mulch (lupin no till) in autumn and 8.5–13.3 for best (Vicia no till) in spring		Duyar et al. (2012)
Tomato/lettuce	Two different legume and broccoli pre-crops	Little effect on tomato but increase in lettuce of 18–27% compared to fallow	N incorporation increased: fallow 41, broccoli 25, faba bean 70, vetch 202 kg/ha/year	Al Chami et al. (2012)
Tomato, zucchini, pepper	Two different legumes and broccoli as pre-crops	Revenue higher for all cover and pre-crops relative to fallow. Highest for broccoli		Bilen et al. (2012)
Zucchini/lettuce	Seven self-seeding winter annual legumes	Trifolium > Medicago Best legume 42.6 Mg/ha compared to 24 fallow for zucchini and 48 compared to 39 for lettuce	Best legume 147 kg/ha N input leading to N uptake of 226 kg/ha in zucchini and 48 in lettuce	Driouech (2012)
Peach	White clover	Peach root density increased by clover. Yield effects small	Total N 17–81 kg/ha with clover, 1–62 with tillage	Parveaud et al. (2012)

ha for clover swards down to 80 kg/ha in beans (Barker 2010). Riesinger and Herzon (2010) measured N fixation in a grass-clover ley as 39 kg N/ha/year for ley growth after the harvest of the nurse crop followed by 185 kg in subsequent cuts and 62 kg in the aftermath.

The legume component has a particular importance when the rotation is stockless. Here, the grass-clover element brings in no direct income and so must benefit succeeding crops in relation to both yield and quality. Martens and Entz (2011) reported that grazing a sward could provide an income of $385–770 through sales of animal products as well as returning at least 75% of the N. A significant element is the impact of such systems on nitrogen loss (Borgen et al. 2012). Continuous cereal undersown with grass-clover winter cover crops gave the best whole rotation grain yield, N yield/input ratio, N loss/yield ratio and the minimum emissions. This rotation had the poorest soil organic matter retention. N use and loss indicators were poor when ley years were consecutive and there was autumn ploughing. Management changes can influence autumn/winter N losses.

9.8 Soil Microbial Populations and Inoculation

The founders of the organic movement were clear about the importance of soil microbiology to plant health. Sir Albert Howard's foundation text (Howard 1945) contains pictures of an apple root with associated fungal hyphae. Lady Eve Balfour's book on the soil (Balfour 1943) establishes infection with arbuscular mycorrhizal fungi (AMF) as the basis of the concept of plant health. Probably the single greatest attempt to increase nutrient supply through the manipulation of soil microbial activity has been through the use of additions of AMF. The impact of this is discussed in Chapter 11 and so the potential impact is merely referred to here for completeness. For example, in a study of AMF on soils being used for the production of chillis, Boonlue et al. (2012) found the presence of 14 different taxa of AMF, with most coming from the genus *Glomus*. Infection with *Glomus clarum*, in particular, increased growth, flowering, fruit production and P uptake compared to non-AMF controls.

9.9 The Impact of Different Soil Nutrients

9.9.1 Nitrogen: How Much N Does a Crop Need?

The soil N cycle is important to the provision of N to crops. Doltra et al. (2011) found that in organic systems, regardless of the crop rotation and so the source of N, a limitation to N availability reduced yields compared to where it was supplied as mineral N. The mechanism for this is discussed in Chapter 8. In all systems, most of the N in soil is present in complexed organic forms. In a given season, only a small fraction of this becomes available (Spargo et al. 2011). The total amounts of N held in soils can be large; a deep peat soil can hold 40 Mg/ha of N in the surface 15 cm (Barker 2010). Making any of this available depends on soil microbial processes.

Commonly, organic N is mineralised to ammonium and then nitrified to nitrate, the preferred form for crop uptake. In its organic forms, N is less at risk of loss but as ammonium it can be lost as ammonia and as nitrate as either N gas or as the very damaging N_2O following denitrification. The needs of a crop for N are thus the product of that needed for crop production plus that lost through these processes. While the amount of N in the soil, especially as NO_3, influences the N status of the crop, N requirement is often diagnosed on the basis of the quantity of N in crop leaves, although the relationship is far from precise. High soil N levels were associated with high concentrations in the leaves but maximum leaf levels could be associated with soil concentrations varying from 30 to 60 ppm while a soil N concentration of 30 ppm could be associated with a leaf N concentration of between 2.5% and 3.1% N. The concentration of NO_3 in soil is also variable over a season. There would seem to be no simple but precise way of establishing the N status of an organic system.

Decisions on the quantity of N required for optimised production will also vary depending upon whether the criteria are driven primarily by production or environmental concerns. Giola et al. (2012) identified that the annual uptake of N by a maize crop was 276 kg/ha while environmental concerns would dictate that applications should not exceed 170 kg/ha N. They investigated the impact of slurry, compared to manure, on N loss through leaching. The highest nitrogen contents found in the soil from slurry and manure were 345 and 627 mg/kg respectively and the lowest 8.9 and 2.5 mg/kg soil. Prior to nutrient applications and the growth of the crop, the soil N content was 1.32 and 1.27 Mg/ha for slurry and manure treatments respectively. After application and cropping, it was 1.37 and 1.23 Mg/ha for slurry and manure treatments respectively. In a 50-year simulation, NO_3 leaching was 11.1 and 9.25 Mg/ha for the two nutrient additions.

Of all the elements needed in significant quantities for crop production, nitrogen probably has the greatest impact in most situations and so is most often associated with variation in yields. Nikolic et al. (2012) concluded, from a study on 30 different wheat cultivars, that:

> One way to enhance productivity, maintain efficient production and minimise environmental impact is to develop specific cropping strategies and to select productive genotypes which can grow under low N conditions.

They assessed the inter-relationships in wheat between grain N content and other indicators of N efficiency. Above-ground N content was strongly related to yield, N accumulation and N reutilisation which had significant effects on grain N. N reutilisation was important to the total grain N supply process.

Improving the efficiency of N supply to the crop is important but different crops and different varieties of a specific crop can have different responses. The demand for N in maize, 18–22 kg/Mg of grain, is lower than that in wheat, commonly 30 kg/Mg. In both crops, potential yield was closely related to N supply (Lorgeou and Prioul 2011). It is within this area of variation that there exists real potential to develop organic systems with yields closer to those produced in fertiliser-managed systems.

The identification of N-efficient varieties is important. In a study using Brussels sprouts as the test species, Fiedler and Stutzel (2012) found that although N accumulation efficiency (NAE) was variable, the relationship between sprout dry matter and total N accumulation explained over half of all the variation in trials with varieties and N supply. Nitrogen use efficiency (NUE) was influenced by both sprout N content and harvest index while the relationship between sprout N and total above-ground N accounted for much variation. Harvest index was linearly related to NUE, suggesting the importance of partitioning. Varieties which performed well under conditions of high N supply also did well when the supply of N was limited.

The supply-orientated model which underpins fertiliser applications in conventional agriculture has tended, even in organic systems, to focus on getting more N into the soil through the use of manures and N fixation by legumes. While this is important, given the extent of N removal in crops (see Tables 9.2, 9.3), it also matters to focus on absorbing N from what is available in the soil and on retaining N within the system. Sapkota et al. (2012) assessed the potential impact of different kinds of catch crops on N retention and its consequences for yield. Chicory, fodder radish and ryegrass functioned in different ways as cover crops. Radish developed a deep root system and depleted N from deeper soil layers to a greater extent than the other species, leading to reduced loss of N. Systems with radish decreased N loss by 79% compared to a non-catch cropped control, resulting in a 2% increase in yield. Chicory and ryegrass reduced N leaching by 71% and 67% respectively but at the cost of a 3% and 5% reduction in yield. Systems with ryegrass had reduced N losses from the top 1 m of soil. Discontinuing the use of a catch crop increased yields but at the cost of a 17% increase in N leaching.

The design of crop rotations has been heavily influenced by the need to optimise the N supply to the main cereal crops in the rotation. Within the cereal component of a rotation, intercropping with a legume can positively affect the N status of the cereal (Tosti and Guiducci 2010). When durum wheat was intercropped with faba beans up to the period when the wheat began to shoot, at which time the beans were incorporated into the soil, this resulted in a positive impact upon yield and a significant increase (10%) in grain protein.

9.9.2 Phosphorus

As with N, crop yields can be related to the amount of P available and to the amount which can be removed from the soil. A study in which a series of rotational treatments were combined with a number of different nutrient additions showed a good relationship between grass yield and the total amount of P removed from the soil by the grass crop.

In soil, P can exist in three major forms: organic P, mineral P and soil solution P. The quantity of P in the soil solution is always low while commonly half of the total P present can be in an organic form. Mineral P exists in chemical forms with substantial variability in their availability to crops (see Chapter 4). Phosphate in solution can also be present in three different ionic forms. A significant proportion of soluble forms of P are fixed by the soil and changed so as not to be immediately available. Nitrogen can be introduced to an organic crop production system by enhancing the impact of fixation from the atmosphere but the supply of P needs an alternative approach to nutrient introduction – the use of materials with very limited immediate crop availability. This is essentially the opposite of the fertiliser philosophy where added materials are intended primarily to have an instant impact.

Organic systems allow the use of rock phosphate as a means of increasing the amount of P within the system and to compensate for the removal of P with harvested crop product. Rock phosphate occurs in a number of forms, of which the most common and the most used in agriculture are calcium phosphate, calcium iron aluminium phosphate and iron aluminium phosphate. Calcium phosphate occurs as apatite and like the other forms is insoluble in water and so not directly useable by a crop. The release of P from rock P is influenced by its mineralogy and particle size but in soil primarily by pH; dissolution is enhanced by acidity, and by the concentration of Ca and P in the soil solution. Fungi have been shown to be effective in increasing the rate of solubilisation of apatite (Dissanayake 1992).

The quantity of P required for good crop production is commonly assessed by soil analysis. Over the years, a range of extraction media have been tested and soil P concentration related to crop response. The variation found in different soils between prevalent forms of P means that different extractants work variably well on different soils. On organically farmed soils, where rock P has been used as a source of P, the relationship between the amount of P removed by a common acid extractant, acetic acid, and an anion resin, which is thought to simulate the ability of roots to remove P from a soil, clearly was different between the two extractants.

Dissanayake (1992) found that on a low pH soil ryegrass was able to absorb more P from apatite (8.6 mg/pot) than from superphosphate (5.8 mg/pot) over a six-month period. On a high pH soil there was no significant effect (Table 9.10). For swedes grown in an acid soil, rock phosphate was almost as effective as triple superphosphate while this was not the case for potatoes grown on the same soil.

Plant species which are known to acidify the soil such as brassicas and legumes seem to be most effective at solubilising rock P (Dissanayake 1992) In addition, Knox et al. (2011) found that green manure crops such as forage rye could increase the availability of P from rock phosphate to following crops.

The characteristics of the plant root system and associated micro-organisms such as AMF have a major impact on the release of P from semi-labile forms in the soil. The mechanism of such an effect was summarised by Atkinson (2006).

Table 9.10 The impact of phosphate type and crop on P uptake (mg/pot) and the impact on soil pH at the conclusion of the study

Character	P source	Oil seed rape	Cabbage	Pea	Sunflower
P uptake	Nil	8.2	8.1	13.9	10.5
	Apatite	26.7	25.2	14.6	10.6
	Super P	24.8	27.2	17.6	14.5
Soil pH	Nil	5.3	5.3	5.4	5.4
	Apatite	5.1	5.1	5.3	5.2
	Super P	5.1	5.2	5.4	5.2

Source: Data represented from Dissanayake (1992).

9.9.3 Potassium

The K content of plant residues is relatively high (1–4%) and is immediately available. The content of farmyard manures is lower than that of the feed sources of the stock, as a consequence of the loss of urine in the field and leaching of the manure during the maturation or composting process. Potassium is found in feldspars and micas and so in sands, such as greensands (5.8%) and rocks, such as granite (4.2%), but these materials are not available to crops. However, the quantities of K in soils are high relative to N or P. The amounts in the surface 15 cm of soil vary from around 3.5 Mg/ha for sandy soils to as much as 80 Mg/ha in clay soils. Comparable values for P are 0.7–2.0 and for N 1.6–8.2 Mg/ha. However, >90% of the K in soils is present as primary minerals with around 1% as available K or in soil solution (Barker 2010).

Analysis of soils across the UK has indicated that significant numbers of soils are deficient in potassium: >30% of arable soils and >40% of grassland soils (Anon 2012a). Potassium, like nitrogen, is needed in large quantities by all crops, commonly 248 kg/ha (300 kg K_2O/ha). Unlike N, it cannot be sourced from the atmosphere and unlike P, there is no sparsely soluble form; all K minerals have release times in hundreds if not thousands of years. Supplying adequate amounts of K to organic crops is thus an important issue.

In conventional crop production, the application of a fertiliser has two distinct purposes. First, it is intended to replace the quantities of the nutrient removed with the harvested crop. Second, it is aimed at eliminating any soil deficiency which could limit the performance of subsequent crops (Anon 2012a). Where soils have a high K status, this amount will be small, around 30 kg/ha/year, but where soils are identified as being at soil test index 0 or 1, correcting this would need larger additions or smaller amounts over a longer period of years. This will need significant total inputs – around 500 kg/ha on many soils and up to 1 Mg/ha on heavy soils. The first element of this approach is equally important to organic production while the second would be seen as of lesser importance when considered against organic principles. The selection of crops to grow is important. Crops with an extensive root system, which exploits a larger soil volume, will cope for longer on low-index soils. However, it is important to recognise that organic crops do need significant quantities of K to be available to them.

9.9.4 Sulphur

The ending of the extensive burning of coal resulted in an increasing need to supply S, especially to crops with a relatively high S content. Sulphur is most commonly applied either as potassium sulphate, which is soluble and immediately available, or as elemental sulphur, usually in a micronised form, or as calcium sulphate. Both of the latter need to be weathered prior to becoming available. Elemental sulphur is effective as a fungicide as well as a source of sulphur. Calcium sulphate, Ca $SO_4 \cdot 2H_2O$, gypsum, like rock phosphate, comes in a crystalline form, which requires to be both weathered and metabolised by soil micro-organisms to make it available. A study of the impact of different particle sizes of gypsum showed that particle size and rate of application affected both crop growth and the levels available in the soil (Yarwood 1988) (Table 9.11).

The addition of S to durum wheat has been shown to increase yield, S and N uptake (Ercoli et al. 2012). The increase was due to an increase in the number of grains per ear. On average, 35 kg/ha S was leached in a season although the amount increased with an increase in the rate of application.

9.10 Conclusion

Nutrient additions to organic production systems centre either on the use of by-products from the rotation and from other on-farm activities or on materials which are waste products from other systems or activities. There are also additions of non-labile sources of nutrients such as rock phosphate and calcium sulphate, both of which require chemical/microbial and physical degradation for nutrient release. In all of these materials, the nutrients needed by the crop are not immediately available and in most cases there is need for microbial decomposition as a prelude to the nutrient being available to the crop. Microbial activity will be greatly influenced by soil temperature, which itself is a product of radiant energy.

The science needed to improve the gearing of crop needs to nutrient supply thus relates to our understanding of these decomposition processes, the way in which they are affected by soil

Table 9.11 The effect of the composition and particle size of gypsum applications on the yield and S content of ryegrass (*Lolium perenne*)

Parameter	Coarse agricultural gypsum	Fine agricultural gypsum	Fractionated fine gypsum	Potassium sulphate	Elemental sulphur
% >52 BS mesh	87	90	100	–	–
% >100 BS mesh	70	72	72	–	–
% S	15.4	18.2	17.6	18.3	80
Grass yield (g/pot)	24.0	21.5	26.0	23.1	25.7
Cumulative S content of grass (mg)	72.0	68.8	85.3	83.3	66.0

Source: Data from Yarwood (1988).
BS, British Standard.

conditions and the scope for better understanding of how systems design may influence these links. There is also evidence that different cultivars of crop species may respond differently to nutrient additions. This seems to be related to variation in the intensity of demand, the period of peak demand and to the crop's ability to exploit resources present throughout the soil profile rather than in a limited part of the soil profile. This suggests that there is scope both to increase organic yields and to improve the gearing of production to a varied range of nutrient additions.

References

Aiyelaagbe, I.O.O., Oshuniyi, A.A., and Adegoke, J.O. (2012). Response of smooth cayenne pineapple to organic fertiliser in south western Nigeria. *Acta Horticulturae* 933: 261–264.

Al Chami, Z., Al Bitar, L., Amer, N. et al. (2012). Evaluation of precrops and organic fertilisation programme on soil chemical properties and on subsequent crop under Mediterranean conditions: case study of south of Italy. *Acta Horticulturae* 933: 313–319.

Anon 1998 Annual Report of the Scottish Standing Committee for the Calculation of Residual Values of Fertilisers and Feeding Stuffs. HMSO, Edinburgh.

Anon 2012a Standard recommendations cannot make a poor soil good overnight. PDA, York. Available at: www.pda.org.uk/standard-recommendations-cannot-make-a-poor-soil-good-overnight

Anon (2012b). *The Farm Management Handbook 2012/13*. Edinburgh: SAC.

Atkinson, D. (2006). Arbuscular mycorrhizal fungi and the form and functioning of the root system. In: *Microbial Activity in the Rhizosphere* (ed. K.G. Muckerji, C. Manoharachary and J. Singe), 199–222. Berlin: Springer.

Balfour, E. (1943). *The Living Soil*. New York: Universe Books.

Barker, A.V. (2010). *Science and Technology of Organic Farming*. Boca Raton: CRC Press.

Bilen, E., Nazik, C.A., Unal, M. et al. (2012). Economic performance of pre-crops in a three-year rotation programme for organic vegetable production. *Acta Horticulturae* 933: 321–327.

Bohme, M., Pinker, I., Gruneberg, H., and Herfort, S. (2012). Sheep wool as fertiliser for vegetables and flowers in organic farming. *Acta Horticulturae* 933: 195–202.

Boonlue, S., Surapat, W., Pukahuta, C. et al. (2012). Diversity and efficiency of arbuscular mycorrhizal fungi in soils from organic chilli farms. *Mycoscience* 53: 10–16.

Borgen, S.K., Lunde, H.W., Lars, R.B. et al. (2012). Nitrogen dynamics in stockless organic clover-grass and cereal rotations. *Nutrient Cycling in Agroecosystems* 92: 363–378.

Brito, L.M., Amaro, A.L., Mourao, I., and Moura, L. (2012). Yield and nitrogen uptake of white cabbage (Brassica oleracea var capitata) with organic and inorganic fertilisers. *Acta Horticulturae* 933: 107–113.

Brito, L.M., Pinto, R., Mourao, I. et al. (2012). Organic lettuce growth and nutrient accumulation in response to lime and horse manure compost. *Acta Horticulturae* 933: 157–163.

Buchan, G. (2001). Soil temperature regime. In: *Soil and Environmental Analysis* (ed. K.A. Smith and C.E. Mullins), 539–594. New York: Dekker.

Cerutti, A.K. and Beccaro, G.L. (2012). Short term effects of manure fertilisation in a hazel nut orchard in northern Italy. *Acta Horticulturae* 933: 245–252.

Corre-Hellou, G., Dibet, A., Hauggaard-Nielsen, H. et al. (2011). The competitive ability of pea-barley intercrops against weds and the interactions with crop productivity and soil N availability. *Field Crops Research* 122: 264–272.

Dissanayake DMAP (1992) Plant and soil factors influencing the availability of phosphorus from natural phosphate sources. PhD thesis, University of Aberdeen.

Doltra, J., Laegdsmand, M., and Olesen, J.E. (2011). Cereal yield and quality as affected by nitrogen availability in organic and conventional arable crop rotations: a combined modelling and experimental approach. *European Journal of Agronomy* 34: 83–95.

Driouech, N. (2012). Annual self-reseeding legumes and their application into Mediterranean organic cropping systems. *Acta Horticulturae* 933: 329–336.

Duyar, H., Tuzel, Y., and Oxtekin, G.B. (2012). Effects of cover crops on yield and quality of organically grown greenhouse tomatoes. *Acta Horticulturae* 933: 307–312.

Ercoli, L., Arduini, I., Mariotti, M. et al. (2012). Management of sulphur fertiliser to improve durum wheat production and minimise S leaching. *European Journal of Agronomy* 38: 74–82.

Fiedler, K. and Stutzel, H. (2012). Nitrogen efficiency of Brussels sprouts under different organic N fertilisation rates. *Scientia Horticulturae* 134: 7–12.

Giola, P., Basso, B., Pruneddu, G. et al. (2012). Impact of manure and slurry applications on soil nitrate in a maize-triticale rotation: field study and long term simulation analysis. *European Journal of Agronomy* 38: 43–53.

Gravel, V., Dorais, M., and Menard, C. (2012). Organic fertilization and its effect on development of sweet pepper transplants. *Hortscience* 47: 198–204.

Howard, A. (1945). *Farming and Gardening for Health or Disease*. London: Faber.

Kayser, M., Muller, J., and Isselstein, J. (2010). Nitrogen management in organic farming: comparison of crop rotation residual effects on yields, N leaching and soil conditions. *Nutrient Cycling in Agroecosystems* 87: 21–31.

Knox, O.G.G., Walker, R.L., Edwards, A.C. et al. (2011). Implication of green manure crop cultivation with rock phosphate in organic rotations. *Aspects of Applied Biology* 113: 101–110.

Llorens, E., Gallardo, A., Garcia-Agustin, P. et al. (2012). Response of tomato crops (*Solanum lycopersicum* Montecarlo) to sewage sludge based compost fertilisation. *Acta Horticulturae* 933: 123–130.

Lorgeou, J. and Prioul, J.-L. (2011). Agrophysiology: light interception, crop photosynthesis, mineral nutrition and crop managing. In: *Advances in Maize* (ed. J.-L. Prioul, C. Thevenot and T. Molnar), 393–410. London: SEB.

Martens, J.R.T. and Entz, M.H. (2011). Integrating green manure and grazing systems: a review. *Canadian Journal of Plant Sciences* 91: 811–824.

Mourao, I., Amaro, A.L., Brito, L.M., and Coutinho, J. (2012). Effects of compost maturation and time of application on the growth and nutrient accumulation by organic cabbage. *Acta Horticulturae* 933: 91–98.

Mourao, I., Pinto, R., Brito, L.M., and Rodrigues, R. (2012). Response of protected organic pea crop to increased application rate of manure compost. *Acta Horticulturae* 933: 181–186.

Muramoto, J., Smith, R.F., Shennan, C. et al. (2011). Nitrogen contribution of legume/cereal mixed cover crops and organic fertilisers to an organic broccoli crop. *Hortscience* 46: 1154–1162.

Nikolic, O., Zivanovic, T., Jelic, M., and Djalovic, I. (2012). Interrelationships between grain nitrogen content and other indicators of nitrogen accumulation and utilisation efficiency in wheat plants. *Chilean Journal of Agricultural Research* 72: 111–116.

Parveaud, C.E., Gomez, C., Bussi, C., and Capowiez, Y. (2012). Effect of white clover (*Trifolium repens* Huia) cover crop on agronomic properties and soil biology in an organic peach orchard. *Acta Horticulturae* 933: 373–380.

Ramirez-Guerrero, H.O., Molina-Aguilera, G., and Moyeja-Guerrero, J.C. (2012). Using phosphogypsum and vinasse for enrichment of ruminant content composting in the tropics. *Acta Horticulturae* 933: 293–296.

Riesinger, P. and Herzon, I. (2010). Symbiotic nitrogen fixation in organically managed red clover-grass leys under farming conditions. *Acta Agriculturae Scandinavica Section B Soil and Plant Science* 60: 517–528.

Roccuzzo, G., Fabroni, S., Allegra, M. et al. (2012). Effects of organic fertilisation on Valencia late orange bearing trees. *Acta Horticulturae* 933: 221–225.

Sanders, H.G. (1944). *Rotations*, Bulletin no 85. London: HMSO.

Sapkota, T.B., Askegaard, M., Laegdsmand, M., and Olesen, J.E. (2012). Effects of catch crop type and root depth on nitrogen leaching and yield of spring barley. *Field Crops Research* 125: 129–138.

Spargo, J.T., Cavigelli, M.A., Mirsky, S.B. et al. (2011). Mineralizable soil nitrogen and labile soil organic matter in diverse long-term cropping systems. *Nutrient Cycling and Agroecosystems* 90: 253–266.

Srek, P., Hejcman, M., and Kunzova, E. (2010). Multivariate analysis of relationships between potato (*Solanun tuberosum* L) yield, amount of applied elements, their concentration in tubers and uptake in a long-term fertiliser experiment. *Field Crop Research* 118: 183–193.

Stockdale, E.A. and Watson, C.A. (2011). Can we make crop rotations fit for a multi-functional future? *Aspects of Applied Biology* 113: 119–126.

Strik, B., Bryla, D., Larco, H., and Julian, J. (2012). Organic high bush blueberry production systems research- management of plant nutrition, irrigation requirements, weeds and economic sustainability. *Acta Horticulturae* 933: 215–220.

Thorup-Kristensen, K., Dresboll, D.B., and Kristensen, H.L. (2012). Crop yield, root growth and nutrient dynamics in a conventional and three organic cropping systems with different levels of external inputs and N recycling through fertility building crops. *European Journal of Agronomy* 37: 66–82.

Tosti, G. and Guiducci, M. (2010). Durum wheat-faba bean temporary intercropping: effects on nitrogen supply and wheat quality. *European Journal of Agronomy* 33: 157–165.

Wickenden MF (1977) Meteorological records 1975–1976. Report of East Malling Research Station for 1976, pp. 29–34

Yarwood V (1988) The impact of gypsum particle size on its use as an agricultural additive. BSc thesis, Liverpool John Moores University.

10

Crop Attributes Facilitating the Use of Soil Resources

David Atkinson

SRUC, Aberdeen, Scotland, UK

10.1   Introduction

In her key 1943 text, reissued in 1976, setting out her views on the needs of the organic farming movement, Eve Balfour (1976) highlighted the critical importance of the crop root system and its associated micro-organisms. Balfour emphasised a number of key ways in which the root system was critical to the management of an organic system.

- The root system of crops which were able to exploit deeper soil horizons were the key to the ability to run closed systems of production.
- Variation in root systems was at the heart of the rotational system, which allowed sequential crops to use different nutrients within the soil.
- The material transferred through the root system to the soil was the key to the creation of soil fertility and structure and so of productivity.
- The micro-organisms, especially mycorrhizal fungi, were the basis of both soil and crop health and at the core of an organic approach to the management of growing systems.

Similarly, in his foundation text, Sir Albert Howard (1945) emphasised the importance of the relationships between roots, mycorrhizal fungi and soils and, commenting on a view of the soil *in situ*, wrote 'on the root systems of fruit trees fungous threads could be seen approaching the young apple roots'. He characterised such relationships as being the key to successful crop production systems.

A more recent study (Thorup-Kristensen et al. 2012) has confirmed much of this and has highlighted the importance to optimised organic production of the ways in which the root system explores the soil volume. From the beginning of a formalised philosophy and approach to organic crop production, the root system has been seen as critical to success. Together with soil micro-organisms, discussed in Chapter 11, it is at the heart of the crop's ability to obtain resources from and maintain those resources in the soil. Understanding the functioning of the root system is vital to appreciating the science underpinning organic production and how it differs from the science base of other forms of farming.

The root system represents the medium of contact between the soil, the source of nutrients and water, and the above-ground parts of the crop. The documentation of the root system, in all types of production systems and especially under field conditions, has lagged behind that of the above-ground components (Franco et al. 2011). Only a few studies have specifically assessed the impact of organic production on the development and functioning of the root system. However, this does not mean that nothing is known about the effect of the features which distinguish organic production on the form, functioning and effectiveness of crop root systems. Essentially, we are asking whether the activities which characterise conventional production, such as the use of fertilisers and pesticides, are greater than other factors such as soil management, soil condition, inherent fertility and crop genetics. The answer seems to be No. We can thus use non-organic field data. Much can be inferred by relating what we know of the development of the root system under field conditions to our knowledge of the impact of organic production on soil processes and condition. Such an approach has much in common with the use of mathematical modelling (Dupuy et al. 2010) and the development of rapid assessment systems as part of crop breeding (Gregory et al. 2009). It emphasises the independence of science fact from the legal and regulatory requirements of organic production and permits the examination of our current knowledge base about the contribution of root systems in all crop production situations.

Many of the limited number of field studies of the development of root systems provide information independent of the production system being assessed. It is important to view the potential impact of the root system from a broad perspective. Re-emphasising one of the major conclusions drawn from the long-running Haughley Experiment (Balfour 1976), Berta et al. (2002) pointed out that it is a mistake just to see the root system as providing stability for the shoot and as the site of water and nutrient uptake. Root systems need to be assessed at community and ecosystem scales as they provide a source of photoautotrophic carbon for the organisms, which make up the rhizosphere and so have a significant influence on the soil in which they grow and on the other species with which they share that environment concurrently and in the future.

As a result, much of importance can be drawn from studies of natural ecosystems as they, like organic production, work as closed systems which evolve in terms of their soil condition and microbiology over time.

The major differences in soil management between organic and other crop production systems relate to specific practices, most relatively well understood, and where the development of the crop root system can have an impact as a result of the underpinning principles (Howard 1945). Howard's key elements were:

- healthy plants have the capacity to resist disease
- a healthy soil with an abundant microflora will produce healthy plants.

The general principles listed above and the specific differences introduced to a production system by the use of a rotation (discussed in Chapter 6), rather than a monoculture, help us to begin to identify the characteristics of the root system which are likely to be important.

In assessing soil resource use, we can ask what is the impact of such factors on the functioning of the root system. Unfortunately, too much of our information base comes from laboratory studies (Russell 1977). Here, the potential for the crop root system to deal with a limiting factor is restricted to performance within those conditions. The plant cannot avoid stress by growing and developing in another part of the soil volume, that is, by exploiting available plasticity. While information from such laboratory studies is thus of use on a comparative basis, it does not always help us to see the response of the root system to the types of challenges found under field conditions.

Here, we assess the requirements of an effective root system in a range of different crop production systems, the principal sources of variation which exist in the root systems of different crops and what is known of the effect of variation in soil management on root system functioning. The literature relevant to this is extensive and so the references selected have been chosen as being illustrative. Although it is likely that all roots of mycotrophic species in organic systems of production will be mycorrhizal, the potential impact of arbuscular mycorrhizas, their effects on root development and their contribution to crop health are discussed in Chapter 11.

10.2 Nutrient Capture and Utilisation

10.2.1 Basic Issues

Plants are sessile organisms and so need roots, and plasticity in those roots, to be able to 'forage' for nutrients. Under N-limited conditions, roots commonly adopt an active foraging strategy characterised by lateral root outgrowth and transcriptase reprogramming. This allows distinct responses to variations in N supply and demand. This is true of all root systems in both natural and cropped situations but is critical in organic production where conditions of limited N availability are relatively common. Plasticity within the root system is important because, regardless of how nutrients are managed, the availability of mineral nutrients is not uniform throughout the soil profile (Table 10.1). Nutrient concentrations do not decrease uniformly with increasing depth so variations in the distribution of roots will affect nutrient availability.

Table 10.1 Variation in the concentration of available nutrients in a Malling series soil with depth

Depth (cm)	% Organic matter	Available P	Available K	NO$_3$-N
0–15	2.74	36	400	7
15–30	1.29	28	219	4
30–45	0.68	12	180	3
45–60	0.58	10	145	4
60–75	0.49	12	118	3
75–90	0.41	13	128	6
90–105	0.49	7	219	7
105–120	0.22	11	137	6
120–135	0.18	8	115	5
135–150	0.11	6	93	5
150–165	0.11	4	102	8

Source: Data from Atkinson (1973).

The concentration of most nutrients tends to be highest close to the soil surface and to decrease with depth. Significant concentrations of nutrients such as NO_3 at depth occur during parts of the season. The significance of this was discussed by Greenwood (1986). He found that many crops could extract significant quantities of nutrients from the subsoil although the potential for this to occur varied between crop species, with cereal crops, such as wheat, able to remove nutrients from much greater depths than vegetables, such as spinach. Much of this variation was a function of size, with larger plants rooting more deeply. Gaiser et al. (2012) showed that similar considerations were important for water extraction by wheat and that the design of the crop rotation could influence both root distribution and water extraction. Where lucerne preceded spring wheat, water extraction from soil at a depth of 90–105 cm was increased, an increase associated with an increase in root length density of around 100%. Palta et al. (2011) found that having a large root system was important to the early-season capture of water and nutrients. These issues are explored later in the chapter in a case study on apple.

10.2.2 Nutrient Availability

In addition to major variations in the quantities of nutrients found in different parts of the soil profile individual nutrients differ greatly in the concentrations found in the soil solution at any depth. This changes the level of root system activity needed to provide the crop with sufficient nutrients (Fried and Broeshart 1967). The variation in the concentrations of different nutrients suggests that the form and density of the root system are likely to be especially important for nutrients. Richardson et al. (2009) demonstrated that root characteristics such as growth rate, specific root length and density had a significant impact on a crop's ability to use P. Matching root distribution to profiles of nutrient availability is important to nutrient capture but nutrient availability is dynamic. Here we examine what root system features seem most likely to influence nutrient capture.

10.3 The Functional Requirements of a Root System

10.3.1 Basic Issues

Although its production requirements are usually unstated, organic agriculture is based on a series of principles and assumptions (Atkinson and Watson 2000a). Obtaining the maximum possible yield is not the sole driver (see Chapter 1). However, producing food and optimising yields is important. As part of its vision, organic production assumes that with the correct management of the soil, substantial, but submaximum, yields can be obtained and sustained while having regard to long-term needs. It is also assumed that through a better understanding of soil processes, yields can be improved without compromising basic principles.

10.3.2 Relation of Root Activity to Soil Processes

If cropping systems are to be developed then improving the ability to make use of lower concentrations of soil nutrients is important. This also helps reduce the loss of nutrients to ground and surface waters. This raises the issue of whether genetic engineering could improve nutrient capture, an issue discussed in Chapter 15. Understanding soil processes and linking the availability of nutrients within the soil to crop demand are critical to optimum production. However, equally important is the means to absorb nutrients and the view that variation in root systems, whether a consequence of crop choice, varietal differences or the impact of soil management, makes a difference.

The relationship between root system functioning and crop nutrient supply was reviewed by Wild et al. (1987) for crops grown in solution culture and by Atkinson (1990) for soil-grown crops. Nutrient uptake and plant growth are interdependent. Growth drives the demand for nutrients (Clement et al. 1978) while growth is dependent on an adequate rate of nutrient supply (Woodhouse et al. 1978). Total and available amounts, chemical speciation and the seasonal availability of nutrients in the soil have a major influence on the nutrient status of the crop, hence the traditional reliance placed on soil nutrient analysis; nutrients must be absorbed to be effective. For a given crop, the quantity of nutrients needed to complete the annual cycle of growth and production defines the total annual requirement. This varies greatly between crops and between different cultivars of the same species. These numbers tend to minimise the importance of the turnover and recycling of nutrients within a single season.

Total nutrient requirement is, however, not the sole issue in relation to supply. Different species require nutrients to be available from the soil at different times during the season and with different intensities of supply, an issue discussed in Chapter 8. In annual crops, the maximum potential rate of nitrogen uptake commonly exceeds the potential of even the most fertile soil to meet crop needs. Although the total nitrogen need of spring barley is relatively low, it commonly absorbs 80% of its total nitrogen uptake of 170 kg/ha during a one-month period, in May, during which time the soil needs to provide and the root system absorb nitrogen at a rate of 4.5 kg/ha/day. Uptake by winter barley, March to June, and swede, July to October, occurs over a longer period (Atkinson 1990).

The ability of the root system to influence nutrient supply thus depends on both the characteristics of the root system and the availability of nutrients in the soil. Where nutrient supply is abundant, as in solution culture, or where large amounts of readily available nutrients have been supplied then the influence of root system characteristics will be small. Where nutrient supply is spatially diverse or limited in availability then a wider range of root system characteristics come into play. This is commonly the situation in organic crop production.

Organic agriculture focuses on providing the nutrients required for crop growth from sources within the soil. As a consequence, such production recognises that crops can grow without fertiliser additions. Hansson (1987) found that, without N additions, barley could produce 55% of the yield obtained with a normal rate of N addition and was able to absorb N from the soil at 58 kg/ha.

At the present time, few crop species have been bred for root system characteristics although variation does occur in different genotypes of the same species (O'Toole and Bland 1987). There is a wider body of information relating to the impact of growing conditions such as soil nutrient status (Grime et al. 1991), competition from other species (Atkinson and White 1980) or soil physical conditions (Goss 1991) on root system development. Here, we explore the variation which is possible within the root system, the effects of both genetics within a species and soil condition on the expression of this variation and the functional significance of these effects within organic production systems.

10.3.3 The Impact of Root System Form

The most basic of assumptions is that the supply of nutrients in an organic system is a function of nutrient availability (capacitance) and the intensity of supply. Where this is adequate then the supply of nutrients to the crop will depend on root length production, the volume of soil being exploited and the timing of root production. Factors such as root branching and specific root length act by influencing the relationship between root length and the volume of soil utilised. Measurable sources of variation were detailed by Atkinson (2000), who identified 29 sets of morphological characteristics and a further seven physiologically based features which could vary and influence functionality. Some of the characteristics most relevant to organic farming situations are summarised in Table 10.2.

The features of the root system (see Table 10.2) of most value in systems receiving their N as mineral fertiliser or as organic manures are identified in Table 10.3.

Where the principal source of N is a soluble fertiliser, good early root growth leading to a significant amount of root close to the soil surface and present at the time of fertiliser application is important. In contrast, with organic manuring, where the supply is dependent on microbial activity and consequently is slower but extended, then root production over a longer period is desirable.

Table 10.2 Key root system parameters and their functional significance in the context of organic production

Parameter and units	Definition	Functional significance/organic importance
Root length (m)	The total length of all root members present	The potential for absorbing water and nutrients from the soil. The basis of functional calculations. Exhibits large temporal variation. Basis of choice for comparing species, plasticity and impact
Root weight (g)	The dry weight of the whole root system	The mass of the root system and the amount of assimilate moved below ground. Basis for estimating carbon flows into the soil
Root length density (Lv. cm/cm^3 or La. cm/cm^2)	The length of root present in a volume of soil (Lv) or under a unit area of soil surface (La)	The intensity of soil exploitation and estimation of the limitations likely to result from soil exploitation. Basis for comparisons of species, systems and expressed plasticity
Specific root length (SRL m/g)	The length of root associated with a unit of dry root weight	The within-root system allocation strategy. An indicator of the importance of soil exploitation. A major element in within-species root system plasticity
Root longevity (d)	The length of time for which an individual root member is alive in the soil	The potential for rapid adjustment of root length and an indicator of soil exploitation strategy. Key to the input of carbon into the soil and to the longevity of AMF in soil
Root distribution (depth m, horizontal m)	The depth to which the root system penetrates and the spread of the roots from the plant stem	Physical stability and anchorage, the depth of soil exploitation and so the potential for use of water and nutrients from the soil. Important to assessing root system plasticity in particular systems

Source: Abbreviated from Atkinson (2000).

Table 10.3 The probable value of root system characteristics leading to optimum use of different forms of N

Root characteristic	Soil N regime	
	Mineral N fertiliser	Organic N supply
Length	Rapid root length production prior to fertiliser addition. High rate of root production	Extended duration of production and root survival. Rate and timing of production closely geared to soil temperature
Weight	Limited below-ground partitioning of resources. Low specific root length	High partitioning of assimilates to roots and AMF. High specific root length
Distribution	High proportion of length near soil surface	Extensive branching. Good distribution down soil profile
Activity	Efficient uptake at high concentrations	Efficient uptake at lower concentrations. The ability to use a variety of chemical species

AMF, arbuscular mycorrhizal fungi.
Source: Modified from Atkinson et al. (1995).

Individual roots vary in diameter, length, the balance between cortical and vascular tissues, relative amounts of vacuolar and cell wall material, longevity and, in perennial crops, the amount of woody tissue. The final form of the root system involves a series of trade-offs in the use of carbon and nutrient resources. Assimilate used to build roots cannot be used for shoot growth although restricting shoot growth does not necessarily lead to an increase in root production (Palta et al. 2007). In addition, the development of large-diameter roots usually results in a reduction in total root length. Roots vary in respect of their microbial associates principally in whether they are infected and to what extent by arbuscular mycorrhizal fungi. The significance of this is discussed in Chapter 11. Root systems can vary in their mass both absolutely and in relation to the above-ground material which they support. At a gross level, they vary in respect of the volume of soil they exploit, the depth of soil penetrated and the density of roots, all functions of the length of individual root members and their branching.

10.3.4 Variation Between Crop Species

The properties listed above are known to vary between species; for example, the roots of graminaceous species tend to be thinner than those of dicotyledonous species. Atkinson (1990) discussed the functional significance of variation in many of the properties listed in Table 10.2. The extent of variation in some of these is shown in Table 10.4, and its consequences in Table 10.5.

Experiments with root pruning have shown that the size of the root system is important. A pruned, thus smaller, root system absorbs less water and had a reduced ability to compete with plants with a larger root system (Ma et al. 2010). Gab and Kacorzyk (2011) demonstrated that key parameters could be altered by changes in growing conditions. They studied species-rich grassland with a range of different nutrient inputs. They characterised the root systems in respect of root length density, mean root diameter and specific root length. Treatment effects

Table 10.4 Variation between species in root system properties with significance for nutrient uptake

Root system property	Common range of reported variation between species
Root length density (cm/cm^2)	2–1000
Root depth (cm)	40–860
Specific root length (m/g)	5–750
Root branching (Rb)	3.5–11.9
P inflow (pmol/cm/s)	0.1–2.8
K inflow (pmol/cm/s)	0.3–0.9

Source: Redrawn from Atkinson (1990).

Table 10.5 Consequences of changes in root properties for water and nutrient uptake

Property	Change in root property	Estimated consequence for nutrient uptake
Root length density	An increase in the length of root per unit soil volume (Lv) from 0.1 cm/cm^3 to 10 cm/cm^3. An increase of Lv from 1 to 3 cm/cm^3	Access to an additional 45 ug P/cm^3 soil. Increased K uptake of 130 mmol per plant
Root depth	Increased by 10 cm	Access to an additional 0.45 μg of P and 20 mm water
Specific root length	Increased from 100 to 200 m/g	An increased K uptake of 60 mmol per plant

Source: Redrawn from Atkinson (1990).

on the root system only occurred in the upper soil layers. Root length density was highest (56 cm/cm^3 compared to 40 cm/cm^3) in unfertilised treatments while root diameter was lowest in these treatments (0.22 compared to 0.26 mm). Gaudin et al. (2011) found that variation in nitrogen supply changed the balance between root number, decreased by low N, and root length, increased by low N, as a result of an acceleration of growth rate rather than an increase in the length of the growing period. Low N increased the proportion of the root system made up of lateral roots and the number and length of root hairs. There was also variation both within and between species in the maximum depth of the root system (Weaver 1920). Most species will root to a depth of around 1 m although in practice the maximum depth of rooting is limited by soil depth. It is possible for different species to show different intensities of soil exploitation within similar depths of penetration.

Variations of this type can have significant impacts on the quantities of nutrients available (see Table 10.5). Selecting the right species or species combinations can significantly change the ability to make use of available nutrients. Christiansen et al. (2006) compared the root systems and nitrogen use of a range of vegetable crops. The maximum rooting depths of beetroot, sweet corn and celeriac were 1.55–1.8, 0.6–0.9 and 0.45–0.6 m respectively, with an associated variation in nitrogen use. Kristensen and Thorup-Kristensen (2007) found the maximum rooting depth of leek, potato, Chinese cabbage, beetroot, summer squash and white cabbage to be 0.5, 0.7, 1.3, 1.9, 1.9 and 2.4 m respectively. Shallow placement of N benefited shallow-rooted crops and deep placement the deeper rooted crops. Similarly, Thorup-Kristensen (2006) found that at harvest onion and lettuce had rooting depths of 0.3 and 0.6 m respectively compared to depths of over 1.1 m in carrot and cabbage. The deeper-rooted crops showed a greater response in N uptake after different forms of leguminous green manuring. In a comparison of three oil-seed crops and three pulses, Liu et al. (2011) found that all crops had most roots in the surface 60 cm of soil, with the highest densities at 0–20 cm depth. Root diameter was greatest in the pulses. In oil seed rape a good supply of water increased root length, number and surface area. Roumet et al. (2008) compared root system characteristics in the Asteraceae, Fabaceae and Poaceae. The Poaceae allocated more C to roots but developed a more sparsely branched system of smaller diameter and higher dry matter content roots.

The extent of variation in root morphology is often greater than that reported for physiological properties. So to increase the ability to absorb more nutrients from soil, it may be easier to increase root length rather than breed for carrier systems able to absorb more rapidly, particularly as this is influenced by mycorrhizal infection (see Chapter 11). Other authors have, however, emphasised the importance of variation in physiological properties. Hirel and Gallais (2011) reviewed the extent of variation in nitrate uptake within maize. They commented on the variation in transporters within the root cell membrane and on our limited knowledge of these compared to knowledge of model species such as *Arabidopsis*. For a range of species other workers, such as Peret et al. (2011), have demonstrated that plants have evolved a range of physiological and morphological modifications to cope with limitations to P supply. *Arabidopsis* adapts to P deficiency by inhibiting primary root growth, increasing lateral root formation and increasing the production of root hairs, all of which aid foraging in surface soil.

In general, however, it seems that morphological responses dominate the plant's response to variation in nutrient supply. In addition, in a number of crops the uptake of nutrients is dominated by the functioning of the hyphae of a mycorrhizal association with the fungus downregulating the plant genes following infection. Fernandez et al. (2009) showed that different crops varied in P efficiency although the differences depended on the definition of efficiency. Maize showed the highest utilisation efficiency in terms of biomass or grain produced per unit P absorbed. Soya and sunflower showed higher acquisition efficiency. Soya had 69% of the root system in the surface 20 cm of soil. P uptake per unit root length was similar in the crops studied but uptake per unit root weight was lowest in maize, a consequence of the higher specific root length: 59, 94 and 34 m/g for soya, sunflower and maize respectively. The dicot species were able to explore more soil with the same below-ground biomass and to absorb more P per unit of C invested below ground. The three species showed similar rates of P uptake per unit root length.

Plasticity is an important feature of a root system. Rose et al. (2009) found that plasticity varied between crop species under conditions of heterogeneous supply of soil P. Oil seed rape and wheat allocated more root length to zones where there was more P than did lupin. Serisena (1994) assessed the ability of different green manure species to remove NO_3 from the soil solution and found big differences between species commonly used as green manures. There was considerable variation between species, with the fastest absorbers removing half of the NO_3 present in the soil solution in a quarter of the time taken by the slowest absorbers. Differences were not explained by differences in the size of the root system.

The activity of the root systems of different crops is important when they are used as green manures. The purpose of a green manure is to absorb mineralised nitrogen and, by incorporating it into the plant, to prevent it from being leached. Green manures are used principally in autumn, after the completion of crop production, and in the spring prior to the establishment of a spring crop. In both cases, the ability of the crop to grow roots and to remove nitrogen from the soil solution rapidly is critical to reducing the environmental impact of leaching and to providing a nutrient resource for the following crop. (Serisena 1994)

10.3.5 Variation Within Crop Species

Few studies to date have assessed the extent of variation in many of the characteristics listed in Table 10.2 for varieties of the same crop species. Russell (1977) lists seven major crop species where variation between genotypes of the same species had been demonstrated. O'Toole and Bland (1987) reviewed data for 14 monocotyledonous crop species and 15 dicotyledonous species. Their principal findings as to the within-species variation in root system properties are summarised in Table 10.6.

There are more data available for some species than for others. A number of papers deal with barley, maize, rice, sorghum, wheat, soybean and cotton while for other species available information is limited. Across all species, most information is available for gross characteristics such as root weight, root length and the ability of the root system to penetrate to depth.

Table 10.6 The principal root system characteristics reported as showing variation in some of the major crop species

Root system feature	Number of crop species (29 assessed) reported as showing genetic variation in this trait	Functional significance reported as associated with this trait
Root weight	17	Efficiency of nutrient and water use
Root length	15	Efficiency of nutrient and water use and ability to resist water stress
Penetration to depth	12	Optimised strategies for water use
Root to shoot ratio	7	Resilience in response to adverse soil biotic and physical factors
Root branching	6	Responses to temperature and water availability
Root and root system growth rates	6	Ability to access water and nitrogen
Root diameter	5	Anchorage and ability to cope with soil compaction
Lateral spread of root system	4	Anchorage
Root length density	4	Ability to extract water

Source: Summarised from O'Toole and Bland (1987).

Table 10.7 The functional significance of some genetically determined root characteristics

Reference	Crop species	Root properties showing genetic variation	Functional impact
Ao et al. (2010)	Soybean *Glycine max*	Root system size, distribution and activity. Root morphology	Efficiency of P extraction
Bauerle et al. (2008)	Grape vine *Vitis vinifera*	Relationship between root system size and plasticity. Periodicity of new growth	Water extraction and drought stress effects
Crush et al. (2009)	Perennial ryegrass *Lolium perenne*	Allocation to roots absolutely and relative to shoots, distribution with depth	Cropping performance under various conditions
Dodd et al. (2011)	Wheat	Root sensing of abiotic stress, root distribution	Water use efficiency
Gao et al. (2010)	Tobacco	Root length and surface area, distribution with depth	Phosphorus use efficiency
Liu et al. (2009)	Maize *Zea mays*	Relationship between root system size and activity. Axial root development. Expression of N transporter genes	Nitrogen use efficiency, response of root growth to nitrogen
Webb et al. (2010)	Red clover *Trifolium pratense*	Root system senescence and N release, changes in gene expression	Adaptation to abiotic stress
Wojciechowski et al. (2009)	Wheat	Impact of dwarfing genes on root growth and root architecture, e.g. root diameter and branching	Seedling establishment

Hirel and Gallais (2011) assessed the impact of phenotypically expressed traits on N use efficiency. They found that these were of most importance when N supply was low because here competition for nutrients was enhanced. They found that in maize, the size and activity of the root system affected N uptake although they commented that little was known of the consequences of genetic variation in the root system. N uptake was related to root density although there was no clear relationship with N use efficiency. Magalhaes et al. (2011) assessed the impact of genotype on P uptake in maize.

A number of recent studies have attempted to look at the genetic control of root characteristics and the relationship of root form to function. These are summarised in Table 10.7.

Ao et al. (2010) assessed the genetic variation in morphological traits and growth and how this related to the ability of soybean to use soil P. Morphological traits seemed to be controlled by quantitative trait loci. Root traits and P efficiency were related. In addition, most root morphological traits were correlated and more heritable than architectural properties. The more P-efficient genotypes established longer and larger root systems, with many roots present in surface soil layers, and maintained more active roots. Bauerle et al. (2008) found that plants with greater vigour had the ability to show greater morphological plasticity in response to variation in soil moisture but without an increased ability to tolerate moisture stress through enhanced root survivorship. Genotypes varied in the relative amounts of root produced in summer and winter. Crush et al. (2009) demonstrated that the variation present in current varieties could be less than that in wild-type populations.

O'Toole and Bland (1987) concluded:

> The fact that so little is known about applications of genetic variation in root parameters to crop improvement should serve as a warning not to follow single trait solutions.

They also suggested:

> Successful genetic improvement of the root function of crop plants requires that the impact of the soil environment on the expression of the phenotype be recognised. The responsiveness of roots to the soil environment (phenotypic plasticity) appears to be of potential value and should be appreciated in its own right.

The extent to which root system characteristics can vary and influence the ability of the crop to interact with the soil is important. It is also important to look at variation within a single species, which we do here for birch and spring barley.

10.4 Case Studies

10.4.1 Case Study 1: *Betula pendula*

With birch as a model species, Lavender (1992) and Lavender et al. (1993) used a series of genotypes to assess the parameters of the root system, which differed between varieties. All of the genotypes used were selected as having relatively similar shoot growth, maximum growth not exceeding 15% more or less than the average. Five of the genotypes used out of 30 had root growth more than 15% lower than the average while four genotypes had growth more than 15% above average.

10.4.1.1 Variation in the Root System

The principal variations in root system parameters are summarised in Table 10.8.

The length of root was correlated with weight for roots with a diameter of <1 mm. For plants with small amounts of woody root, this correlation held but for plants with a larger quantity of woody root, length and weight were unrelated, indicating variation in allocation either to length or to thickening. Across all of the genotypes, the ratio of root to shoot weight was positively correlated with root weight but unrelated to shoot weight. Over a series of harvests there were significant differences between clones in root/shoot ratio, root weight, the weight of new and of woody roots, root length, the proportion of the system present as thin roots and distribution with depth. There were no significant differences between the specific root length of any kind of root or the weight of woody roots. In this species, there was variation in partitioning of assimilate into the root system and of partitioning into distribution with depth and within the root system into the length of woody roots but not the extent of thickening. There were also significant differences in the life of individual roots between clones but not of the proportion which survived to make up the permanent root system.

Table 10.8 The range of variability recorded for a series of root system parameters in *Betula pendula*

Root system parameter	Range of variation measured	Variation (%) relative to the population mean	Suggested functional significance
Length of new root	14–50 m	128	New absorbing potential generated at a point in time
Specific root length of new roots	25–54 m/g	83	Allocation strategy to a given length of root through either cell size or cell number
Root weight	4.9–10.4 g	75	Absolute allocation of assimilate to the root system
Weight of woody roots	2.8–5.8 g	74	Anchorage and the potential for new root production in later years
Length of fine roots	343–704 m	67	Base absorption potential, intensity of soil exploitation
Total root length	377–747	64	As for fine root length but related to total distribution with depth
Root to shoot ratio	0.28–0.48	54	Partition strategy between photosynthetic and nutrient and water acquisition capacities

Source: Data from Lavender et al. (1993).

10.4.1.2 Plasticity Under Different Growing Conditions

In the study with *B. pendula*, Lavender (1992) and Lavender et al. (1993) assessed whether the differences between genotypes, discussed above, affected the ability of the clones to cope with conditions of low phosphate, water or nitrogen supply. Although various degrees of restriction in water supply affected the growth of both the root system and the canopy and modified root distribution, there were no differences in the ability of any of the clones to cope with water stress. Ranges of different root distributions were able to cope equally well with a restricted water supply.

In contrast, the different clones coped in different ways with limitations to nutrient supply. With a low supply of nitrogen, the weight and length of root were related but this was less the case with a higher supply of nitrogen. With a higher supply of nitrogen the weight of woody roots was unrelated to length. Root/shoot ratio was higher with a restricted nitrogen supply and was independent of either root or shoot weight. With a high rate of nitrogen supply, root/shoot ratio was relatively constant regardless of the size of the root or shoot system. In this study, there were significant interactions between clone and nitrogen supply for the weight of both fine and woody roots, the specific length of new roots and the proportion of the root system made up by woody roots. Most of these effects were a result of variation in the ability of clones to make use of high nitrogen rather than an enhanced ability to cope with a restricted supply.

Key characteristics and features for which there was a significant interaction between clone and level of N supply are summarised in Table 10.9.

Similar trends were apparent as a consequence of variation in phosphorus supply. Total root weight and the weight of both fine and woody roots showed a significant interaction between

Table 10.9 The effect of different levels of N supply on the root development of four clones of *B. pendula*

Root character	N treatment	Clone 11	Clone 14	Clone 23	Clone 29
Root length (m)	H	100	98	139	148
Treatment and clone significant P < 0.001 Interactions non-significant	L	39	47	58	52
Fine root weight (mg)	H	403	468	676	738
Treatment, clone and interactions significant P < 0.002	L	111	172	220	207
SRL of new roots (m/g)	H	28	35	29	30
Interactions significant P < 0.05	L	39	31	31	33
R/S	H	0.40	0.32	0.43	0.45
Treatment significant P < 0.001	L	0.67	0.72	0.78	0.69

Source: Data from Lavender (1992).

clone and response to phosphorus supply. It is important to note that while clone type can affect a particular root characteristic, environmental changes, such as variation in nutrient supply, often resulted in a degree of plasticity in response, which mitigated the advantage which might have been associated with a particular characteristic of the root system.

10.4.2 Case Study 2: Spring Barley

Atkinson et al. (1995) detailed the range of variation in the root system of 26 different varieties of spring barley grown under identical conditions. The principal sources of variation detected and their probable significance for organic systems of production are detailed in Table 10.10 and the variation recorded in key attributes at final harvest in Table 10.11.

The clones showed the greatest variation in early root growth and least variation in root survival. Partitioning to the root system, the plasticity of the root system and the rate of root production were relatively similar.

Root volume and weight were the most variable characteristics, suggesting that the greatest potential variance was in the total amount of assimilate partitioned into the root system (root weight) and its use to make cell wall material and to expand the cells created (root volume). Root length was substantially less variable than either of these, potentially indicating variation in cell expansion and root diameter in the use of resources partitioned into the root system. Root length was, however, more variable than was specific root length, the amount of length created with a unit amount of assimilate.

Table 10.10 Variation in the characteristics of 26 different spring barley variety root systems grown in observation tubes (Atkinson and Dawson 2001) and with root development characterised as intersections with a grid system

Functional characteristic	Recorded measurement	Range of values measured intersections	Ratio of highest to lowest
Rapid early root growth	Length at 4 weeks	33–89	2.7
High partitioning to the root system	Total production	2205–4337	2.0
Plasticity	Length at 8 weeks	203–428	2.1
Rate of production	Ratio of length at 7 and 5 weeks	1.58–2.81	1.8
Root survival	Ratio of length at 18 and 9 weeks	1.11–1.71	1.5

Source: Modified from Atkinson et al. (1995).

Table 10.11 Variation in the range of root system features of spring barley varieties at harvest

Root system volume mL	Root length m	Root diameter mm	Root weight mg	Specific root length m/g
0.23–3.38	11–60	0.15–0.32	30–290	133–635
Highest as % of Lowest 1469%	545%	213%	966%	477%

The five varieties with the highest root weights (>180 mg) had substantially lower specific root lengths (mean 188 m/g) than did the five varieties with the lowest (<70 mg) root weights (mean 403 m/g), suggesting that varieties with low assimilate supply compensate by generating a relatively greater root length. Similarly, the five varieties with the greatest root length (>43 m) produced roots of higher diameter (mean 0.23 mm) compared to the five varieties with the lowest (<22 m) root length (0.19), suggesting that the potential to contact the greatest volume of soil, a function of length, is of greater importance than the additional layers of cortical cells which are the usual product of an increased root diameter.

Variation existed in a range of characters as did the magnitude of variation which was highest for rapid early root growth, potentially of value to crops receiving soluble N and least for root survival previously identified as important to organic crops. Plasticity and partitioning to the root system, both of potential value in organic systems, showed significant variation.

10.5 Root Dynamics and Carbon Inputs to the Soil

10.5.1 Root Dynamics

Root production does not occur uniformly over the duration of the growing season. This is true of both annual cropping species, where there is a need to develop root length from a zero base, and perennial crops where new growth can be initiated from the permanent root system.

10.5.2   Root Longevity

The root system represents one of the major mechanisms whereby atmospheric carbon fixed by photosynthesis is transferred into the soil to become soil organic matter. We are used to crop leaves surviving from initiation until the end of production. There is an assumption that the root system must behave similarly. While some roots do survive for the duration of the season, most roots of both annual and perennial crops survive for short periods. For example, Atkinson and Watson (2000b) found that the life of 50% of the roots of red clover, pea, mustard, white clover and oats was respectively 20, 6, 12, 40 and 6 days. Black et al. (1998) found similar variation in the longevity of the roots of trees: 40% of the roots of *Prunus avium* but only 6% of the new roots of *Picea sitchensis* survived for more than 14 days. Survival values for *Acer pseudoplatanus* and *Populus* were intermediate. Survival times varied for cohorts of roots produced at different times of the year. This differed from the results of Head (1966) and Atkinson (1985) who found the survival of the roots of apple and strawberry to be relatively uniform across the growing season.

The process of root death usually begins with the loss of the cortical tissues followed by the loss of the rest of the root tissues in the roots of annual crops and in most of the roots of perennial crops. Some roots in perennials survive to form the permanent woody root system. When the root cortical tissue dies, associated arbuscular mycorrhizal fungi (AMF) are also lost. Together these processes are the basis of the transfer of plant carbon into the soil (Figures 10.1–10.3).

Survival is influenced by temperature and by infection by AMF. Forbes et al. (1997) found for *Lolium perenne* that after 35 days growth at temperatures of 15, 21 or 27 °C, around 65%, 40% and >20% of roots produced remained. Watson et al. (2000) compared root longevity for *Trifolium repens* and *L. perenne* at sites in Italy and the UK. In Italy, 73% of *T. repens* roots survived for under 21 days while in the UK the comparable figure was 29%. The comparable figures for *L. perenne* were 84% and 38%. As a result, much of the carbon which is transferred to the root system remains briefly before becoming soil organic matter. The importance of this has been discussed by Goss and Watson (2003) who identified this; roots require significant assimilate for basic respiratory requirements, as concentrating roots in zones with optimum resources. Soil organic matter is usually recorded as that present at the end of a season. The periodicity of root growth suggests more carbon cycles through the root organic matter pool than later determined by chemical analysis. Using data from mini rhizatron measurements, Black (1997) calculated the amount of carbon and associated N and P which flowed into dead root material during a single field season (Table 10.12).

These calculations indicate that the input of organic material to the soil during a season can be substantial and that this is associated with the recycling of N and P so crop needs during a season can be met by the reuse of nutrients from the root system. Data for both changes in the organic matter content of soil and the input of root material are uncommon. Data for studies on apple are shown in Table 10.13.

A number of recent studies have characterised the impact of different levels of carbon input via the root system on soil conditions. These are summarised in Table 10.14. It is clear that both the amount of carbon input and where that input occurs are significant. Input at depth is important as roots can transfer C to depths which are inaccessible by any method of incorporating manure (Calonego and Rosolem 2010). The amount of material incorporated is influenced by crop management. Multicropping may adversely affect crop yield.

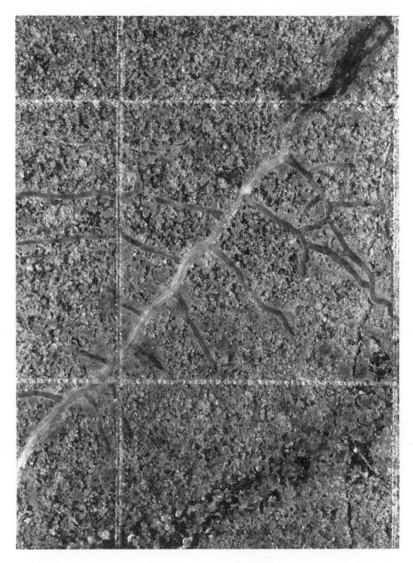

Figure 10.1 A new apple root in the process of turning brown and beginning the process of decomposition. The etched lines represent 12 mm.

10.6 Variation in Root Systems in Practice

10.6.1 Variation in Root Systems with Functional Significance

Under laboratory conditions, a range of properties have been shown to be important. Under field conditions many of these have no impact. This is a consequence of the greater variation which crops experience under field conditions. Under laboratory conditions, the aim is to standardise experimental conditions so plants are grown with a standard temperature, light and soil water potential. Under field conditions, all of these vary. In addition, under field conditions

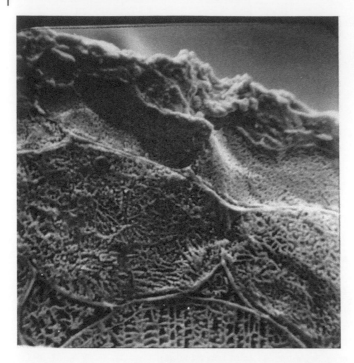

Figure 10.2 A stereoscan electron micrograph of a freeze fracture across a young apple root showing epidermal and cortical cells. Micrograph by WJ McHardy.

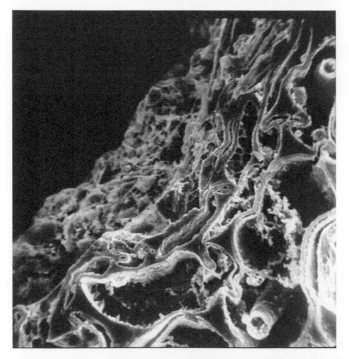

Figure 10.3 A similar picture of an apple root where the cells have begun to collapse. Micrograph by WJ McHardy.

Table 10.12 The quantities of carbon, nitrogen and phosphorus released to the soil as a result of root death by *Lolium perenne* and *Trifolium repens* over a 20-week season for crops grown at field sites in either Aberdeen, Scotland (highest mean summer air temperature 17 °C), or Santa Sofia, Italy (highest mean air temperature 24 °C)

Crop species and site	Carbon released to soil by root death Mg/ha/year	Nitrogen released to soil by root death kg/ha/year	Phosphorus released to soil by root death kg/ha/year
Lolium UK	120	858	292
Lolium Italy	150	1156	129
Trifolium UK	11	229	39
Trifolium Italy	27	623	31

Source: Data reworked from Black (1997).

Table 10.13 Estimates of the contribution of root production to soil organic matter levels

Crop and assumptions	Organic matter mg/g	Estimated change in organic matter (OM) content of soil Mg/ha/year	Root La cm/cm^2	Estimated root production Mg/ha
High-density apple *Change in OM based on changes over 5-yr period adjusted for changes in uncropped soil*	17.4	3.4	30	5.6
Grass sward in an orchard *Change data based on increased OM when bare soil is grassed. Root data from a parallel experiment*	30.3	5.7	90	4.1

Source: Organic matter data derived from Atkinson et al. (1978) and root production data from Atkinson (1985).

Table 10.14 The influence of the root system on soil condition

Root system properties	Change in soil condition	Crop species	Reference
Deep root growth	Increase in biopores and reduced penetration resistance resulting from rotational cover crops improved growth more than chisel ploughing	Soya/triticale, sunflower, millet, sorghum, sunn hemp	Calonego and Rosolem (2010)
Root density	Competition for soil resources by living mulches reduced crop performance	Maize/Italian ryegrass	Faget et al. (2012)
Root length density, root diameter, specific root length, root weight	Fertiliser and grazing treatments impacted on root system distribution and functioning	Grass pastures	Gab and Kacorzyk (2011)

Table 10.15 The effect of root system characteristics on the contribution of rotation components to nutrient and water supply

Rotational feature and root impact	Impact on soil condition and nutrient supply	Crop	Reference
Impact of green manure on rooting depth	Increased depth of rooting and uptake of N	Beetroot, sweet corn, celeriac	Christiansen et al. (2006)
Intercropping; impact on root depth	Faster root growth in cereals gives access to more N	Pea/barley	Corre-Hellou et al. (2007)
Impact of catch crops root system properties	Catch crops exhibiting early root growth reduced N leaching	Spring wheat/ phacelia, sunflower, brassica	Herrera et al. (2010)
The impact of catch crops on retention of N	The root systems varied in depth of rooting associated with differences in N capture. Chicory and radish were best at reducing N leaching from depth	Spring barley/ chicory, fodder radish, perennial ryegrass	Sapkota et al. (2012)
The effect of rooting depth on the use of green manure N	Deep-rooted vegetables, increased N uptake after a legume cover. Shallow-rooted vegetables did better after rye	Carrot, cabbage, onion, lettuce/winter-hardy and non-winter-hardy legumes, rye	Thorup-Kristensen (2006)
The impact of crops with different rooting characteristics	Deep-rooted crops were important to the retention of N	Crop rotations with different crops	Thorup-Kristensen et al. (2012)
The impact of fodder crops on root distribution	Lucerne improved the ability of wheat to absorb water	Spring wheat/lucerne	Gaiser et al. (2012)
The impact of intercrops on root development	Intercropping maize with wheat gave greater root exploitation	Wheat/maize/faba beans	Li et al. (2011)

crops can employ avoidance strategies, such as compensating for the absence of water in one part of the soil by exploiting other areas (Atkinson et al. 2003). Below we examine the functional significance of variation in root system properties through case studies.

10.6.2   The Ability of the Crop Plant to Extract Nutrients from the Soil

The functioning of the root system is related to the crop demand. In spring cereals, demand is intense for short periods. In perennial crops, demand is spread over a longer period. This requires different patterns of root development for optimised performance, something important to the design of the rotation (see Chapter 6). Functioning is affected by root properties (Table 10.15).

10.7   Case Study 3: Apple

Apples have long been grown grafted onto a root system, the rootstock, which differs genetically from the cropping variety. The use of the compound tree allows the scion, the cropping component, to be bred for production characteristics and the rootstock, the element interacting

Table 10.16 The effect of rootstock and soil management on the exploitation of the soil to a depth of 50 cm (La) cm/cm^2 by the apple root system

			Rootstock		
Grass cover	M27	M9	M26	MM106	MM111
Beside tree (50 cm)	30.6	29.9	30.9	49.4	55.5
Away from tree (150 cm)	5.9	15.8	10.5	13.1	16.7
Weighted mean	12.1	19.3	15.6	22.2	26.4
Bare soil					
Beside tree (50 cm)	13.8	14.5	16.5	22.7	29.6
Away from tree (150 cm)	11.1	19.5	12.3	19.8	30.6
Weighted mean	11.8	18.3	13.4	20.5	30.4

Source: Data recalculated from Atkinson (1990).

with the soil, to be developed for its ability to control the size of the tree and perhaps to optimise the use of soil conditions. As part of the development process, rootstocks are tested on a variety of sites. Differences in susceptibility to soil-borne diseases and environmental features such as low temperatures have been found. In trials, many rootstock comparisons are compromised by the planting of stocks of different vigour at different distances so that spacing and the genetic effects of rootstock interact. To overcome such effects, Atkinson (1990) assessed the effect of rootstock on root development when trees on rootstocks of varying vigour were planted at a common wide spacing. Root length density (La) was assessed for 11-year trees (Table 10.16).

The results showed that all major factors, soil depth, distance from the trunk, the way in which the soils was being managed, that is, bare or with a grass cover, and rootstock affected measured root density. Root length density was generally greater close to the tree and near the soil surface. It varied with rootstock vigour, generally being highest with the more vigorous rootstocks. It was higher for those trees grown with a grass cover. Although rootstock did not interact with any of the other factors, it did have a major effect on root density and as such would be significant where soil resources were limiting, particularly where the rootstock produced a vigorous tree with a significant demand for soil nutrients and water.

A key question which follows from this is whether these differences have functional significance and if so, under what conditions and whether these conditions occur sufficiently often in organic production to be important. For the trees detailed in Table 10.16, the size of the root system can be related to the length of new shoots produced in a season, the nutritional status of the tree and their use of soil water (Table 10.17).

The amount of root associated with a given length of shoot was highest for the smallest trees, that is, those on the most dwarfing rootstocks, but was relatively constant for larger trees, suggesting relatively greater soil exploitation by the dwarf trees. The seasonal water use generally followed tree size except that trees on M9 used rather more water than an assessment based only on shoot length would have suggested. Trees on this stock were good at absorbing 32 P injected into the soil and had higher concentrations of P in leaves than did trees on M27.

Table 10.17 The relationship between the size of the root system and its functional impact

Property	Rootstock				
	M27	M9	M26	MM106	MM111
Total root length per tree (km)	18.8	29.3	21.4	32.8	48.6
Annual shoot production per tree (m)	5.1	29.0	80.7	141.1	166.9
Root length per unit shoot length (km/m)	3.7	1.0	0.26	0.23	0.29
Nutrient status *P concentration in leaves (mg/g)* (SE 0.10)	2.04	2.11	2.13	2.10	2.10
Nutrient uptake potential *Uptake of 32P from 25 to 30 cm depth near tree (cpm/g)*		4391	1724		2779
Water use *Mean soil water deficit (cm) averaged over the growing season* (SE 1.00)	2.56	3.33	3.04	3.79	4.46

The association of root length with root system functioning under field conditions is complex.

In addition to the effects of genetics, the development of the root system is greatly influenced by the conditions under which the crop is being grown. Soil factors such as temperature and water potential can significantly influence the rate of growth of individual roots, their branching and their survival. The example discussed above (see Table 10.16) indicated that root length density is affected by whether a tree is grown with the soil maintained bare or with a grass cover. This indicated the inherent plasticity within the root system of trees with identical genetics. In addition, Gurung (1979) and Atkinson and White (1980) showed that when trees were grown with part of their root system in bare soil and part where there was competition with other species, they adjusted their distribution between the two soil types. These effects on root distribution affect where the tree obtains its nitrogen (Atkinson et al. 1978). Where trees were surrounded by uniform bare soil then they made use of the whole volume of soil. Where trees were grown in a bare area with grassed inter-rows then uptake was concentrated in the bare areas.

The density at which trees are planted is also known to have a major impact upon how they distribute root growth within the soil volume Atkinson (1978) (Figure 10.4). The excavation of the root systems of five-year apple tress of Golden Delicious/M9 showed that closely spaced trees made more use of deeper soil and that there were major changes in the form of the root system from one where the preferred direction of growth was horizontal to one where the predominant direction of growth was vertical. In this study, the depth of root penetration was limited by available soil depth, which was around 120 cm. While genetics has a significant impact on root system properties, many root systems show great plasticity and so how the plant is grown will significantly affect its ability to use soil resources.

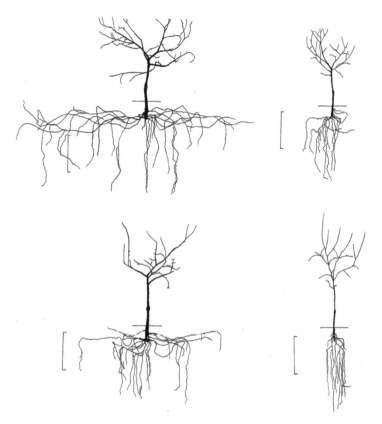

Figure 10.4 The excavated root systems of apple trees grown at low or high planting densities. *Source:* Modified from Atkinson (1978).

10.8 So How Much Root Does a Plant Need?

The amount of root needed by a plant to function has exercised the attention of breeders and agronomists. While breeding for root characteristics has rarely been a predominant aim, it often contributes to resistance to pathogens or maximising assimilate transfer to crop. With some difficulty it is possible to estimate the amount of root which should be needed, under idealised conditions, to absorb the nutrient needed for growth. Inflow rates per unit root length can be amalgamated to estimate the total root length needed to supply the quantity of nutrient required over the period of active growth. Such conditions can be approximated for crops growing in solution culture. Although fertiliser and irrigation technologies aspire to achieve the same result, variations induced by climate and soil mean that allowance has to be made for suboptimal root system functioning and for additional root length to be generated. With organic production, the availability of nutrients is limited further.

While estimating the root system needed for nutrient uptake is difficult, other functions are equally hard to quantify. Water supply is dependent on radiant energy and rainfall. When the need for the root system to be part of a soil biological community and to provide material for

Table 10.18 Seasonal variation in nutrient inflow rates mol/cm/s $\times 10^{-13}$

Element	June	July	August	October
N	1.7	11	42	30
P	0.23	0.7	2.1	0.16
K	0.25	3.7	15	5.9

Source: Modified from Asamoah (1984).

soil structure is added then even good estimates are difficult. It is possible, however, to estimate the length of root likely to be needed under good conditions.

Nye and Tinker (1977) summarised data on the recorded inflow rates of N, P and K and estimated rates of $1.2–13 \times 10^{-13}$, $0.12–6 \times 10^{-13}$ and $1.5–10.7 \times 10^{-13}$ mol/cm/s respectively. The range of variation in recorded nutrient uptake is rather small, usually of the order of one order of magnitude. Inflows of N and K are similar on a length basis at around 5×10^{-12}, with P being more than an order of magnitude less. Asamoah (1984) assessed the extent to which such inflow rates could be approached by young field-grown apple trees (Table 10.18). Measured root lengths in June, July, August and October were 27.2, 26.5, 31.2 and 37.9 m respectively. Total demand to supply new growth rather than root length was responsible for much of the variation in inflow rate with the root system responding to increased demand. In perennial crops, much of the nutrient stored in perennial organs can be reused in later years. The potential contribution of this was quantified by Blasing et al. (1990) who found that around 60% of the N in bark could be remobilised together with a similar proportion of the N previously located in leaves.

10.9 Conclusion

The availability of nutrients such as N and P is critical to the productivity of all agricultural systems. Unlike natural vegetation, where most nutrients are conserved within the system, all agricultural systems will lose nutrients as a consequence of crop removal. These losses, but especially losses of N, need to be replaced either by the nitrogen fixed by legumes within the rotation or by additions of nutrients as manures, themselves the product of either legume N or N released from soil organic matter. This N will not be concentrated in a small part of the soil volume nor will it all be immediately available and so it is to meet this need that the root system becomes the critical link between soil resource and crop need.

We have shown above the various ways in which crop root systems can be varied. The choice of crop species, the selection of a particular variety of a species but most of all the way in which that species/variety is grown will affect the crop's ability to use the soil resources which are made available. In organic systems, this is the key means of influencing the need for nutrients. It also represents a key approach to developing soil microbial communities, which are critical to making nutrients available both from added manures and from the nutrients stored within the soil (see Chapter 11).

Organic systems are largely about managing soil resources and in such systems roots (and AMF, discussed in Chapter 11) are both a tool in that management and the basis of the key management needs. It is here that the ecological science underpinning organic systems gives potential avenues for progress in achieving sustainable yield increases. Most research to date has focused either on nutrient supply linked to fertiliser technology or on the improvement of the capture of photosynthetic energy and its partition into food product. Moving nutrient from soil to the needs of the crop has been less well studied beyond assessments of the quantity of root needed to absorb well-targeted and immediately available nutrient additions. Using current and new knowledge of how root systems might be better configured or engineered to deal with the very different supply conditions found in organic agriculture represents one of the next opportunities to increase production while working with the environment.

References

Ao, J., Fu, J., Tian, J. et al. (2010). Genetic variability for root morph-architectural traits and root growth dynamics as related to phosphorus efficiency in soybean. *Functional Plant Biology* 37: 304–312.

Asamoah TEO 1984 Fruit tree root systems: effects of nursery and orchard management and some consequences for growth nutrient and water uptake. PhD Thesis, University of London.

Atkinson D (1973). The root system of Fortune/M9. Annual Report of East Malling Research Station for 1972, pp. 72–78.

Atkinson, D. (1978). The use of soil resources in high density planting systems. *Acta Horticulturae* 65: 79–90.

Atkinson, D. (1985). Spatial and temporal aspects of root distribution as indicated by the use of a root observation laboratory. In: *Ecological Interactions in Soil* (ed. A.H. Fitter, D. Atkinson, D.J. Read and M.B. Usher), 43–65. Oxford: Blackwell.

Atkinson, D. (1990). Influence of root system morphology and development on the need for fertilisers and the efficiency of use. In: *Crops as Enhancers of Nutrient Use* (ed. V.C. Baligar and R.R. Duncan), 411–451. San Diego: Academic Press.

Atkinson, D. (2000). Root characteristics: why and what to measure. In: *Root Methods: A Handbook* (ed. A.L. Smit, A.G. Bengough, C. Engels, et al.), 1–32. Berlin: Springer.

Atkinson, D., Black, K.E., Forbes, P.J. et al. (2003). The influence of arbuscular mycorrhizal colonisation and environment on root development in soil. *European Journal of Soil Science* 54: 751–757.

Atkinson, D. and Dawson, L.A. (2001). Root growth: methods of measurement. In: *Soil and environmental Analysis* (ed. K.A. Smith and C.E. Mullins), 435–498. New York: Marcel Dekker.

Atkinson, D., Johnson, M.G., Mattam, D., and Mercer, E.R. (1978). The effect of orchard soil management on the uptake of nitrogen by established apple trees. *Journal of the Science of Food and Agriculture* 30: 129–135.

Atkinson D, Watson CA (2000a) The research needs of organic farming: distinct or just same as other agricultural research? Proceedings BCPC Conference Pests and Diseases. pp 151–159.

Atkinson, D. and Watson, C.A. (2000b). The beneficial rhizosphere: a dynamic entity. *Applied Soil Ecol* 15: 99–104.

Atkinson, D., Watson, C.A., Hooker, J.E., and Black, K.E. (1995). Nutrient conservation in organic crop production systems: Implications for root systems. In: *Soil Management in Sustainable Agriculture* (ed. H.F. Cook and H.C. Lee), 54–60. Wye College Press.

Atkinson, D. and White, G.C. (1980). Some effects of orchard soil management on the mineral nutrition of apple trees. In: *Mineral Nutrition of Fruit trees* (ed. D. Atkinson, J.E. Jackson, R.O. Sharples and W.M. Waller Butterworths), 241–254. London.

Balfour, E.B. (1976). *The Living Soil and the Haughley Experiment.* New York: Universe Books.

Bauerle, T.L., Smart, D.R., Bauerle, W.L. et al. (2008). Root foraging in response to heterogeneous soil moisture in two grape vines that differ in potential growth rate. *New Phytologist* 179: 857–866.

Berta, G., Fusconi, A., and Hooker, J.E. (2002). Arbuscular mycorrhizal modification to plant root systems: scale, mechanism and consequences. In: *Mycorrhizal Technology in Agriculture* (ed. S. Gianinazzi, H. Schuepp, J.M. Barea and K. Haselwandter), 71–85. Basel: Birkhauser.

Black KE (1997) Root longevity as affected by biotic and abiotic factors. PhD thesis, University of Aberdeen

Black, K.E., Harbron, C.G., Franklin, M. et al. (1998). Differences in root longevity of some tree species. *Tree Physiology* 18: 259–264.

Blasing, D., Atkinson, D., and Clayton-Green, K. (1990). The contribution of roots and reserves to tree nutrient demands: implications for the interpretation of analytical data. *Acta Horticulturae* 274: 51–69.

Calonego, J.C. and Rosolem, C.A. (2010). Soybean root growth and yield in rotation with cover crops under chiselling and no-till. *European Journal of Agronomy* 33: 242–249.

Christiansen, J., Thorup-Kristensen, K., and Kristensen, H. (2006). Root development of beetroot, sweet corn and celeriac and soil nitrogen content after incorporation of green manure. *Journal of Horticultural Science and Biotechnology* 81: 831–838.

Clement, C.R., Hopper, M.J., and Jones, L.H.P. (1978). The uptake of nitrate by Lolium perenne from flowing nutrient solution: II. Effect of light, defoliation, and relationship to co2 flux. *Journal of Experimental Botany* 29: 453–464.

Corre-Hellou, G., Brisson, N., Launay, M. et al. (2007). Effect of root depth penetration on soil nitrogen competitive interactions and dry matter production in pea-barley intercrops given different soil nitrogen supplies. *Field Crops Research* 103: 76–85.

Crush, J., Nichols, S., Easton, H. et al. (2009). Comparisons between wild populations and bred perennial ryegrasses for root growth and root/shoot partitioning. *New Zealand Journal of Agricultural Research* 52: 161–169.

Dodd, I.C., Whalley, W., Ober, E.S., and Parry, M. (2011). Genetic and management approaches to boost UK wheat yields by ameliorating water deficits. *Journal of Experimental Botany* 62: 5241–5248.

Dupuy, L., Gregory, P.J., and Bengough, A. (2010). Root growth models: towards a new generation of continuous approaches. *Journal of Experimental Botany* 61: 2131–2143.

Faget, M., Liedgens, M., Feil, B. et al. (2012). Root growth of maize in an Italian ryegrass living mulch studied with a non-destructive method. *European Journal of Agronomy* 36: 1–8.

Fernandez, M., Belinque, H., Gutierrez boem, F., and Rubio, G. (2009). Compared phosphorus efficiency in soybean, sunflower and maize. *Journal of Plant Nutrition* 32: 2027–2043.

Forbes, P.J., Black, K.E., and Hooker, J.E. (1997). Temperature-induced alteration to root longevity in *Lolium perenne*. *Plant and Soil* 190: 87–90.

Franco, J., Banon, S., Vicente, M. et al. (2011). Root development in horticultural plants grown under abiotic stress conditions – a review. *Journal of Horticultural Science and Biotechnology* 86: 543–556.

Fried, M. and Broeshart, H. (1967). *The Soil Plant System*. New York: Academic Press.

Gab, T. and Kacorzyk, P. (2011). Root distribution and herbage production under different management regimes of mountain grassland. *Soil & Tillage Research* 113: 99–104.

Gaiser, T., Perkons, U., Kuepper, P.M. et al. (2012). Evidence of improved water uptake from subsoil by spring wheat following lucerne in a temperate humid climate. *Field Crops Research* 126: 56–62.

Gao, J., Deng, B., Zeng, X. et al. (2010). Genotypic variation in phosphorus efficiency of tobacco in relation to root morphological characteristics and root architecture. *Xibei Zhiwu Xuebao* 30: 1606–1613.

Gaudin, A.C., Mcclymont, S.A., Holmes, B.M. et al. (2011). Novel temporal, fine-scale and growth variation phenotypes in roots of adult-stage maize (*Zea mays* L.) in response to low nitrogen stress. *Plant, Cell and Environment* 34: 2122–2137.

Goss, M.J. (1991). Consequences of the activity of roots on soil. In: *Plant Root Growth* (ed. D. Atkinson), 171–186. Oxford: Blackwell.

Goss, M.J. and Watson, C.A. (2003). The importance of root dynamics in cropping systems research. *Journal of Crop Production* 8: 127–155.

Greenwood, D.J. (1986). Prediction of nitrogen fertiliser needs of arable crops. *Advances in Plant Nutrition* 2: 1–62.

Gregory, P.J., Bengough, A., Grinev, D. et al. (2009). Root phenomics of crops: opportunities and challenges. *Functional Plant Biology* 36: 922–929.

Grime, J.P., Campbell, B.D., Mackey, J.M.L., and Crick, J.C. (1991). *Plant Root Growth* (ed. D. Atkinson), 381–398. Oxford: Blackwell.

Gurung HP (1979) The influence of soil management on root growth and activity in apple trees MPhil thesis, University of London

Hansson, A.C. (1987). *Roots of arable crops: Production, growth dynamics and nitrogen content Report no 28*. Uppsala: Swedish University of Agricultural Sciences.

Head, G.C. (1966). Estimating seasonal changes in the quantity of white unsuberised roots on fruit trees. *Journal of Horticultural Science* 41: 197–206.

Herrera, J.M., Feil, B., Stamp, P., and Liedgens, M. (2010). Root growth and nitrate-nitrogen leaching of catch crops following spring wheat. *Journal of Environmental Quality* 39: 845–854.

Hirel, B. and Gallais, A. (2011). Nitrogen use efficiency-physiological, molecular and genetic investigations towards crop improvement. In: *Advances in Maize* (ed. J.L. Prioul, C. Thevenot and T. Molnar), 285–310. London: SEB.

Howard, A. (1945). *Farming and Gardening for Health or Disease*. London: Faber.

Kristensen, H. and Thorup-kristensen, K. (2007). Effects of vertical distribution of soil inorganic nitrogen on root growth and subsequent nitrogen uptake by field vegetable crops. *Soil Use and Management* 23: 338–347.

Lavender EA 1992 Genotypic variation in the root system of *Betula pendula* Roth. PhD thesis, University of Aberdeen

Lavender, E.A., Atkinson, D., and Dawson, L.A. (1993). Variations in root development in genotypes of *Betula pendula*. *Aspects of Applied Biology* 34: 183–192.

Li, L., Sun, J., and Zhang, F. (2011). Intercropping with wheat leads to greater root weight density and larger belowground space of irrigated maize at late growth stages. *Soil Science and Plant Nutrition* 57: 61–67.

Liu, J., Chen, F., Olokhnuud, C. et al. (2009). Root size and nitrogen-uptake activity in two maize (*Zea mays*) inbred lines differing in nitrogen-use efficiency. *Journal of Plant Nutrition and Soil Science-Zeitschrift fur Pflanzenernahrung und Bodenkunde* 172: 230–236.

Liu, L., Gan, Y., Bueckert, R., and van Rees, K. (2011). Rooting systems of oilseed and pulse crops. II: vertical distribution patterns across the soil profile. *Field Crops Research* 122: 248–255.

Ma, S.C., Li, F.M., Xu, B.C., and Huang, Z.B. (2010). Effect of lowering the root/shoot ratio by pruning roots on water use efficiency and grain yield of winter wheat. *Field Crops Research* 115: 158–164.

Magalhaes, P., de Souza, T., and Cantao, F. (2011). Early evaluation of root morphology of maize genotypes under phosphorus deficiency. *Plant, Soil and Environment* 57: 135–138.

Nye, P.H. and Tinker, P.B. (1977). *Solute Movement in the Soil-Root System*. Oxford: Blackwell.

O'Toole, J.C. and Bland, W.L. (1987). Genotypic variation in crop plant root systems. *Advances in Agronomy* 41: 91–146.

Palta, J.A., Chen, X., Milroy, S.P. et al. (2011). Large root systems: are they useful in adapting wheat to dry environments? *Functional Plant Biology* 38: 347–354.

Palta, J.A., Finery, I.R., and Rebetzke, G.J. (2007). Restricted-tillering wheat does not lead to greater investment in roots and early nitrogen uptake. *Field Crops Research* 104: 52–59.

Peret, B., Clement, M., Nussaume, L., and Desnos, T. (2011). Root developmental adaptation to phosphate starvation: better safe than sorry. *Trends in Plant Science* 16: 442–450.

Richardson, A.E., Hocking, P.J., Simpson, R.J., and George, T.S. (2009). Plant mechanisms to optimise access to soil phosphorus. *Crop & Pasture Science* 60: 124–143.

Rose, T.J., Rengel, Z., Ma, Q., and Bowden, J.W. (2009). Crop species differ in root plasticity response to localised P supply. *Journal of Plant Nutrition and Soil Science-Zeitschrift fur Pflanzenernahrung und Bodenkunde* 172: 360–368.

Roumet, C., Lafont, F., Sari, M. et al. (2008). Root traits and taxonomic affiliation of nine herbaceous species grown in glasshouse conditions. *Plant and Soil* 312: 69–83.

Russell, R.S. (1977). *Plant Root Systems*. London: McGraw Hill.

Sapkota, T.B., Askegaard, M., LaegdsmanD, M., and Olesen, J.E. (2012). *Effects of Field Crops Research* 125: 129–138.

Serisena H (1994) Role of cover crops in nitrogen cycling in farming systems. MPhil thesis, University of Aberdeen.

Thorup-Kristensen, K. (2006). Root growth and nitrogen uptake of carrot, early cabbage, onion and lettuce following a range of green manures. *Soil Use and Management* 22: 29–38.

Thorup-Kristensen, K., Dresboll, D.B., and Kristensen, H.L. (2012). Crop yield, root growth, and nutrient dynamics in a conventional and three organic cropping systems with different levels of external inputs and N re-cycling through fertility building crops. *European Journal of Agronomy* 37: 66–82.

Watson, C.A., Ross, J.M., Bagnaresi, U. et al. (2000). Environment-induced modifications to root longevity in *Lolium perenne* and *Trifolium repens*. *Annals of Botany* 85: 397–401.

Weaver, J.E. (1920). *Root Development in the Grassland Formation*. Washington: Carnegie Institution of Washington.

Webb, K., Jensen, E., Heywood, S. et al. (2010). Gene expression and nitrogen loss in senescing root systems of red clover (*Trifolium pratense*). *Journal of Agricultural Science* 148: 579–591.

Wild, A., Jones, L.H.P., and Macduff, J.H. (1987). Uptake of mineral nutrients and crop growth: the use of flowing nutrient solutions. *Advances in Agronomy* 41: 171–220.

Wojciechowski, T., Gooding, M., Ramsay, L., and Gregory, P. (2009). The effects of dwarfing genes on seedling root growth of wheat. *Journal of Experimental Botany* 60: 2565–2573.

Woodhouse, P.J., Wild, A., and Clement, C.R. (1978). The uptake of nutrient by plants from flowing nutrient solution. I Control of phosphate concentration in solution. *Journal of Experimental Botany* 29: 885–894.

11

Mycorrhizal Activity, Resource and Microbial Cycles

David Atkinson

SRUC, Aberdeen, Scotland, UK

11.1   Introduction

Other than those brought into soils with rain, all the nutrients absorbed by crops in organic systems depend on soil microbial processes for their supply. Even the nutrient additions given in organic farming systems are provided either as complex organic manures, such as farmyard manure and other animal waste products, or as complex mineral materials such as rock phosphate (see Chapter 9). In both cases the ability of crop roots to absorb such nutrients is influenced by the effects of soil condition and factors such as temperature and a range of soil microbial processes. Detailed information on the extent of such processes is given in Chapters 4 and 6. It is not possible to review here all the microbial presses which occur in soils and so we have been selective. A major group of soil microbes important to nutrient cycling in organic systems are the arbuscular mycorrhizal fungi (AMF). In considering the role of soil microbes, emphasis is usually given to bacteria. Here, we place significant emphasis on the role of this group of fungi within organic processes but also as an example of the types of effects which soil microbes can have. Their impact was highlighted by the founders of the organic farming movement (Balfour 1976; Howard 1945). Unlike most soil microbes, which are capable of an independent existence, mycorrhizal fungi are obligate symbionts and depend on their association with crop roots for both survival and functioning. It is appropriate to consider the role of AMF in terms of both their direct effects and those which are a consequence of the impact of mycorrhizal infection on the form and functioning of the root system.

11.2   Mycorrhizal Establishment

Mycorrhizal fungi form mutualistic associations with around 240 000 plant species, with AMF being the most common, associated with the roots of around 80% of plant species (Berta et al. 2002). The nature of the association has been documented by van Tuinen et al. (1994). The AMF are believed to have originated from a common ancestor between the Ordovician and

The Science Beneath Organic Production, First Edition. Edited by David Atkinson and Christine A. Watson.

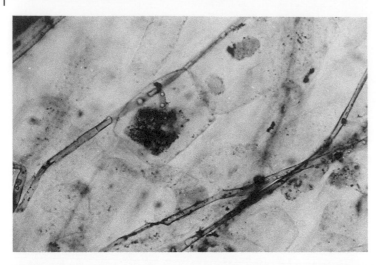

Figure 11.1 A stained onion root showing AMF hyphae and an arbuscle within a root cortical cell.

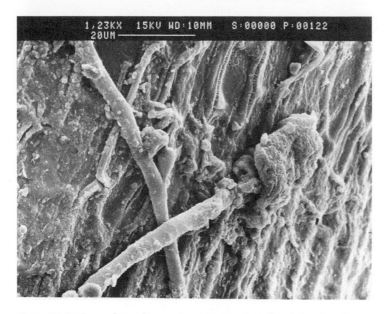

Figure 11.2 The surface of an apple root grown in soil and showing the presence of AMF on the root surface.

Devonian periods and have resulted in three families, six genera and around 130 species. Sexual reproduction is not known and AMF have an aseptate multinucleate mycelium, which shows abundant mycelia development in root tissue (Figure 11.1) and the soil (Figure 11.2) when the symbiosis has been established.

Integration of plant and fungus occurs at a series of levels. Structurally, the fungus modifies cells so as to form a dual host/fungal plasmalemna structure – the arbuscles (Figure 11.3).

Figure 11.3 A stereoscan electron micrograph of cortical cells where a cell on the upper right is infected by AMF which are present as unbranched arbuscles. Micrograph by WJ McHardy.

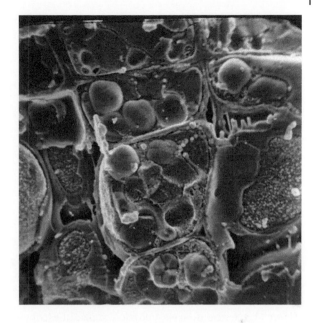

Physiologically, the nutrient uptake properties of the roots change when colonised with a rearrangement of the enzymatic activities of the root (Azcon-Aguilar and Bago 1994). Benefits to the plant include improved nutrition, protection from pathogens, enhanced tolerance of heavy metals and partial resistance to drought. With the exception of brassicacious and cruciferous crops, all crop species are associated with AMF. The level of infection varies greatly, from close to 100% of roots infected in many broad-leaved crops to relatively low levels of infection in cereals. This has led to a view that in cereal crops AMF are of little or no importance. Such a view has, however, to be modified by information on the relative root density in the soil (see Chapter 10). Root length density in cereals can be two orders of magnitude higher than that in broad-leaved crops and so even with low rates of infection, the density of AMF in a volume of soil can be high.

11.3 Mycorrhizal Effects

Under laboratory conditions, a range of root system properties have been shown to be important (see Chapter 10). It is now clear that AMF change the form of the crop root system (Hooker et al. 1998). However, under field conditions many such properties have no impact. The consequences of this emphasis for mycorrhizal research have been discussed by Read (2002). A result of this is that while we know the potential of the mycorrhizal–plant symbiotic relationship, the extent to which this potential matters under field conditions is less documented. For organic production systems, this represents a major gap in our functional understanding. For example, while the infection may help the uptake of P, it does so much else besides, including influencing resistance to disease and the ability to cope with drought stress, and in a very real way affects the fitness of the infected plant (Howard 1945). AMF have been shown to have a significant impact on plant nutrient uptake (Jeffries and Barea 1994), on soil structure (Figure 11.4) (Rillig

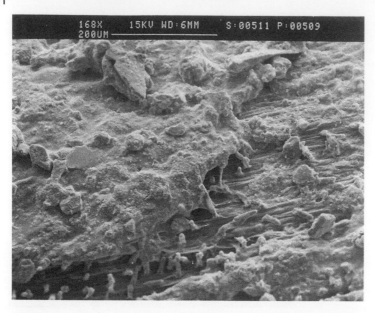

Figure 11.4 A stereoscan electron micrograph of the root surface of an apple root grown in soil and showing the layer of soil at the surface held together by organic materials and fungi and the short root hairs found in apple.

and Mummey 2006) and on the crop's interactions with its physical environment (Atkinson et al. 2005; Sanchez-Diaz and Honrubia 1994) and with pathogenic fungi (Hooker et al. 1994).

Arbuscular mycorrhizal fungi are thus important to root system and crop functioning. They seem to be more prevalent and active in organic production systems (Lumini et al. 2011). The role of AMF in sustainable cropping systems was reviewed by Harrier and Watson (2003) who emphasised their critical importance for crop production systems focused on linking production to biological cycles. They concluded that AMF could have significant effects as a result of an improved supply of P, N, S and B and trace elements but also had a significant role in relation to crop protection, water supply, partition to grain, soil structure and soil microbial populations.

The process of AMF infection of roots is adversely affected by key conventional agriculture practices such as the use of fertilisers and fungicides. In situations where nutrient supply is dominated by fertiliser chemistry and crop protection is delivered by pesticides, the activities of AMF are of lesser importance. For organic crops, they play a key role in both nutrient uptake and protection from diseases, especially soil-borne disease. In addition, Berta et al. (2002) suggest that they have the ability to facilitate co-existence between species and promote stability through enhancing complementarity between root systems of different crops. Here we briefly review the infection process and assess some of its effects on plant performance.

11.4   The AMF Association

While AMF are ubiquitous in terrestrial ecosystems, including crops, only around 150 mycorrhizal species have been identified. Low taxonomic diversity is combined with a broad geographic range and it is possible to find as many as 10–30 spore types present in a single soil.

Current studies suggest substantial diversity at a molecular level (Clapp et al. 2002). The relationship between plant and fungus is an ancient one, at least 400 million years old (the nitrogen-fixing symbiosis with *Rhizobium* is of much more recent establishment, about 65 million years), and involves much molecular interaction between the two partners. The persistence of the plant Glomeromycota fungus symbiotic relationship through evolutionary time is due to a widespread molecular dialogue between the two partners. AMF can germinate spontaneously but hyphal branching, appressoria differentiation, root penetration and interradical development, which are needed for the establishment of the symbiosis, are under the control of an interaction between the two partners (Gianinazzi-Pearson et al. 2009).

A number of the genes which are induced in response to the presence of the appressoria and prior to root penetration belong to protein-encoding families, which are associated with defence strategies against pests or pathogens. The establishment of the symbiosis seems to require the partial suppression of salicylic acid-dependent plant responses. A significant number of differentially regulated genes are involved, such as those which affect in wall constituent synthesis. The fungus affects the functioning of plant membrane transporters (Liu et al. 1998). The genes involved in the formation and functioning of AMF have been reviewed by Gollette et al. (2002). The response of plants to AMF is polygenic in wheat involving genes on six of 21 chromosomes. At least 12 of the genes involved in mycorrhizal colonisation are also involved in the association with N-fixing bacteria although others are unique to the AMF symbiosis. It is clear that both plant and fungal genes regulate the symbiosis. The development of the arbuscles in the cortical cells (see Figure 11.3) of the root causes a modification of both the cell contents and transcriptional activation of a number of major metabolic genes.

11.5 Effects on Plant Nutrition: Basic Mechanisms

Crops are affected by AMF, with infected plants being less responsive to nutrient additions (Jakobsen et al. 1994). AMF rely for their growth and activity on carbon provided by the host plant, up to 20% of plant-assimilated C, and in exchange they improve the mineral nutrition of the plant, particularly by the acquisition of P and to a lesser extent N and micronutrients (Ferrol and Perez-Tienda 2009). There is a bidirectional transfer of nutrients and carbohydrates. To facilitate this, AMF develop an extended contact area with root cells, which change structurally. The fungus penetrates the cell wall but fungal material remains outside the plant plasma membrane.

This process is so effective that it can become responsible for 100% of the supply of P to the plant, negating the use of the direct pathway through plant epidermal cells. Plant P transporter genes become downregulated as a consequence of the development of symbiosis. H+ ATPases generate proton motive forces and drive the uptake of P across the fungal plasma membrane. These transporter genes show maximal expression at the P concentrations usually found in soil solutions. The fungus controls the amount of P delivered to the plant. Infection with mycorrhizal fungi has long been known to influence the uptake of P and through this mechanism to increase plant growth (Atkinson 2006). The increased P acquisition by AMF-infected plants is principally due to the extension of extraradical mycelium into the soil over distances of several centimetres from the root surface (Jakobsen et al. 1992). Despite the number of recent studies of AMF, there is much which remains unknown and which needs resolution if organic and other systems are to be developed so as to optimise the functioning of the mycorrhizas.

11.6   Impact on Crop Nutrition

Reviewing the literature on AMF nutrient interactions, Jakobsen et al. (2002) found that mycorrhizal fungal hyphae could be almost as effective as roots and fungi combined. While the uptake of nitrogen, as both NH_4 and NO_3, could be enhanced to the same extent as that of phosphate, the limited mobility of P in soils made the latter attribute of greater functional significance. Under adverse soil conditions, the fungus may take over the supply of P to the plant, although this varies between plant species, while different fungi may elicit different responses.

Jakobsen et al. identified a range of root system characteristics which would make functioning AMF advantageous. These included a short root length, a low root to shoot ratio, a lack of root branching, large-diameter roots and a low nutrient influx capacity. Crop plants whose root systems show such features are likely to benefit from infection with AMF and to be disadvantaged by practices which reduce AMF functioning or infection. They also identified a range of environmental factors which could enhance the importance of AMF. These included a low soil nutrient availability and the presence of root pathogens (the importance of AMF in crop protection is discussed in Chapter 12). In addition, in dry soils where nitrogen movement by mass flow becomes limited and diffusion becomes of greater importance, AMF are likely to have a significant role in the uptake of N.

Studies using mutant plants which are unable to develop a normal AMF association have helped our understanding of the role of AMF. Using such plants, it has been shown that AMF can have a significant effect on the potential losses of NO_3 and NH_4 by leaching (Asghari and Cavagnaro 2012). They do not, however, seem to have an effect on the emission of N_2O from the soil (Cavagnaro et al. 2012) Further, there is some evidence from studies using N14/15 ratios that plants infected with mycorrhizal fungi can use nitrogen from organic compounds (Aerts 2002).

While AMF have been associated with an increased ability to absorb trace elements, they have also been associated with the ability to enhance tolerance of heavy metals (Gonzalez-Guerrero et al. 2009). This can involve the binding of the metal, commonly Cu, Cd, Zn or Fe, to the fungal wall, the production of chelators and a reduction in the impact of free radicals.

11.7   The Impact of AMF on Soil Structure

The AMF mycelium extends out from the root for several centimetres and can spread from plant to plant to form a linked nutrient-absorbing and soil-stabilising network. Tillage can disrupt such networks. AMF have been shown to be involved in the formation of stable soil aggregates (Miller and Jastrow 1992). Estimates of the amount of mycelium produced by AMF suggest values of 2.6–54 m/g. They seem in addition to have a role in encapsulating organic material within an aggregate and even in the weathering of minerals (Jefffries and Barea 1994).

11.8   Carbon Flows into the Soil

There are few estimates available of the quantity of carbon in the soil as mycorrhizal hyphae and fewer still of the amount of carbon passing through this soil compartment during a season. Miller and Jastrow (1990) estimated the amount of mycelium produced under grassland as

Table 11.1 The impact of AMF infection on the release of carbon, nitrogen and phosphorus as a result of root death for poplar and clover

Species	Treatment	Carbon returned by root death Mg/ha	Nitrogen recycled by root death kg/ha	Phosphorus recycled by root death kg/ha
Poplar	AMF	133	952	324
	Uncolonised	64	458	156
Clover	AMF	11	221	38
	Uncolonised	17	349	59

Source: Data reworked from Black (1997).

varying between 2.6 and 54 m/g soil. Dodd (1994) gave values for hyphal diameter varying between 2 and 20 μm, with runner hyphae having diameters of 10–15 μm. On this basis, if we assume an average value of 10 μm for diameter and an average length of 30 m/g then the volume of hyphae in the soil will be 2.4 mm^3 and, assuming a density of 1, a weight of 2.4 mg. Atkinson and Dawson (2001) estimated that 50% of hyphae arose and died in a seven-day period and most of the remainder over a further period of 14 days and so over a typical season of 200 days, we can estimate a hyphal production rate of around 45 mg/g soil. If we assume that most AMF are to be found at this density in the surface 0.3 m of soil and a soil density of 1.2 g/mL, then a m^2 of soil surface to this depth should over a season have 16.2 kg of hyphae as an input to soil organic matter. This approximates to an input of 162 Mg ha/year.

In addition, whether or not roots are infected with AMF influences root survival and so the transfer of carbon from atmosphere to soil and the recycling of nutrients such as N and P from the root pool to the soil pool. Atkinson et al. (2003) found that AMF infection changed the proportion of *Trifolium repens* roots, which survived for given periods of time; 20% of control roots survived longer than 42 days compared to 37% of AMF-infected roots. AMF infection had no effect on root survival in *Lolium perenne*. This effect was not a consequence of effects on root branching. The impact of AMF on carbon and nutrient transfers was estimated by Black (1997) whose results are summarised in Table 11.1.

11.9 The Impact of AMF on Adaptation to the Soil Physical Environment

Safir et al. (1971, 1972) assessed the potential impact of AMF infection on water transport in soya. They found that AMF reduced resistance to water transport, which was associated with an increase in shoot growth and an enhanced ability to recover from the adverse impact of a reduced water supply. Consistent with this, Dunsiger (1999) found that infection with AMF appeared to make the plant more responsive to a decreasing supply of soil water. Plants seemed to be better linked to changing environmental conditions, as if the AMF were acting as a bio-sensor (Table 11.2).

Sanchez-Diaz and Honrubia (1994) discussed the impact of AMF on plant water relations. They distinguished between effects occurring when plants were suffering water stress, when

Table 11.2 The relationship between water use and soil water potential for AMF

Treatment	High water use Transpiration (mL/plant/day)	Soil water potential (-MPa)	Low water use Transpiration (mL/plant/day)	Soil water potential (-MPa)
Control	11	0	7	0.55
Glomus mosseae	15	0	0	0.55
Glomus intraradices	15	0	3	0.55

Glomus mosseae and *G. intraradices* infected and control plants during periods of high and low water demand.

the issue is whether AMF improve the plants' ability to function with a limited supply of water, and when they were growing with an adequate water supply, where the issue is whether they improve the plants' ability to absorb water. Under non-stress conditions, AMF infection commonly results in increased transpiration. These effects have been attributed to a direct water flow via the hyphae, the impact of improved P supply and an impact on hormonal balances. Some of these may also be important under conditions of water stress.

AMF may interact with the hormonal signals which are part of normal root–shoot communication, as in altered levels of the hormone ABA. Dunsiger et al. (2003) and Atkinson et al. (2005) suggested that the extraradicular hyphae of AMF acted as biosensors warning the host plant of an impending period of low water supply. This is an alternative approach to being able to reduce the impact of drought by making extra resources available which is the role in respect of mineral nutrients. AMF seem able to influence both the supply and the demand sides of a plant's water balance. A key element in this is the ability of the AMF-infected plant to reduce transpiration more than control plants in response to an increasing soil water deficit. A premature downregulation of water use as a result of the sensing of the soil environment beyond that close to the root indicates a very different contribution of AMF to the functioning of a crop plant. This would allow available soil water to be used over a longer period of time.

11.10   The Impact of AMF on Plant Pathogens

The impact of infection by AMF on both resistance to and tolerance of plant diseases is discussed at length in Chapter 12 and so is only mentioned here for the sake of completeness. The subject has been reviewed by Pozo et al. (2002, 2009) and by Azcon-Aguilar et al. (2002). This was, however seen by the founders of the organic movement as a major element in both plant protection and the design of organic systems.

11.11   Impact of AMF on Roots

The ability of the plant's root system to absorb immobile nutrients such as P is dependent on the root surface area, which is a consequence of branching, and length. Different species show different patterns of branching (Black et al. 1998). Plants with a sparsely branched system seem most likely to become dependent on AMF. Root system geometry is not the only major factor influencing the degree of mycorrhizal dependency. Growth rate and plasticity, the ability to respond to localised changes in the soil environment, both influence responsiveness (Azcon-Aguilar and Bago 1994). While for most plant species the mycorrhizal association is normal, its value to the crop will depend on the conditions which the crop is experiencing.

Ectomycorrhizal fungi, which are commonly associated with the root systems of forest tree species, have long been known to affect the form of tree roots and root systems. For long, it was assumed that AMF did not have an impact on the roots which they infect. Although AMF effects are less dramatic, it is now known that infection with AMF influences the branching of the root system principally by increasing the number of orders of roots produced and, as a result, the longevity of individual roots (Atkinson et al. 1994). Mycorrhizal root systems thus tend to be more branched and with a higher proportion of small-diameter, higher-order roots, some <0.1 mm in diameter (Berta et al. 1990, 2002; Hooker et al. 1992; Atkinson 2006). Infection with AMF can alter all the primary characteristics of the root system identified above, including size, structure, spatial distribution and temporal presence, as detailed by Atkinson (2000).

Berta et al. (1995, 2002) report the results of a number of studies in which the effect of AMF infection on root architecture has been assessed. There are reports where infection has increased branching, in *Allium porrum* (145%), *Vitis vinifera* (140–260%), *Plantago lanceolata* (145%) and poplar (181–716%). There are also reports where infection has had no effect or even produced a reduction. The impact of AMF infection on root architecture is shown in Table 11.3.

Berta et al. (1993) found that infected onion root systems were more highly branched, containing roots of larger diameter and lower specific root length. For grape, Schellenbaum et al. (1991) obtained similar results but with the magnitude of the difference between control and AMF plants increasing with root order. Hooker et al. (1992) found increases in branching for every root order, with effects greatest for higher-order laterals.

These effects seem to be a consequence of the effect of colonisation on the development of root apices, with the proportion of active apices on primary roots decreasing more rapidly in infected roots. The production of a greater number of higher-order roots has an effect on root longevity and so on the input of carbon to the soil where it is available to be used as a resource by other soil organisms (Atkinson 2006).

As the pattern of branching is important to accessing soil nutrients and especially those with limited mobility such as phosphates, AMF effects on root morphology are potentially impor-

Table 11.3 The effect of AMF infection on root branching (% of control)

Reference and species	Length of individual roots			No of Branches on		
	Primary	Secondary	Tertiary	Primary	Secondary	Tertiary
Berta et al. (1993) Onion	77	81			164	
Hooker et al. (1995) Poplar	92	200	219	98	181	717
Schellenbaum et al. (1991) Grape vine	95	89	93	140	200	266

Source: Data from Atkinson et al. (1994).

tant. Branching patterns can influence the exploration of a heterogeneous soil environment. Berta et al. (2000) discussed the relative value of determinate, as in woody plants, and indeterminate, as in herbaceous species, patterns of root growth and development. Optimal exploitation of the soil can occur either by producing a large root system which persists (indeterminate) or a dynamic root system which is largely ephemeral (determinate) which thus allows a large area of soil to be exploited in a shorter time and with the use of fewer resources than would be the case with an indeterminate system of a similar total length. Eissenstat et al. (2000) showed that roots were not equally efficient at all ages and that there were considerable costs in relation to simply keeping a root in existence. Infection with AMF changes the character of the exudates produced by plants and as a result can reduce the susceptibility of AMF-infected plants to subsequent infection by pathogens such as *Phytophthora* (Norman et al. 1996, 2000).

Table 11.4 The effect of AMF on the functioning of roots in organic systems

AMF effect	Root system modification	Impact on cropping system
Dynamic presence in soil and roots	Modifying the form and longevity of individual roots Ability to infect a range of plant species Switches on some plant genes and downregulates others such as those related to nutrient absorption	Provides an additional element in the matching of a crop species to its place in a rotation Does not limit the design of a rotational sequence. Increases the potential for the root system to make use of low concentrations of available nutrients in the soil solution
Variable longevity of infection	A component of root plasticity A potential regulator of root presence and its gearing to crop resource demands	Systems which allow flexibility to have impact One of the few tools available to modify the root system during the cropping cycle
Enhanced C flow into the soil	Modifies the life of individual roots and root systems and increases the requirement for more C flow to the underground part of the plant	Increased soil organic matter content and flow of C to soil microbial communities
Improved penetration into the whole soil volume	Increases the ability of the root system to absorb soil resources, so increasing the functional significance of any given root length density	More effective gearing of soil resource use to crop demands Greater potential use of the available soil volume by roots
Penetration into small-diameter soil pores	Permits roots to access parts of the soil which are not accessible directly to them Signals a need for a change in root activity in response to changing soil available resources	A mechanism to allow the crop to better relate its needs to temporal variation in soil resources, especially water
Interactions with other soil micro-organisms	Enhances the resistance of plant roots to pathogenic organisms	The ability to both resist and tolerate pathogenic organisms

11.12 Arbuscular Mycorrhizal Fungi and the Management of Soils

The development of appropriate systems of soil management is essential in order for AMF to have an optimal impact on crop production (Harrier and Watson 2003). AMF colonisation of cereals was higher in early season when the cereals were grown in a rotation rather than in monoculture. The sequence of crops in a rotation also influenced colonisation, with infection being higher after soybean than barley. The inclusion of non-mycorrhizal crops in a rotation reduced infection rates and early-season P uptake. Varietal selection, cultivation method, nutrient management, the use of crop protection chemicals and grazing management have all been shown to affect crop infection with AMF and their contribution to nutrition and production.

11.13 Conclusions: AMF and Root Functioning

Arbuscular mycorrhizal fungi are a normal part of a healthy root system. Their impact is minimised by the application of chemicals as occurs in non-organic systems. In systems which encourage their functioning, they modify the working of the root system and provide an additional resource by making available an increased supply of nutrients, by improving the gearing of the root system to the environment in which it has to function and by providing an additional means of minimising the impact of plant pathogens. The impact of AMF is summarised in Table 11.4.

References

Aerts, R. (2002). The role of various types of mycorrhizal fungi in nutrient cycling and plant competition. In: *Mycorrhizal Ecology* (ed. H. van der MGA and I. Sanders), 117–133. Berlin: Springer Verlag.

Asghari, H.R. and Cavagnaro, T.R. (2012). Arbuscular mycorrhizas reduce nitrogen loss via leaching. *PLoS One* 7: 151–155.

Atkinson, D. (2000). Root characteristics: why and what to measure. In: *Root Methods: A Handbook* (ed. A.L. Smit, A.G. Bengough, C. Engels, et al.), 1–32. Berlin: Springer.

Atkinson, D. (2006). Arbuscular mycorrhizal fungi and the form and functioning of the root system. In: *Microbial Activity in the Rhizosphere* (ed. K.G. Mukerji, C. Manoharachary and J. Singh), 199–222. Berlin: Springer.

Atkinson, D., Berta, G., and Hooker, J.E. (1994). Impact of mycorrhizal colonisation on root architecture, root longevity and the formation of growth regulators. In: *Impact of Arbuscular Mycorrhizas on Sustainable Agriculture and Natural Ecosystems* (ed. S. Gianinazzi and H. Schuepp), 89–99. Switzerland: Birkhauser.

Atkinson, D., Black, K.E., Dawson, L.A. et al. (2005). Prospects advantage and limitations of future crop production systems dependent on the management of soil processes. *Annals of Applied Biology* 146: 203–215.

Atkinson, D., Black, K.E., Forbes, P.J. et al. (2003). The influence of arbuscular mycorrhizal colonization and environment on root development in soil. *European Journal of Soil Science* 54: 751–757.

Atkinson, D. and Dawson, L.A. (2001). Root growth: methods of measurement. In: *Soil and Environmental Analysis* (ed. K.A. Smith and C.E. Mullins). New York: Marcel Dekker.

Azcon-Aguilar, C. and Bago, B. (1994). Physiological characteristics of the host plant promoting an undisturbed functioning of the mycorrhizal symbiosis. In: *Impact of Arbuscular Mycorrhizas on Sustainable Agriculture and Natural Ecosystems* (ed. S. Gianinazzi and H. Schuepp), 47–60. Switzerland: Birkhauser.

Azcon-Aguilar, C., Jaizme-Vega, M.C., and Calvet, C. (2002). The contribution of arbuscular mycorrhizal fungi to the control of soil borne plant pathogens. In: *Mycorrhizal Technology in Agriculture* (ed. S. Gianinazzi, H. Schuepp, J.M. Barea and K. Haselwandter), 187–197. Basel: Birkhauser.

Balfour, E.B. (1976). *The Living Soil and the Haughley Experiment*. New York: Universe Books.

Berta, G., Fusconi, A., and Hooker, J.E. (2002). Arbuscular mycorrhizal modification to plant root systems: scale, mechanism and consequences. In: *Mycorrhizal Technology in Agriculture* (ed. S. Gianinazzi, H. Schuepp, J.M. Barea and K. Haselwandter), 71–85. Basel: Birkhauser.

Berta, G., Fusconi, A., Sampo, S. et al. (2000). Polyploidy in tomato roots as affected by arbuscular mycorrhizal colonisation. *Plant and Soil* 226: 37–44.

Berta, G., Fusconi, A., and Trotta, A. (1993). VA-mycorrhizal infection and the morphology and function of root system. *Environmental and Experimental Botany* 33: 159–173.

Berta, G., Fusconi, A., Trotta, A., and Scannerini, S. (1990). Morphological modifications induced by the mycorrhizal fungus Glomus strain E3 in the root system of Allium porrum. *New Phytologist* 114: 207–215.

Berta, G., Trotta, A., Fusconi, A. et al. (1995). Arbuscular mycorrhizal induced changes to plant growth and root system morphology in Prunus cerasifera. *Tree Physiology* 15: 281–294.

Black KE (1997) Root longevity as affected by biotic and abiotic factors. PhD thesis, University of Aberdeen

Black, K.E., Harbron, C.G., Franklin, M. et al. (1998). Differences in root longevity of some tree species. *Tree Physiology* 18: 259–264.

Cavagnaro, T., Barrios-masias, F., and Jackson, L. (2012). Arbuscular mycorrhizas and their role in plant growth, nitrogen interception and soil gas efflux in an organic production system. *Plant and Soil* 353: 181–194.

Clapp, J.P., Helgason, T.J., Daniell, T.J., and JPW, Y. (2002). Genetic studies of the structure and diversity of arbuscular mycorrhizal fungal communities. In: *Mycorrhizal Ecology* (ed. H. van der MGA and I. Sanders), 201–220. Berlin: Springer Verlag.

Dodd, J.C. (1994). Approaches to the study of the extra radicular mycelium of arbuscular mycorrhizal fungi. In: *Impact of Arbuscular Mycorrhizas on Sustainable Agriculture and Natural Ecosystems* (ed. S. Gianinazzi and H. Schuepp), 147–166. Basel: Birkhauser.

Dunsiger Z, Atkinson D, Watson CA 2003 Arbuscular mycorrhizal fungi: their role in the ability of crops to cope with stress. Proceedings of the BCPC International Congress on Crop Science and Technology, pp. 433–439.

Dunsiger Z (1999) The influence of arbuscular mycorrhizas on the water relations of trees. PhD thesis, University of Edinburgh.

Eissenstat, D.M., Wells, C.E., Yanai, R.D., and Whitbeck, J.L. (2000). Building roots in a changing environment: implications for root longevity. *New Phytologist* 147: 33–42.

Ferrol, N. and Perez-Tienda, J. (2009). Co-ordinated nutrient exchange in arbuscular mycorrhiza. In: *Mycorrhizas – Functional Processes and Ecological Impact* (ed. C. Azcon-Aguilar, J.M. Barea, S. Gianinazzi and V. Gianinazzi-Pearson), 73–87. Berlin: Springer.

Gianinazzi-Pearson, V., Tollot, M., and Seddas, P.M.A. (2009). Dissection of genetic cell programmes driving early arbuscular mycorrhizal interactions. In: *Mycorrhizas – Functional Processes and Ecological Impact* (ed. C. Azcon-Aguilar, J.M. Barea, S. Gianinazzi and V. Gianinazzi-Pearson), 33–45. Berlin: Springer.

Gollette, A., Brechenmacher, L., Weidmann, S. et al. (2002). Plant genes involved in arbuscular mycorrhiza formation and functioning. In: *Mycorrhizal Technology in Agriculture* (ed. S. Gianinazzi, H. Schuepp, J.M. Barea and K. Haselwandter), 87–102. Basel: Birkhauser.

Gonzalez-Guerrero, M., Benabdellah, K., Ferrol, N., and Azcon-Aguilar, C. (2009). Mechanisms underlying heavy metal tolerance in arbuscular mycorrhizas. In: *Mycorrhizas – Functional Processes and Ecological Impact* (ed. C. Azcon-Aguilar, J.M. Barea, S. Gianinazzi and V. Gianinazzi-Pearson), 107–122. Berlin: Springer.

Harrier, L.A. and Watson, C.A. (2003). The role of arbuscular mycorrhizal fungi in sustainable cropping systems. *Advances in Agronomy* 20: 185–225.

Hooker, J.E., Berta, G., Lingua, G. et al. (1998). Quantification of AMF-induced modifications to root system architecture and longevity. In: *Mycorrhizal Manual* (ed. A. Varma), 515–531. Berlin: Springer.

Hooker, J.E., Black, K.E., Perry, R.L., and Atkinson, D. (1995). Arbuscular mycorrhizal fungi induced alteration to root longevity of poplar. *Plant and Soil* 172: 327–329.

Hooker, J.E., Jaizme-Voga, M., and Atkinson, D. (1994). Bio control of plant pathogens using arbuscular mycorrhizal fungi. In: *Impact of Arbuscular Mycorrhizas on Sustainable Agriculture and Natural Ecosystems* (ed. S. Gianinazzi and H. Schuepp), 191–200. Basel: Birkhauser.

Hooker, J.E., Munro, M., and Atkinson, D. (1992). Vesicular-arbuscular mycorrhizal fungi induced alteration in poplar root system morphology. *Plant and Soil* 145: 207–214.

Howard, A. (1945). *Farming and Gardening for Health or Disease*. London: York Publishing Services.

Jakobsen, I., Abbott, L.K., and Robson, A.D. (1992). External hyphae of vesicular arbuscular mycorrhizal fungi associated with Trifolium subterraneum L 2 hyphal transport of 32P over defined distances. *New Phytologist* 120: 509–516.

Jakobsen, I., Joner, E.J., and Larsen, J. (1994). Hyphal phosphorus transport, a keystone to mycorrhizal enhancement of plant growth. In: *Impact of Arbuscular Mycorrhizas on Sustainable Agriculture and Natural Ecosystems* (ed. S. Gianinazzi and H. Schuepp), 133–146. Switzerland: Birkhauser.

Jakobsen, I., Smith, S.E., and Smith, F.A. (2002). Function and diversity of arbuscular mycorrhizae in carbon and mineral nutrition. In: *Mycorrhizal Ecology* (ed. H. van der MGA and I. Sanders), 75–92. Berlin: Springer Verlag.

Jefffries, P. and Barea, J.M. (1994). Biogeochemical cycling and arbuscular mycorrhizas in the sustainability of plant soil systems. In: *Impact of Arbuscular Mycorrhizas on Sustainable Agriculture and Natural Ecosystems* (ed. S. Gianinazzi and H. Schuepp), 101–116. Switzerland: Birkhauser.

Liu, C., Muchhal, U.S., Uthappa, M. et al. (1998). Tomato phosphate transporter genes are differentially regulated in plant tissues by phosphorus. *Plant Physiology* 116: 91–99.

Lumini, E., Vallino, M., Alguacil, M.M. et al. (2011). Different farming and water regimes in Italian rice fields affect arbuscular mycorrhizal fungal soil communities. *Ecological Applications* 21: 1696–1707.

Miller, R.M. and Jastrow, J.D. (1990). Hierarchy of root and mycorrhizal fungal with soil aggregation. *Soil Biology and Biochemistry* 22: 579–584.

Miller, R.M. and Jastrow, J.D. (1992). The role of mycorrhizal fungi in soil conservation. In: *Mycorrhizae in Sustainable Agriculture* (ed. G.J. Bethlenfalvay and R.G. Linderman). Madison: ASA.

Norman, J.R. and Hooker, J.E. (2000). Sporulation of *Phytophthora fragariae* shows greater stimulation by exudates of non-mycorrhizal than by mycorrhizal strawberry roots. *Mycological Research* 104: 1069–1073.

Norman, J.R., Atkinson, D., and Hooker, J.E. (1996). Arbuscular mycorrhizal fungal induced alteration to root architecture in strawberry and induced resistance to the root pathogen *Phytophthora fragariae*. *Plant and Soil* 185: 191–198.

Pozo, M.J., Slezack-Deschaumes, S., Dumas-Gaudot, E. et al. (2002). Plant defense responses induces by arbuscular mycorrhizal fungi. In: *Mycorrhizal Technology in Agriculture* (ed. S. Gianinazzi, H. Schuepp, J.M. Barea and K. Haselwandter), 103–111. Basel: Birkhauser.

Pozo, M.J., Verhage, A., Garcia-Andrade, J. et al. (2009). Priming plant defense against pathogens by arbuscular mycorrhizal fungi. In: *Mycorrhizas – Functional Processes and Ecological Impact* (ed. C. Azcon-Aguilla, J.M. Barea, S. Gianinazzi and V. Gianinazzi-Pearson), 123–135. Berlin: Springer.

Read, D.J. (2002). Towards ecological relevance-progress and pitfalls in the path towards an understanding of mycorrhizal functions in nature. In: *Mycorrhizal Ecology* (ed. H. van der MGA and I. Sanders), 3–29. Berlin: Springer Verlag.

Rillig, M.C. and Mummey, D.L. (2006). Mycorrhizas and soil structure. *New Phytologist* 171: 41–53.

Safir, G.R., Boyer, J.S., and Gerdeman, J.W. (1972). Nutrient status and mycorrhizal enhancement of water transport in soybean. *Plant Physiology* 49: 700–703.

Safir, G.R., Boyer, J.S., and Gerdemann, J.W. (1971). Mycorrhizal enhancement of water transport in soybean. *Science* 172: 581–583.

Sanchez-Diaz, M. and Honrubia, M. (1994). Water relations and the alleviation of drought stress in mycorrhizal plants. In: *Impact of Arbuscular Mycorrhizas on Sustainable Agriculture and Natural Ecosystems* (ed. S. Gianinazzi and H. Schuepp), 167–178. Switzerland: Birkhauser.

Schellenbaum, L., Berta, G., Ranirina, F. et al. (1991). Influence of endomycorrhizal infection on root morphology in a micropropagated woody plant species (*Vitis vinifera* L). *Annals of Botany* 68: 135–141.

Van Tuinen, D., Dulieu, H., Zeze, A., and Gianinazzi-Pearson, V. (1994). Biodiversity and characterisation of arbuscular mycorrhizal fungi at the molecular level. In: *Impact of Arbuscular Mycorrhizas on Sustainable Agriculture and Natural Ecosystems* (ed. S. Gianinazzi and H. Schuepp), 13–24. Basel: Birkhauser.

12

Crop Protection and Food Quality

Challenges and Answers

David Atkinson and Robin L. Walker

SRUC, Aberdeen, Scotland, UK

12.1   Introduction

Organic agriculture began to diverge in a major way from conventional agriculture in the first half of the twentieth century. They diverged initially in relation to the supply of mineral nutrients. Conventional agriculture met the need for nutrients with inorganic fertilisers while organic production continued to rely on crop rotations and recycling materials from activities such as animal husbandry. For the past 50 years, the major divergence has been in relation to crop protection materials, especially organic pesticides. The two approaches to food production came to rely on a different science base.

The two divergences interacted. Supplying nutrient needs with large applications of nitrogen put crops at increased risk of damage from fungal pathogens. If the yield increases expected from fertiliser use were to be achieved then crops had to be protected from any factors which could reduce yield.

One of the most important is the leaf area available to intercept radiant energy. In most leaves, the rate of photosynthesis is related to the photon flux density received (see Chapter 8). This relationship is changed if leaf functioning is affected by disease as this reduces the duration of effective leaf activity. In most crops, leaf production ends just before flowering and thereafter photosynthesis depends on leaves produced before that time. A shortened period for the production of leaves, as a result of damage by pests, diseases or weeds, results in a reduced leaf area with consequences for yield. In conventional agriculture, pests, diseases and weeds are controlled using chemical pesticides. The ability of pests and diseases to adapt their genomes to chemicals requires a continuing supply of new pesticides. In a very real sense, chemical crop protection depends on sustained warfare against biological entities which compete with the crop. The science which underpins this approach is dominated by an understanding of how unwanted biological organisms might be disrupted by chemicals introduced to and maintained within the system.

The Science Beneath Organic Production, First Edition. Edited by David Atkinson and Christine A. Watson.
© 2020 John Wiley & Sons Ltd. Published 2020 by John Wiley & Sons Ltd.

The approach within an organic system is wholly different. Maintaining an effective leaf area remains important but organic producers accept that organisms which cause disease, compete for resources or damage the crop will always be present. Controlling their impact so as to optimise crop performance is critical but eliminating them is not seen as achievable. Pests, diseases and competing vegetation are of course present in all natural systems. Here, particular species come to dominate the composition of the vegetation or the insect and fungal populations. This dominance is a feature of genetic characteristics. Organic systems base their crop protection strategies on the approaches of ecological science. The relationship between CO_2 absorption and intercepted radiation varies between different plant species (Eckhardt 1977). Organic crop protection makes use of this variation.

Here we explore how crop protection is delivered in organic systems and its basis in ecological science. We also assess the impact of this and of organic systems of production on crop and food quality as a means of answering the question as to whether food produced organically differs in any measurable way from food produced in other systems.

12.2   Crop Protection Against Pests, Weeds and Diseases

Crop protection issues exist in temperate and tropical crops (Burke 2016). The founders of the organic movement believed that healthy organisms were able to resist disease. Howard (1945) saw pests (which include diseases) as an indication that something was wrong with the plant and as a sign of poor vitality. He identified the soil as a major component in plant health. This lead to the approach of feeding the soil as a means of developing vitality. The use of composts was at the heart of his approach. Similarly, Balfour (1943) focused on the importance of an appropriate trace element content within crops and the role of soil fungi, especially arbuscular mycorrhizal fungi (AMF) (see Chapter 11), in delivering both resistance to disease and the ability to function in the presence of disease-causing organisms. This remains at the heart of the organic approach to protecting crops, an approach which aims to develop co-existence with organisms which conventional production describes as pests, weeds or diseases.

At times, this idealised approach of co-existence fails in a major way. When this occurs, to preclude total crop loss, the rules of organic farming exceptionally permit the use of traditional crop protection materials. This is, however, an exception, not the total approach, although the provision has led some commentators to see this as the basis of organic crop protection. The availability of these materials is noted here but our emphasis, in this chapter and in Chapter 13, is on the main approaches to organic crop protection. Permitted substances vary between countries both in what is acceptable and whether it needs explicit authorisation (Table 12.1). The materials are ones with a long history of use in crop protection.

The rotation (see Chapter 6) was and remains a key element in organic crop protection strategies (Sanders 1944). In conventional agriculture, most actions are targeted, such as the application of a specific herbicide to deal with a specific weed. In organic production, most activities have a range of effects. The cover crop in a rotation will affect weed populations but will also influence soil condition and nutrient availability. Defining crop protection in isolation from other parts of crop production is unhelpful. The change of crops in a rotation, from year to year, affects weed populations. This is partly due to different crops having different tillage requirements; root crops reduce weeds because of the amount of tillage they require and

Table 12.1 Traditional pesticides available for use as a last resort in a range of organic certification schemes

Product	Common purpose
Copper	Control of foliar fungal or bacterial diseases
Fatty acids	General control
Potassium permanganate	Control of fungal diseases
Pyrethroids	Insect control
Rotenone	Insect control
Soap	Reduction of aphid numbers and some leaf-borne diseases
Sulphur	Control of foliar fungal diseases and improvement of sulphur content
Acetic acid	Weed control
Micro-organisms (e.g. *Bacillus thuringiensis*)	Control of insect pests

through the impact of shading and allelopathy. Organisms which attack a particular crop need that crop to be present for them to multiply. The absence of that crop reduces the levels of crop-specific pests and diseases. Crop-derived materials are known to have adverse effects on pathogens (Grevsen 2012).

In organic systems, the more holistic approach to crop production makes it less helpful to focus on weed control as opposed to disease control, as the intention is often to reduce weeds and diseases through the same practices. Nevertheless, to allow comparison with approaches in other systems, we consider how weeds, diseases and pests are dealt with in organic production.

12.3 Weed Control

The control or management of weeds is the greatest challenge to all crop production but a particular challenge in organic systems. The ability of the crop to adversely affect weeds is significant, especially in cereals. However, cereals vary in their susceptibility to weed pressures. Oats and rye are competitive crops and are able to perform well under conditions of limited fertility. The habit of the crop also has an affect and so varieties which are long strawed or with a spreading growth habit are likely to compete better with weeds. Both crops and weeds need access to light to grow and so crop varieties able to outgrow the weed species will minimise competition.

Despite crop protection, it is variously estimated that weeds alone commonly cause an average loss of 10–15% of yield in cereals although in some situations and other crops, losses can be greater than this. In addition, in some crops weed seeds contaminate the food crop, so increasing the cost of production. In apple, Atkinson and White (1981) showed that the presence of weeds in the tree row, which is normally kept weed free, resulted in a reduction in yield of 16–49% and an even greater reduction in financial returns. All non-crop vegetation will compete with the crop as a consequence of their need for water and mineral nutrients and perhaps

Table 12.2 Major ecological processes which influence the composition of natural vegetation, which provide the basis for practices designed to manage weed populations in organic crops

Ecological process/ concept	Significance to weed management in organic crop production
Competition/ dominance	Use of crop through, e.g., species selection, planting time, to allow the crop to suppress weed growth
Disturbance	Use of cultivation or chemicals to disrupt weed growth and reduce the weeds' competitive effect
Dispersal	Value of rotational design in limiting the spread of weed species
Stress tolerance	Crop and cultivation effects on the stresses to which weed seedlings and propagules are exposed
Life history	Rotational design to exploit weaknesses in weed life histories
Ruderal habit	Rotations and cultivation patterns to reduce the conditions needed by annual weeds to establish and contribute to the seedbank
Pathogen impact	Rotations designed to increase the impact of soil pathogens on weed species
Allelopathy	Use of crops known to produce chemicals with inhibitory effects on weed species and use of rotations to minimise the impact of such materials produced by weeds on crops

as a result of an impact on soil micro-organisms. In the absence of chemical control, organic systems rely on the management of the same processes which influence the composition of vegetation in natural situations (Table 12.2).

Grime (1979) discussed the various strategies used by plants which affect the composition and structure of vegetation. He identified stress and disturbance as particularly important. Stress, such as that consequent on a shortage of light, water or mineral nutrients, restricts photosynthesis. Disturbance, such as those resulting from cultivation, results in the partial or total destruction of plant biomass. The management of weeds in organic crops is thus a balance between the competitive potential of crop and weed, their relative abilities to cope with environmental stresses and their responses to disturbance. Other ecological factors which influence the balance of advantage between crop and weed include competitive ability, which is affected by the distribution through space and time of leaf area. The relative speed of growth is important; some weed species such as *Arrenatherum elatius* ($115\,\mathrm{g/m^2}$/week) can grow faster than barley ($64\,\mathrm{g/m^2}$/week), making early sowing so as to establish crop dominance important. Plant height is critical to competitive advantage (Boysen Jensen 1929). Small differences in the interception of radiation combined with relative growth rates influence the balance between crop and weed and affect the crop plant's ability to establish dominance. Sowing at high seed rates, particularly when associated with correctly timed cultivations, can also increase crop competition.

Competition has been defined in a number of ways (Firbank 1991). Essentially, it represents an interaction consequent upon a shared need for resources which are in limited supply and which results in a reduction in the reproductive potential of the competing individuals. In the context of crop–weed dynamics, we tend only to be interested in the consequences of the interaction for the crop even though much of the organic strategy for weed control is focused on reducing the competitive pressure of the weed (Firbank 1991). At low densities of plants, there are sufficient resources to go around but at higher densities, because available resources can only support so much total growth, the yield of crop plants will be reduced. Crops are commonly sown at high densities so as to maximise their competitive advantage. However, inevitably any weeds present will reduce the amount of resource available to the crop, thus potentially affecting crop yield. The extent of the weed presence will control the scale of the impact.

The starting position in the competitive race is critical. Competition for light is asymmetrical. Plants with a small initial advantage grow more quickly than shaded neighbours, making early emergence, even of a few hours at the time of germination, a significant advantage. Crop yield is most affected by early weed growth so there are real gains for varieties which rapidly achieve high levels of ground cover, unless this results in exposure to harsher environmental conditions.

Dominance has two components: the way in which the dominant plant becomes greater in size and the deleterious effects which the dominant plant will exert on its neighbours, usually in the form of nutrient or water stresses, the release of phytotoxic compounds or through impacts on soil micro-organisms. One of the key objectives of cereal farming is to create conditions in which the cereal will function as the dominant entity, especially by synchronous germination from a high density of relatively large seeds followed by the rapid development of a dense cover composed of large numbers of plants of comparable maturity.

This aim can be adversely influenced by the differential effects of insect predation or disease (Grime 1979): 'Environments differ in the extent to which they allow the competitive characteristics of the plant to be expressed'.

Many weed species are ruderals and so their ability to recover from weed control operations is important to any strategy. The control of ruderals depends on timing, particularly of cultivations. The period immediately following crop emergence, when the presence of weeds can have a large impact on future yields, is critical. When weeds are eliminated during this period then future weed populations seem to have a smaller impact, probably because of the ability of the developed crop to dominate late-emerging weeds. Weeds are able to use a variety of regeneration strategies such as seeds which persist in the soil or which travel within the field, or rhizomes (Grime 1979). Plants with more than one regenerative strategy tend to occur in more places and so have the potential to be more troublesome as weeds.

The weed pressures in organic production differ from that where control is based on herbicides. Crop varietal selection (see Chapter 13) is a significant factor in the design of the weed control strategy. Recent cereal varieties have been bred with the expectation of significant applications of fertilisers and pesticides, resulting in varieties which respond differently under organic production to the varieties of the past. Grain yields of winter wheat prior to 1950 were 6 Mg/ha, currently they are over 9. This has been achieved with little change in total dry matter, which remains at around 16 Mg/ha. Straw length has fallen from around 150 cm before 1900 to around 80 currently. As a result of both of these trends, harvest indices, the ratios of crop to total dry matter production, have risen over the same period from a little over 0.3 to around 0.5. One consequence of this is that crop plants are less competitive with weeds as a result of reduced height. The achievement of a high harvest index requires rapid development by the crop both to generate sufficient grains and to suppress weeds. Both are more difficult with modern varieties in an organic context.

Fryer (1981) identified that prior to the use of herbicides, the most important weeds were broad-leaved species. In the herbicide era, weed problems have come to be dominated by grass weeds. A consequence of this is that the weed species most likely to be found in organic production may not have been well researched in recent years. Additionally, farms which convert from conventional to organic production may have weed problems that have not been studied in an organic context because they were not a problem in an earlier era.

Where the same crop is grown in successive years and the same methods of preparation repeated then the selective pressures on weeds are less than in a rotation. In addition, if weed control has previously been through the use of herbicides then selective pressures will have resulted in a weed flora whose characteristics mimic the main features of the crop, thus reducing options for weed control. Weeds and crops show parallel features in their evolution, resulting in major similarities in life history and growth forms; hence many of the worst weeds of arable production are now grass weeds able to survive in a disturbed habitat. The production of reproductive propagules by weeds is a function of size. Future weed densities and crop yield losses are thus affected by the emergence and growth of weeds in the current season. Methods available to organic growers are summarised in Table 12.3.

Stockdale and Watson (2011) assessed the impact of rotations on weed control and identified effects due to tillage and changes in the crop life cycle as important to reducing weed seedling establishment. The impacts of various approaches to cultivation have been widely reported. Mays et al. (2014) found that ground covers selected for weed control in organic orchards could also improve soil quality, with benefits for the crop. Mechanical weeding is important. Many devices have been developed although with some differences in the preferred approaches for vegetables, commonly spring tines and brushes and between-row cultivations. Koller and Vieweger (2012) assessed the impact of a number of weed control treatments on both the efficacy and cost in onions. They found that directly sown onions required between 272 and 756 h/ha while the use of planted seedlings reduced this to 65–79 h/ha and the use of sets to 24–61 h/ha. Flame weeding reduced weeding hours by 30–53%. A fabric cover stimulated weed growth and created options for control with cultivation or flame weeding.

Table 12.3 Approaches to the management of unwanted plants (weeds) in organic production systems

Control method	Mechanism of action
Tillage, including hoeing	Physical destruction of weed and stimulation of seedling germination at unsuitable times, e.g. frost periods
Cropping system including selection of crop species and rotation design	Impact of rotation on the timing of cultivations, sowing, etc. aimed at giving advantage to the crop through effects such as shading, nutrient competition
Burning	Physical destruction
Chemical control	Options very limited but soaps can be used
Biological control	Options available for enclosed systems but uncommon elsewhere
Plastic or fabric cover	Black plastic raises soil temperatures while simultaneously depriving germinating seedlings of light and is effective against annual weeds
Mulch, including the use of species for their allelopathic effects	Plant residues are able to work in a similar way to plastic mulches and may simultaneously have an adverse impact as a result of chemicals produced in the decomposition of the mulch

A range of mulch treatments aimed at depriving weeds of access to light have been devised. Solomakhin et al. (2012) assessed the use of sawdust mulch on the control of weeds in an apple orchard. They found that compared to a weed population of 167–273 weeds/m^2 with manual weeding, sawdust numbers were 0–3. Similar numbers were obtained with a grass clipping mulch and a pine bark mulch. In addition to its contribution to weed control, the wood mulch provided around 4 kg/ha of N which was around 10–12% of the requirement of a young tree and caused an increase in the colonisation of the surface soil by roots which would have allowed the use of much of the 60–75 kg/ha/N released by the mineralisation of organic matter. Boydston et al. (2011) assessed the impact of using mustard seed meal (MSM) as a weed control material in onion production. The concentration at which the meal was used and the timing of its use affected both the level of weed control achieved and the extent of crop damage. High concentrations and early applications damaged onion stands. Appropriate applications reduced red root pigweed by 90%, especially when the meal was derived from mustard cultivars with a high glucosinolate or ionic thiocyanate content. Varieties with high content achieved 90% control with application as low as 3.2 g/m^2 while low-content varieties required 128 g/m^2. In field trials applications of MSM at 2.2–4.5 Mg/ha reduced weed emergence by 68–91% although in some years this had an adverse effect on onion yields.

12.4 Living with Crop Diseases

Plant diseases are disturbances which affect the normal functioning of the plant (Barker 2010). In discussing weeds, we commented on the similarities between crops and weeds. In a parallel context, fungi found on crops are also commonly found on adjacent natural vegetation, which is a major issue when considering long-term control (Wheeler 1981).

Three major issues are important to an understanding of the relationship between fungi and their plant hosts (Wheeler 1981)

- Fungi exploit a range of niches on or in plants, with some being found in the roots, others on leaves and still others specialised to colonise flowers or fruits. A crop plant may play host to a number of fungi.
- There are a range of mechanisms by which fungi parasitize their hosts with some causing the immediate death of infected cells and others, while directing assimilate to the fungi, not causing immediate death.
- There is great variation in the ways in which fungi spread and the length of time that spores remain infective.

The development of the fungal population will be influenced by the structure of the host population. Where crop plants are widely spaced, the spread of a pathogen will be reduced. Where a crop is composed of large numbers of the same genotype, which develop synchronously, this aids infection and spread. This is clearly different from the situation in adjacent wild vegetation. Genetic mixtures also reduce spread and so the intensity of infection. The importance of wild plants as a source of inocula is high initially but once a pathogen has become established in a crop, most new infection is from within. Some pathogenic fungi, such as rusts, require two hosts to complete their life cycle and here an awareness of the relationship between wild vegetation and crop is critical to control (Wheeler 1981).

Table 12.4 The means available for the control of plant diseases in organic systems

Approach	Mechanism of effect
Use of disease-resistant cultivars	Cultivar resists infection by pathogen or tolerates normal levels of infection
Selection of appropriate site	Site selection can avoid the presence of a soil pathogen
Rotation design	Changing the crop reduces the possibility of the build-up of pathogens, especially soil pathogens
Maintenance of appropriate soil condition	Pathogen activity is commonly affected by factors such as soil pH and water content
Use of AMF	AMF reduce the impact of pathogens, especially soil pathogens
Other biological controls	Reduce pathogen populations (Adesegun et al. 2012)
Chemical application	Few chemicals are available for use in organic systems but Cu and S reduce the levels of several foliar-acting pathogens. A number of chemicals derived from biological sources are in use for specific crops

AMF, arbuscular mycorrhizal fungi.

In the absence of chemicals which either protect against infection from plant diseases or eliminate infections when they have occurred, organic systems rely for their protection from diseases on a combination of crop genetics and cultural practices which move the balance of advantage to the crop plant. The principal mechanisms available for protection from attack by plant pathogens are summarised in Table 12.4.

12.4.1   The Impact of AMF on Plant Pathogens

The founders of the organic movement were clear that soil microbiology, especially infection with AMF, was at the heart of resistance to disease. Other aspects of the impact of AMF are discussed in Chapter 11. A healthy soil has a diverse soil microflora and a healthy plant, as a consequence of AMF infection, may reduce pathogen infection or impact. The impact of AMF on the plant's ability to cope with the effects of pathogens has been reviewed by Hooker et al. (1994) and Pozo et al. (2002, 2009). Colonisation by AMF can improve resistance to pathogens. Resistance can have a number of components such as improved P or N nutrition, increased tolerance to drought or to heavy metals or as a result of an improvement to soil structure. All of these aid crop development. However, independent of this, infection with AMF can enhance the ability of the plant to both resist infection and tolerate the effects of infection. It is thus being seen as an approach to biocontrol. Major elements relate to the competition between AMF and pathogens for both photosynthates and infection sites. Changes to the morphology of the root system may alter the dynamics of infection.

The controlled defence reaction which seems to be induced when AMF infection occurs is a component in the enhanced disease resistance – systemic acquired resistance (SAR), resulting from AMF infection (Pozo et al. 2002). In addition, it has been demonstrated that arbuscule-containing cortical cells are more resistant to infection by pathogens such as *Phytophthora parasitica* and that lytic activity against the pathogen is induced.

During interactions with micro-organisms, plants are able to recognise microbe-derived molecules and to tailor their response to the micro-organism encountered. The molecular dialogue between organisms determines the outcome of the relationship, which ranges from parasitism through mutualism to a true symbiotic relationship based on significant modifications to cellular processes (Pozo et al. 2009). Molecular exchanges between plant and AMF occur at the start of the infection process and indicate that this is a symbiotic relationship. It is a normal developmental stage for the plant. The obligate nature of the fungal response indicates the extent of the fungus's adaptation (Gollotte et al. 2002). Studies of mutant plants which block AMF development indicate both the detail of the infection process and much about the response of plants to infection by fungal pathogens. The initial plant response does not inhibit spore germination or initial hyphal development or the production of appressoria. However, at the point of contact between appressoria and the plant cell wall, wall deposits containing phenolic compounds, callose and PR1 proteins are produced. In a normal AMF infection, this does not occur but such a response is part of the plant's approach to pathogenic fungi. A common response of a plant to the presence of a pathogen, which results in the production of an elicitor, is the development of specific receptors on the plant cell surface which can lead to responses such as the production of reactive oxygen species and the activation of genes coding for enzymes in the phenylpropanoid pathway which lead to the production of antifungal molecules.

A well-established mycorrhizal association is important to systemic resistance but the principal effects are thought to be due to the modulation of plant response defence mechanisms as a consequence of the AMF infection. The modulation results in a mild and systemic activation of the plant immune system. Quantitative rather than qualitative differences seem to determine resistance. AMF-infected plants can display an enhanced defence reaction when challenged by pathogens such as *Fusarium*. Priming for callose deposition seemed to account for the protection given by *Glomus intraradices* against *Colletotrichum*. AMF provide systemic protection of the root system against pathogens such as *Phytophthora*. Mycorrhizal plants formed papilla-like structures around the sites of pathogen infection, so preventing pathogen spread. Plants recognise pathogens as a consequence of the elicitors which are secreted during the early phase of pathogen attack. Defence-related genes become activated. This includes genes which code for enzymes in the phenylpropanoid pathway, which leads to the production of antifungal or signal molecules, to the production of reactive oxygen species and those coding for pathogenesis-related proteins such as chitenases and glucanases.

As a result, it has been suggested that AMF have the potential to function as biological control agents (Azcon-Aguilar et al. 2002). AMF reduce the impact of root rots such as *Rhizactonia* and *Fusarium* and wilts such as *Verticilium* as well as diseases caused by *Phytophthora*, *Pythium* and *Aphanomyces* (Whipps 2004). Additionally, AMF infection seems to confer some resistance to both nematodes and phytophagous insects.

Norman et al. (1996) assessed whether the modifications to root architecture which resulted from AMF infection (see Chapter 11) were a component of the enhanced resistance shown by strawberry plants to *Phytophthora fragariae*. They found that in non-AMF-infected plants, the extent of root necrosis caused by *P. fragariae* was correlated with the proportion of the root system made up of high-order roots. In AMF-infected plants, necrosis was negatively correlated, suggesting that changes in root architecture were not important to resistance in themselves but seem to be related to factors expressed concurrently (Table 12.5). Norman et al. (1996) found

Table 12.5 The impact of AMF on root necrosis due to *Phytophthora fragariae* in three strawberry varieties

Variety	Control	Infected with *Glomus fasciculatum* (55–71%)	Infected with *Glomus etunicatum* (65–67%)
Cambridge Favourite	75	30	32
Elsanta	39	35	30
Rhapsody	7	10	7

Source: Data from Norman et al. (1996).

that prior inoculation with AMF reduced root necrosis due to *P. fragariae* by 60% and 30% in the strawberry varieties Cambridge Favourite and Elsanta. Different species of AMF gave different levels of control of the pathogen.

Arbuscular mycorrhizal fungi have thus been shown to enhance resistance to major pathogen groups such as to *Rhizactonia*, *Fusarium*, *Verticilium*, *Phytophthora*, *Pythium* and *Aphanomyces*. In AMF-colonised plants, root system necrosis due to the pathogen was negatively correlated with the development of higher-order roots.

Effects on foliar diseases are less clear although there is evidence that AMF-infected plants show better growth and resistance when infected by pathogens such as mildew and rust (Pozo et al. 2009). On occasions, AMF infection has led to increased levels of infection by rusts and mildews but with the pathogens having a level of effect below that which would have been anticipated for this level of infection (Pozo et al. 2009).

12.4.2 Plant Varietal-Based Resistance

Selecting a crop variety is important. While genetic resistance rarely provides complete resistance to all diseases, some varieties are significantly more resistant. It is important to select a variety with resistance to the diseases most likely to be found. The importance of genotype to all aspects of crop performance, including disease resistance, is dealt with in Chapter 13. Its impact on crop protection means that it is necessary to comment on some aspects here.

While resistance to pathogens induced by AMF and other microbes is important, so is natural genetic-based resistance. Villegas-Fernandez et al. (2011) assessed the potential for developing resistance to fungal diseases in faba bean and identified a number of cultivars with resistance to both rust (*Uromyces viciae-fabae*) and chocolate spot (*Botrytis fabae*). The absence of a genotype–environment interaction for these organisms suggested that the varieties should be resistant in a range of settings and that the resistance should be stable with time. Gragera-Facundo et al. (2012) found that disease resistance to mildew in melon was correlated with fruit weight and sensory quality while Castejon et al. (2012) found that there were large differences in the susceptibility of plum cultivars to the rust *Tranzschelia pruni-spinosae*. Against a five-point scale where 5 represented 80% plus of the leaf surface showing evidence of infection and 0 no sign of infection, the best cultivar when grown under organic conditions and in a year when infection was prevalent recorded a score of 0.5 and the worst one of 4.4. Conventional

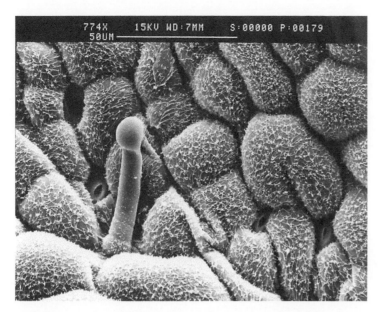

Figure 12.1 Scanning electron micrograph of the germination of a mildew spore on the surface of a strawberry leaf. *Source:* Micrograph taken by DS Skene, East Malling Research.

crop protection treatments maintained disease levels in most cultivars under 1. In other studies on plum, Arroyo et al. (2012) found that the cultivar showing the greatest tolerance to rust diseases was the most susceptible to mildew. Mildews have a significant impact on many organic crops (Figures 12.1 and 12.2). Applications of sulphur, permitted under most organic standards, controlled mildew but not rust.

The pattern of planting also affects the development of disease. Crop disease can be reduced by the use of crop mixtures, especially where one of the varieties would otherwise show significant fungal colonisation. Different species of cereals such as barley and oats can be mixed. Fernandez-Aparicio et al. (2011) showed that the incidence of chocolate spot (*B. fabae*) on faba beans was reduced when it was intercropped with a cereal such as barley or wheat but not when the intercrop was a legume. The effects were attributed to a combination of host biomass reduction, changed microclimate and the presence of a physical barrier to spore dispersal.

The soil environment influences the infectivity of many soil-borne pathogens. Accordingly, management is the basis of the control of a number of pathogens. Many horticultural crops, such as strawberry, have traditionally been grown on ridges so as to reduce the impact of *Phytophthora*, which benefits from wet conditions. Soil pH provides another means of control as pH affects a number of pathogens. Brown and Jennings (2012) assessed the effect of a preplanting soil treatment with calcium hydroxide on the impact of apple replant disease. Under their conditions, they found CaOH to be as effective as dazomet or methyl bromide, which are used to minimise the effect of replant disease in conventional orchards. The use of CaOH gave a cumulative yield of 450 Mg/ha over a 10-year period compared to 325 in an untreated control.

Thus, at the heart of the protection of crops from fungal and bacterial diseases is the maintenance of healthy growing conditions. Nevertheless, attempts have been made to parallel

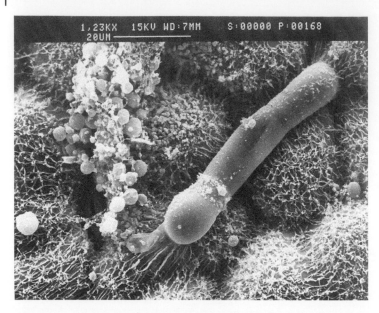

Figure 12.2 Scanning electron micrograph of the beginning of the development of appressoria by a germinated mildew spore on a strawberry leaf. *Source:* Micrograph taken by DS Skene, East Malling Research.

conventional approaches to crop protection by using materials acceptable within organic standards and biological control approaches developed for use in enclosed environments. Wei et al. (2012) found that extracts of garlic bulbs gave up to 80% control of *Fulvia fulva* on tomato in controlled trials. Similarly, Adesegun et al. (2012) found that extracts of spices gave 51–77% control of *Sclerotium rolfsii* on tomato in *in vitro* and pot culture studies. Attempts to use plant-derived 'pesticides' have rediscovered what has been long known in the agrochemical industry, that few of the compounds which work well in culture or in the glasshouse perform as well or as consistently under the more varied conditions found in the field. Grevsen (2012) found that a biofumigant treatment derived from brassicacious materials, while effective under controlled conditions, did not have significant effects on disease development in field-grown stored carrots.

12.5 Pest Control

The various ways of protecting organic crops from insect pests have been reviewed by Barker (2010). The approaches he details are summarised in Table 12.6.

As with fungal pests, control of insects depends primarily on the design of the rotation and the integration of appropriate cultivars and physical protection (Figure 12.3).

Chemicals would be used only when all else had failed. Protecting crops against insect pests without agrochemicals requires the development of approaches which depend on natural processes and biological interactions and synergies. Much depends on combinations of enhancing crop vigour and enhancing the role of beneficial arthropods and antagonists (Alteri and

Table 12.6 Potential approaches to the control of insect pests

Approach	Mechanism	Materials available
Chemical treatment	Physical harm to insect wings or cuticle	Water, soaps, oils, Neem, nicotine, alcohol, diatomaceous earth, sulphur, whitewash
Dusts	Damage to nervous system	Pyrethrum, rotenone, ryania, sabadilla, quassia, lime sulphur
Biological control	Competition, predation, disease	Lacewings, parasitic wasps, mites, nematodes
Microbial pesticides	Bacterial or fungal diseases of insects, presence of proteinaceous toxins	*Bacillus thuringiensis*, other *Bacillus* species, *Verticilium* species
Crop management	Timing of crop establishment, tillage, removal of crop residues	
Physical controls	Prevention of insect contacts with plant surfaces and removal by hand or machine	Barriers such as sticky bands, traps including pheromone traps, trap plants such as mustard, nasturtium and radish
Cultivar selection	Physical barriers to insect attacks and inhibitory chemicals	
Rotation design	Gaps in the presence of target host species, effects of other crops in the rotation and of companion crops planted to modify insect behaviour	

Source: Derived from Barker (2010).

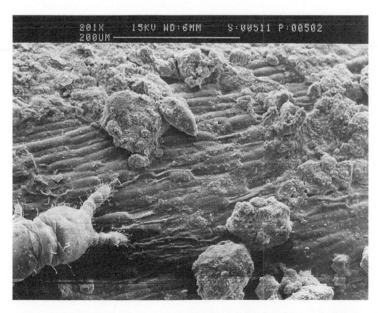

Figure 12.3 Scanning electron micrograph of a collembolan grazing the surface of an apple root. *Source:* Micrograph taken by DS Skene, East Malling Research.

Nicholls 2012). The key is enhancement of the immunity of the system through natural pest control systems and regulatory processes, which increase functional complexity. Increasing crop diversification reduces insect pest pressure by a reduced pest colonisation rate, reduced pest tenure time, oviposition interference and increased pest mortality as a result of predators and parasitoids. Intercropping and the application of compost decreased pest abundance in broccoli. With some intercrops there was an increase in parasitism. In vineyards with diversified cover crops, pest populations were reduced through enhanced numbers of parasitoids and predators. Summer cover crops provided a consistent food source for a diverse community of natural enemies (Alteri and Nicholls 2012). The structure and physiology of an apple tree have been shown by Simon et al. (2012) to affect the ease of colonisation by rosy apple aphid.

As with resistance to diseases, AMF infection affects a plant's ability to resist the attack of phytophagous insects (Pozo et al. 2009). Infection with AMF seems to inhibit attacks of root-feeding insects but to be more variable in its effects on leaf-feeding insects. In general, AMF reduce the impact of generalist chewing insects but increase the performance of sap-feeding or specialist insects. Despite this, infection with *Glomus mosseae* has been shown to reduce the performance of potato aphid on tomato.

Insects like most organisms have pests and diseases to which they are susceptible and an approach to organic pest control has been the identification of such diseases and methods which can be used as a means of protecting the crop. Ganga Visalakshy and Krishnamoorthy (2012) sought to control onion damage from thrips by the use of microbial pathogens such as *Metarrhizium anisopliae*, *Verticillium lecanii* and *Beauveria bassiana* and entomophilic nematodes. This approach gave control at a similar level to chemical control. Chemicals which disrupt mating have also been effective in reducing damage by codling moth to pome fruits (Kutinkova et al. 2012).

12.6 The Quality of Organic Crops and Crop-Based Foods

Perhaps the single most contentious question in relation to food produced by organic systems is whether it differs in quality from food produced in other ways. Is organic food better for you? This also links to whether pesticides at the concentrations likely to be found in foods are harmful to consumers. These questions come up in debates over the importance of yield as the key question in relation to food production. That food has major effects on health has long been accepted. More detailed relationships between food and health are less well understood. Some effects are as simple as the provision of adequate amounts of carbohydrates, amino acids and lipids. Others relate to the delivery of specific metabolites. If organic production were better at delivering important elements of the diet, this would change the debate around whether organic production can feed the world.

Food, although traded as a commodity on world markets, nevertheless has a unique position in communities and cultures, a fact attested to by the number of food laws associated with faith traditions (much of the Book of Leviticus, which is common to the Hebrew Scriptures and the Christian Bible, deals with issues related to food). The impact of ethically based approaches to food has been reviewed by Atkinson et al. (2012) who showed that the term *quality* has sociological and theological components which go beyond chemistry or microbiology.

Food quality is, however, commonly defined in relation to its physical, chemical and microbial content. The presence or absence of chemicals and of micro-organisms and the significance of these for a healthy diet is at the heart of the debate. The list of substances found within foods which have a significant impact upon health has increased, with studies assessing the impact of a wider range of compounds beyond the basic mineral elements and major components such as carbohydrates, proteins and lipids. The addition of substances such as antioxidants and free radical binders such as phenolics is of very recent lineage. It seems likely that the identification of products and molecules with positive and negative impacts on health will continue. An important question thus becomes the ways in which the production of crops using an organic approach could affect the quality of the food produced. Important sources of variation include the following.

12.6.1 Varietal Selection

Cultivars of a crop species are known to vary in their ability to both absorb nutrients from the soil and produce secondary metabolites. Varietal selection is discussed in Chapters 6, 8 and 13. Above we discuss the importance of variety to crop protection but the choice of variety can have a major impact on compositional attributes as diverse as quantities of mineral nutrients, the amino acid composition of proteins and the amounts of secondary metabolites. Varieties are selected for a range of reasons, the most common being the ability to grow in a particular location (see Chapter 5) and the ability of a cultivar to resist the impact of potential pathogens, but ultimately this choice can also influence the quality of the food.

One of the major components of the western diet is bread and so it is important to assess whether organically produced wheat is appropriate for bread making and whether it confers any additional attributes. Casagrande et al. (2009) assessed the field factors which could lead to a low grain protein content and so to poorer baking quality. They found that for successful baking, it was necessary for grain to have a protein content above 10.5%. Variation in grain protein content was most commonly caused by selection of a variety with poor baking quality, crop N status and weed density at flowering. To be suitable for bread making, wheat must have a high protein content and an appropriate balance of gliadins and glutenins. In organic systems, the low and variable N status of crop during grain filling makes the achievement of high protein levels more difficult. Low total protein can be compensated by a high percentage of high-quality protein. Achieving this requires cultivars with optimised gliadins and glutenins, effective N uptake and an effective N translocation capacity from the vegetative component into the grain and the genetic ability to turn this N into high-quality protein (Osman et al. 2011). Leithold and Becker (2011) found that modifying row spacings in organic production could significantly influence protein content and other features important to baking quality. The selection of crop variety impacted on nutrient content. The varietal effect differs between years and sites although Murphy et al. (2011) found the concentration of Cu, Fe and P to be broadly adapted relative to other nutrients and that temporal stability was more common than spatial stability.

Compounds with antioxidant activity such as phenols have recently been recognised as being significant in assessing food quality. Heimler et al. (2012) studied polyphenol content and antiradical content in lettuce and found that although yields were higher in conventional production systems, the content of polyphenols were proportionately higher in both organic and biodynamic crops (1.74 and 1.85 mg/g respectively compared to 1.36 mg/g). Anti-free radical activity was correlated with the flavanoid and hydroxycinnamic acid content. Similarly, Kalinova and Vrchotova (2011) found that organic production methods lead to an enhanced content of several flavanoids in buckwheat.

12.6.2 The Production System

Interactions with other organisms, tillage and spacing have all been shown to affect the composition of plant tissues used as foods. The impact of the system used to produce crops has been assessed for major characteristics such as dry matter content, mineral content and vitamins in a number of studies. Most commonly, but not universally, foods from organic systems have higher dry matter, mineral and vitamin contents (Table 12.7).

The way in which apple trees are grown has significant effects on the mineral composition of their fruit. Where trees are grown with a grass cover, there are adverse effects on fruit size but often positive effects on composition. Adverse impacts are commonly the result of competition for water which may be alleviated by irrigation. Atkinson (1984) found that apples from trees grown under grass with irrigation contained 152 mg/P/g fresh weight compared to 112 mg in fruit from a chemically managed treatment. In a study comparing the composition of Cox apples grown on a number of rootstocks (apple trees are grown with a scion variety grafted onto a rootstock which provides the root system) under either grass or a bare soil surface, the concentrations of a number of nutrients were found to be higher with the grass treatment. While rootstock influenced the nutrient content of the fruit, there were no significant rootstock–soil management interactions (Table 12.8).

The source of nitrogen applied to potatoes has been shown to affect their processing quality (Zaman et al. 2011), with organic sources having a positive impact on processing quality.

12.6.3 Inputs Used as Part of the Cultural System

Crop protection chemicals have come under the greatest focus. In an area which is controversial, the impact and importance of the residues of pesticides are amongst the most commonly discussed subjects. Prior to their release for use on crops, all pesticides are tested and safe levels of residues determined. However, despite such evaluations, concern remains over issues of long-term and cumulative effects and the impact of combinations of chemicals. Advances in analytical chemistry have meant that it is now possible to detect the presence of residues which would not have been possible in earlier years. As a result it is common for residues of

Table 12.7 The influence of growing system on the content of conventionally and organically grown fruit and vegetables found in a number of studies

Component	Higher in organic	Inconsistent or no significant difference	Higher in conventional
Dry matter	10	8	1
Mineral content	7	6	1
Vitamin C	7	6	0

Table 12.8 The effect of soil management and rootstock on the composition of Cox apples in 1983

Rootstock	Dry matter	(% DW)	P	mg/kg	K	g/kg
			Treatment			
	Bare	Grass	Bare	Grass	Bare	Grass
M27	13.9	15.8	88	102	1.21	1.31
M9	14.5	15.1	105	128	1.37	1.50
M26	16.9	17.7	131	172	1.50	1.80
MM106	15.1	15.2	115	146	1.40	1.55
MM111	15.4	15.6	122	154	1.45	1.63

DW, dry weight.

Table 12.9 The frequency with which residues were detected in horticultural produce in the USA

Commodity	Pesticide	Type of material	Frequency found (%)
Orange	Thibendazole	Fungicide	64
Potato	Chlorpropham	Growth regulator	59
Apple	Thibendazole	Fungicide	57
Peach	Iprodione	Fungicide	54
Grapefruit	Thibendazole	Fungicide	54
Peach	Dicloram	Fungicide	47
Celery	Permethrin	Insecticide	39
Banana	Thibendazole	Fungicide	38
Celery	Chlorothalonil	Fungicide	32
Apple	Azinphos-methyl	Insecticide	31

Source: Data from Atkinson et al. (2007).

insecticides, fungicides and growth regulators to be detected in fresh produce. Table 12.9 shows the frequency with which residues of commonly used pesticides were found in studies carried out by the United States Department of Agriculture (USDA).

The most commonly detected pesticides are fungicides as these are most commonly applied close to the time of harvest. Similar pesticides can be found with different frequencies in different crops and residue frequencies vary between seasons. The mere presence of a residue does not mean that presence at the concentration detected is either harmful or in excess of permitted levels (Table 12.10).

While produce from organic systems should be free from pesticides, drift from surrounding areas using conventional crop protection, residues of long-surviving materials used in earlier eras (e.g. DDT) and mixing of produce in pack houses and shops have resulted in organic

Table 12.10 Pesticide residues found in food samples in UK

Crop	% with residues	Number of pesticides found	Number with residues >MRL
Strawberry	67	33	5
Celery	66	30	11
Carrot	64	12	3
Lettuce	58	37	30
Potato	37	15	6

Source: Data from Atkinson et al. (2007).
MRL, maximum residue level.

Table 12.11 The percentage of samples of conventional (26 571 samples), integrated pest management (IPM) and organic (127 samples) foods containing pesticides

Sample	Organic	IPM	Conventional
All fruit	23	49	82
Fruit excluding banned materials	23	49	82
All vegetables	23	45	65
Vegetables excluding banned materials	9	44	61

Source: Data from Baker et al. (2002).

produce not being wholly free from pesticides (Baker et al. 2002) (Table 12.11). The occurrence of pesticide residues was higher in conventional than in organic production with 73% of conventional and 23% of organic production containing residues of at least one pesticidal material. Thirteen percent of the small number of organic samples tested contained residues of products not allowed in organic systems, including banned organochlorine pesticides, which it is assumed remain from an earlier era.

12.6.4 Inherent Attributes

For many consumers, the fairness of the system to producers and the environment has significant impacts on their perception of 'quality'.

As discussed in Chapter 1, the decision to produce food using organic methods goes beyond yield or even environmental protection. The agricultural system selected commonly relates to values and ethics. Atkinson (2009) concluded that approaches which centred on the use of the soil's microbial resources were likely to have the lowest externalities. The organic philosophy aims to reduce negative externalities whether resulting from environmental pollution, the destabilisation of rural communities or the unemployment of agricultural workers. The input of people is a key factor throughout the whole food chain. Those who see the importance of fairness to the environment, communities and individuals feel that food which has been produced so as to exploit the natural environment or producers cannot be of high quality, regardless of its chemical composition or its microbial status.

These issues have been reviewed by Atkinson et al. (2012). They devised the term *inherent quality* to cover a range of ethically related characteristics which would reduce the value of food in the eyes of consumers who held these views. Related issues are discussed in Chapter 3. Many of these issues are at the heart of the debate on the acceptability of genetically modified (GM) crops, issues discussed by Bruce and Bruce (1998). These values would cause some consumers to reject GM crops on the basis of them having an unacceptable inherent quality. This view may derive from the removal of a gene which is part of the characteristics of one species and its transfer to another species being viewed as identity theft. It could be a consequence of the patenting of genetic constructs, which preclude these genes being used by others and may prevent farmers using self-saved seeds. It may be a consequence of the economic imbalance between small producers in developed countries and multinational companies who control the costs of inputs to farming systems and so influence indebtedness. Those using conventional systems of production would probably not see issues of this type as significant but for organic producers they can be important.

12.6.5 Microbial Content and Chemical Contamination

The presence of microbes or the products of their metabolism has long been an important determinant of food safety. Prange (2012) discussed the impact of preharvest and postharvest technologies on the quality of horticultural produce. Developments in controlled atmosphere storage, such as the use of very low concentrations of O_2 with the monitoring of chlorophyll levels to preclude damage resulting from such low concentration, allow improvements to both the maintenance of product quality and reduction in decay. All food is susceptible to contamination by toxic materials introduced accidentally. Where soil and grassland are contaminated, substances may enter food materials over an extended period and this is likely to be similar in both organic and conventional production systems.

12.7 Conclusion

Crop protection in organic production systems will always be difficult. The degree of difficulty will be related to the importance placed on maximising yields. Where the aim is a yield similar to that which could have been achieved in a conventional system then the difficulties will be pronounced. Where other aspects are relatively more important, more options become available. The key is that organic 'crop protection' depends on a set of very varied processes, from the design of the rotation through to the selection of appropriate cultivars. The particular issues affecting a range of common arable crops are summarised in Table 12.12.

These examples illustrate the importance of the design of the whole system in seeking to control the interactions between the crop plant, other vegetation (weeds) and other biological organisms (pests and diseases) yet still with the aim of optimising the performance of the crop and its ability to produce food.

Table 12.12 The impact of crop protection issues on the production of a range of common crops

Crop	Advantages	Problems
Wheat	Provides good returns for good yields Disease-resistant varieties available Straw of value in mixed farming systems	Weed control challenging Needs good fertility Spring varieties can be late Accumulation of weed and disease issues reduces yield as second cereal crop
Barley	Harvested early Disease-resistant spring varieties available Spring varieties can be undersown Acceptable as second cereal	Winter varieties susceptible to disease Weed control difficult Significant fertility requirements Susceptible to low pH
Oats	Yields well under low fertility Premium milling markets Disease-resistant varieties Spring varieties can be undersown Tolerant of low pH	Weed control difficult Winter varieties affected by cold Late harvested
Beans	Nitrogen fixing Limited number of pests Winter and spring varieties exist	Poorly competitive with weeds Some difficult diseases, e.g. chocolate spot
Potatoes	Premiums for organic available Weed control easier than in many crops Good break crop in a rotation	Potato blight a major issue Requirement for significant breaks between crops

References

Adesegun, E.A., Adebayo, O.S., and Akintokun, A.K. (2012). Antifungal activity of spice extracts against Sclerotium rolfsii. *Acta Horticulturae* 931: 415–419.

Alteri, M.A. and Nicholls, C.J. (2012). Agro ecological diversification strategies to enhance biological pest regulation in horticultural systems. *Acta Horticulturae* 933: 35–41.

Arroyo, F.T., Castejon, M., Garcia-Galavis, P.A. et al. (2012). Keys to improve the Japanese plum organic farming in Mediterranean climate conditions. *Acta Horticulturae* 931: 469–475.

Atkinson D (1984) The effect of orchard soil management on crop yield, fruit size distribution colour and phosphorus nutrition. Proceedings of a Workshop on Pome Fruit Quality, Universitat Bonn, Germany, pp. 7–18.

Atkinson, D. (2009). Soil microbial resources and agricultural policies. In: *Mycorrhizas, Functional Processes and Ecological Impact* (ed. C. Azcon-Aguilar, J.M. Barea, S. Gianinazzi and V. Gianinazzi-Pearson), 1–16. Berlin: Springer.

Atkinson, D. and White, G.C. (1981). The effects of weeds and weed control on temperate fruit orchards and their environment. In: *Pests Pathogens and Vegetation* (ed. J.M. Thresh), 415–428. Boston: Pitman.

Atkinson, D., Burnett, F., Foster, G. et al. (2007). Pesticide use and pesticide residues. In: *Pesticides in Food* (ed. A. Chakraborty), 158–200. Hyderabad: Icfai Press.

Atkinson, D., Harvey, W., Leech, C. et al. (2012). Food security: the approach of the Scottish churches. *Rural Theology* 10: 27–42.

Azcon-Aguilar, C., Jaizme-Vega, M.C., and Calvet, C. (2002). The contribution of arbuscular mycorrhizal fungi to the control of soil borne plant pathogens. In: *Mycorrhizal Technology in Agriculture* (ed. S. Gianinazzi, H. Schuepp, J.M. Barea and K. Haselwandter), 187–197. Basel: Birkhauser.

Baker, B.P., Benbrook, C.M., Groth, E., and Benbrook, K.L. (2002). Pesticide residues in conventional, integrated pest management (IPM) grown and organic foods: insights from three US data sets. *Food Additives and Contaminants* 19: 427–446.

Balfour, E. (1943). *The Living Soil*. New York: Universe Books.

Barker, A.V. (2010). *Science and Technology of Organic Farming*. Boca Raton: CRC Press.

Boydston, R.A., Morra, M.J., Borek, V. et al. (2011). Onion and weed response to mustard (Sinapsis alba) seed meal. *Weed Science* 59: 546–552.

Boysen Jensen, P. (1929). Studier over Skovtracerres Forhold til Lyset. *Dansk Skovforen Tidssk* 14: 5–31.

Brown, G. and Jennings, D. (2012). The effect of calcium hydroxide pre-plant soil treatment on apple replant disease. *Acta Horticulturae* 931: 505–512.

Bruce, D. and Bruce, A. (1998). *Engineering Genesis: The Ethics of Genetic Engineering*. London: Earthscan.

Burke, M. (2016). Tropical organic farming. *Chemistry and Industry* 7: 11.

Casagrande, M., David, C., Valantin-Morison, M. et al. (2009). Factors limiting the grain protein content of organic winter wheat in south eastern France: a mixed model approach. *Agronomy for Sustainable Development* 29: 565–574.

Castejon, M., Arroyo, F.T., Garcia-Galavis, P.A. et al. (2012). Susceptibility of Japanese plum cultivars to Tranzschelia pruni-spinosae under organic and conventional management in southern Spain. *Acta Horticulturae* 931: 463–467.

Eckhardt, F.E. (1977). Physiological behaviour in relation to the environment. A comparison between a crop and various types of natural vegetation. In: *Environmental Effects on Crop Physiology* (ed. J.J. Landsberg and C.V. Cutting), 157–171. London: Academic Press.

Fernandez-Aparicio, M., Shtaya, M.J.Y., Emeran, A.A. et al. (2011). Effects of crop mixtures on chocolate spot development on Faba Bean grown in Mediterranean climates. *Crop Protection* 30: 1015–1023.

Firbank, L.G. (1991). Interactions between weeds and crops. In: *The Ecology of Temperate Cereal Fields* (ed. L.G. Firbank, N. Carter, J.F. Darbyshire and G.R. Potts), 209–232. London: Blackwell.

Fryer, J.D. (1981). Weed control practices and changing weed problems. In: *Pests Pathogens and Vegetation* (ed. J.M. Thresh), 403–414. Boston: Pitman.

Ganga Visalakshy, P.N. and Krishnamoorthy, A. (2012). Comparative field efficacy of various entomopathic fungi against Thrips tabaci: prospects for organic production in India. *Acta Horticulturae* 931: 433–437.

Gollotte, A., Brechenmacher, L., Weidmann, S. et al. (2002). Plant genes involved in arbuscular mycorrhizal formation and functioning. In: *Mycorrhizal Technology in Agriculture* (ed. S. Gianinazzi, H. Schuepp, J.M. Barea and K. Haselwandter), 87–102. Basel: Birkhauser.

Gragera-Facundo, J., Daza-Delgardo, C., Gill-Torralvo, C.G. et al. (2012). Comparing the yield of three pepper cultivars in two growing systems, organic and conventional in Extramaduras (Spain). *Acta Horticulturae* 933: 131–135.

Grevsen, K. (2012). Biofumigation with Brassica juncea pellets and leek material in carrot crop rotations. *Acta Horticulturae* 931: 427–431.

Grime, P. (1979). *Plant Strategies and Vegetation Processes*. Chichester: Wiley.

Heimler, D., Vignolini, P., Arfaioli, P. et al. (2012). Conventional, organic and biodynamic farming: differences in polyphenol content and anti oxidant activity of Batavia lettuce. *Journal of the Science of Food and Agriculture* 92: 551–556.

Hooker, J.E., Jaizme-Voga, M., and Atkinson, D. (1994). Bio control of plant pathogens using arbuscular mycorrhizal fungi. In: *Impact of Arbuscular Mycorrhizas on Sustainable Agriculture and Natural Ecosystems* (ed. S. Gianinazzi and H. Schuepp), 191–200. Basel: Birkhauser.

Howard, A. (1945). *Farming and Gardening for Health or Disease*. London: Faber.

Kalinova, J. and Vrchotova, N. (2011). The influence of organic and conventional crop management, variety and year on the yield and flavonoid level in common buckwheat groats. *Food Chemistry* 127: 602–608.

Koller, M. and Vieweger, A. (2012). Weed management in organic onion production: optimising cultivation technique and mechanical weed control to reduce hand labour. *Acta Horticulturae* 931: 391–397.

Kutinkova, H., Dzhuvinov, V., Samirtz, J., and Casagrande, E. (2012). Mating disruption of codling moth cydia pomonella L by applications of the microencapsulated formulation checkmate CM-F in Bulgaria. *Acta Horticulturae* 931: 485–490.

Leithold, G. and Becker, K. (2011). The cultivation methods wide row spacing for winter wheat in organic farming: its possibilities to improve the baking quality in nitrogen limiting conditions. *Archives of Agronomy and Soil Science* 57: 455–475.

Mays, N., Brye, K.R., Rom, C.R. et al. (2014). Groundcover management and nutrient source effects on soil carbon and nitrogen sequestration in an organically managed apple orchard in the Ozark highlands. *Hortscience* 49 (5): 637–644.

Murphy, K.M., Hoagland, L.A., Yan, L. et al. (2011). Genotype x environment interactions for mineral concentrations in grain of organically grown spring wheat. *Agronomy Journal* 103: 1734–1741.

Norman, J.R., Atkinson, D., and Hooker, J.E. (1996). Arbuscular mycorrhizal fungal induced alteration to root architecture in strawberry and induced resistance to the root pathogen Phytophthora fragariae. *Plant and Soil* 185: 191–198.

Osman, A.M., Struik, P.C., and van Bueren, E.T. (2011). Perspectives to breed for improved baking quality wheat varieties adapted to organic growing conditions. *Journal of the Science of Food and Agriculture* 92: 207–215.

Pozo, M.J., Slezack-Deschaumes, S., Dumas-Gaudot, E. et al. (2002). Plant defence responses induced by arbuscular mycorrhizal fungi. In: *Mycorrhizal Technology in Agriculture* (ed. S. Gianinazzi, H. Schuepp, J.M. Barea and K. Haselwandter), 103–112. Basel: Birkhauser.

Pozo, M.J., Verhage, A., Garcia-Andrade, J. et al. (2009). Priming plant defences against pathogens by arbuscular mycorrhizal fungi. In: *Mycorrhizas: Functional Processes and Ecological Impact* (ed. C. Azcon-Aguilar, J.M. Barea, S. Gianinazzi and V. Gianinazzi-Pearson), 123–135. Berlin: Springer.

Prange, R.K. (2012). Pre-harvest, harvest and post-harvest strategies for organic production of fruits and vegetables. *Acta Horticulturae* 933: 43–50.

Sanders, H.G. (1944). *Rotations*, Bulletin no 85. London: HMSO.

Simon, S., Morel, K., Defrance, H., and Hemptinne, J.L. (2012). Development of the rosy apple aphid within its habitat: some structural and physiological aspects in apple trees. *Acta Horticulturae* 931: 491–498.

Solomakhin, A.A., Trunov, Y.V., Blanke, M., and Noga, G. (2012). Organic mulch in apple tree rows as an alternative to herbicide and to improve fruit quality. *Acta Horticulturae* 931: 513–521.

Stockdale, E.A. and Watson, C.A. (2011). Can we make crop rotations fit for a multi-functional future? *Aspects of Applied Biology* 113: 119–126.

Villegas-Fernandez, A.M., Sillero, J.C., Emeran, A.A. et al. (2011). Multiple-disease resistance in Vicia faba: multi-environment field testing for identification of combined resistance to rust and chocolate spot. *Field Crops Research* 124: 59–65.

Wei, T.T., Cheng, Z.H., Ma, Q., and Han, L. (2012). The inhibitive effects of garlic bulb crude extract on Fulvia fulva of tomato. *Acta Horticulturae* 931: 407–414.

Wheeler, B.E.J. (1981). The ecology of plant parasitic fungi. In: *Pests, Pathogens and Vegetation* (ed. J.M. Thresh), 131–142. Boston: Pitman.

Whipps, J. (2004). Prospects and limitations for mycorrhizas in Bio control of root pathogens. *Canadian Journal of Botany* 82: 1198–1227.

Zaman, A., Sarkar, A., Sarkar, S., and Devi, W.P. (2011). Effect of organic and inorganic sources of nutrients on productivity, specific gravity and processing quality of potato (Solanum tuberosum). *Indian Journal of Agricultural Sciences* 81: 1137–1142.

13

Plant Breeding and Genetics in Organic Agriculture

Thomas F. Döring[1] and Martin S. Wolfe[2]

[1] *Department of Agroecology and Organic Farming, Institute of Crop Science and Resource Conservation, University of Bonn, Bonn, Germany*
[2] *The Organic Research Centre, Wakelyns Agroforestry, Eye, UK*

13.1   Introduction

In an issue of a British farmers' newspaper from early August 2012, a cartoon depicted two people standing in front of a flooded field, with rain coming down in torrents. One of them says: 'It's a good idea – providing they can find anywhere dry enough to test it!' The other one is holding a copy of a report entitled 'Drought-proof crop research' (Wignall 2012). It is difficult to imagine a more succinct way of capturing the dilemma of modern plant breeding. Only a few years earlier, climate change research had consistently predicted droughts in the UK under climate change scenarios (Richter and Semenov 2005; Semenov 2007). In response to such predictions, major research funds had been set aside to develop drought-resistant cereal varieties. Yet 2012 saw the wettest April–June in England since records began in 1910. In July 2012, rainfall was also above the long-term average, with floods occurring in several areas.

However, the contrast between predicted trend and single weather events reflected a further result of climate change, namely the greater variability of the weather and increased frequency of extreme weather events (Easterling et al. 2000; Rosenzweig et al. 2001; Arnell 2003; Schär et al. 2004). It is now clear that increased weather variability can have dramatic consequences for crop yields (Urban et al. 2012). So how should a crop breeding strategy respond to this situation? Should it aim for crop varieties that are better adapted to drier or to wetter conditions? Evidently the answer is that, as a safeguard against the inherent uncertainty about future climate and weather, *both* directions need to be pursued simultaneously. Rather than selecting drought-tolerant genotypes and discarding the rest, both functions need to be kept in the active pool of genotypes. In other words, this functional diversity can buffer against environmental disturbances in both directions (Reusch et al. 2005).

A similar question occurs when aiming for better plant protection. As with climate, plant breeding could optimise the response of the plant to a *specific* stress, such as a specific virus strain or race of pathogenic fungus, or it could aim to buffer against variable, evolving stresses

The Science Beneath Organic Production, First Edition. Edited by David Atkinson and Christine A. Watson.
© 2020 John Wiley & Sons Ltd. Published 2020 by John Wiley & Sons Ltd.

and ultimately the unpredictable emergence of different novel forms or races of different pathogens or pests. Because of the ability of pathogens to evolve rapidly and overcome specific resistance mechanisms, it is evident that the latter approach will result in greater durability (McDonald and Linde 2002; Parlevliet 2002; McDonald 2010; Döring et al. 2012b).

Indeed, as crops face multiple stresses, both biotic and abiotic, the variability of climatic conditions and plant diseases are only two of many issues in plant breeding. Good nutrient use efficiency (DoVale et al. 2012; Parentoni et al. 2012; Blair 2013; van Bueren et al. 2014), tolerance to nutritional stress (Fritsche-Neto and DoVale 2012) and high competitiveness against weeds (Hoad et al. 2012; Worthington and Reberg-Horton 2013) are all on the wishlist of plant breeders and farmers alike. Yet, even if plant breeders are successful in creating plant genotypes for each of these problem areas separately, it is unlikely that a single genotype would be able to deal well with all stresses simultaneously. Traits, and thus genes, that are ideal for coping with one stress may have a negative effect on the ability to deal with another. Another important reason is the outcome of trade-offs for resource allocation to different functions in the plant (Denison 2012). Such questions are particularly challenging when it is unknown – as in most cases – which specific stress factors will predominate in a given year and location.

For dealing with biotic and abiotic stresses, non-organic agriculture often relies on non-genetic measures, such as using herbicides against weeds, pesticides against pests and pathogens, and mineral fertilisers for plant nutrition. Under organic management, these measures are banned or restricted (Lampkin 1994; Lotter 2003; European Commission 2007). Therefore approaches based on crop genetics and diversity are of great importance in the attempt to achieve high and stable yields under the specific conditions found in organic agriculture (Lammerts van Bueren et al. 2002, 2010). Meanwhile, with increased costs of nitrogen fertiliser and constraints expected for other major agricultural inputs such as phosphorus (Cordell et al. 2009), fuel and pesticides, it is likely that non-organic agriculture will also pay more attention to crop genetics (Alfred et al. 2014).

In this chapter we explore the potential as well as the limitations of plant genetics and plant breeding for organic crop production, highlighting the importance of plant diversity at various levels of biological organisation. Following and expanding an approach recently brought forward in a chapter from the textbook *Organic Crop Breeding* (Wilbois et al. 2012), we use as a structural framework the four principles of organic agriculture. These are health, ecology, fairness and care as published by the International Federation of Organic Agriculture Movements (IFOAM 2005; Luttikholt 2007). We show how plant genetics relates to each of these principles and how plant breeding can contribute to their attainment through the use of greater functional diversity in plants (Figure 13.1). To do this, we first look at the various biological levels at which diversity can be observed.

13.2 Plant Diversity in Agro-Ecosystems

Most generally, diversity refers to the degree of variation in a given system. In agro-ecosystems, it is useful to distinguish between planned and associated diversity, referring to elements deliberately introduced versus elements occurring spontaneously (Costanzo and Bàrberi 2014). In this chapter we concentrate on planned diversity of crops.

Biological diversity (or biodiversity) generally comprises diversity at the ecosystem, species and genetic level (Hawksworth 1995). We will deal first with genetic diversity, as it has the most

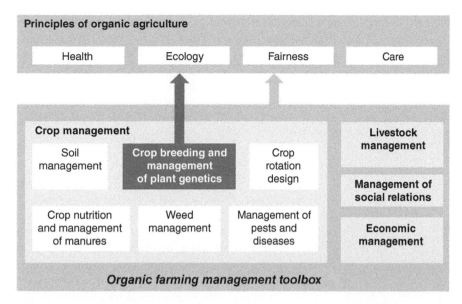

Figure 13.1 Heuristic framework of relationship between organic crop breeding and the principles of organic farming. The dark grey box and arrow show the main topic covered by this chapter, i.e. the ways in which breeding and genotype selection can contribute to the principles of health, ecology, fairness and care. However, crop breeding and management of plant genetics are also intricately linked with other components of crop management, and are also interacting with livestock, social and economic issues at the farm level.

obvious relevance for plant breeding, although diversity at the higher levels is also of critical importance, particularly in the context of organic agriculture.

13.2.1 Genetic Diversity

Genetic diversity can be defined as the variation of genetic characteristics present within a biological species. Let us consider the following example to explore some ways to describe genetic diversity. On the sandy soils of Brandenburg, as in many other regions of Germany, crops are frequently damaged by wild boars (*Sus scrofa* L.) that find shelter in the surrounding woods. Some farmers have resorted to a simple but apparently effective trick to protect their wheat crops from boar damage: they use awned wheat varieties that the animals appear to find unattractive, probably because the plants are just too irritating. The majority of wheat varieties in Europe are awnless but many have fully developed awns, such as the variety 'Goldblumenweizen' which was developed for sandy soils under organic management (Heyden 2004).

Awnedness in cereals (Figure 13.2) has been intensively studied because it can also contribute to increased grain yield, especially under drought conditions (Grundbacher 1963). Genetically, the presence of awns in wheat is controlled, largely, in a relatively simple way, namely by three genes. Each of these genes has two variants, or alleles. The dominant alleles, named *Hd*, *B1* and *B2*, are awn inhibitors, while the three corresponding recessive alleles, *hd*, *b1* and *b2*, do not inhibit awn formation. Wheat plants that have two copies of each of the three recessive alleles (homozygous genotypes) are fully awned. Awnlessness results from the allele combinations that contain either *Hd B2* or *B1 B2* (Sourdille et al. 2002). Now consider a large group of plants,

Figure 13.2 Phenotypic diversity with respect to awnedness in two plants from a composite cross-population of winter wheat. *Source:* Image courtesy of Thomas Döring.

such as all wheat plants in a farmer's field. When more than one alternative allele is found amongst the plants in such a population, this population is said to be polymorphic with respect to this character.

In general, the components of diversity can be summarised as (i) the number of elements (such as alleles); (ii) the evenness of the distribution of these elements; and (iii) their relatedness, whereby diversity increases as relatedness decreases. A simple way to describe genetic diversity with respect to awnedness is to count the awned and awnless wheat varieties in a given area. Here, diversity increases with the overall number of varieties, but if all of them are awnless, the observed diversity with respect to this trait would be zero. For a given number of varieties, diversity would be maximal if the proportion of awned and awnless varieties was 50% each, that is, deviations from an even distribution reduce diversity. However, even if all varieties in the area are awnless, there might still be genetic diversity of awnedness genes amongst them. This is because the phenotype of awnlessness can be generated by a number of different allele combinations, for example *Hd b1 B2* or *Hd B1 B2* or *hd B1 B2*.

At the allele level, there are several measurements of genetic diversity such as the percentage of polymorphic loci, the mean number of alleles per polymorphic locus, the expected proportion of heterozygous loci per individual, Nei's G_{ST} or genetic distance (Berg and Hamrick 1997). Further metrics of genetic diversity are given by Hughes et al. (2008) and some statistical tools to describe and compare genetic diversity are summarised by Mohammadi and Prasanna (2003). Similar to the example above, where diversity is highest when proportions of awned and awnless varieties are evenly distributed, genetic diversity indices referring to alleles use evenness in the distribution of alleles. Diversity decreases when the distribution of alleles within a population or species becomes more skewed, such as when one allele dominates a population (van de Wouw et al. 2010).

It is also important to note that the expression of any gene can be influenced to a greater or lesser extent by the environment, either internally or externally. In this context, the internal environment refers to the remainder of the genotype apart from the gene in question. The expression of specific genes may be influenced in different ways by the presence or absence of different alleles of many other genes, depending on the gene function and the sensitivity of observation or analysis. Gene expression can also be affected by, and interact with, the external physical or biotic environment.

Through selection, the plant breeder attempts to ensure maximal expression of the desired genes, although this will be limited by the range of genotypes available and the sites and seasons used for selection. Under non-organic breeding, there is usually a fixed rotation with specified fertility and the application of herbicides to try to reduce or limit environmental variation; organic breeding, without such inputs, progresses under more variable conditions. Under centralised, or conventional, breeding, selections are made repeatedly at only a few sites, which further limits environmental variation. However, there has been increasing interest in decentralised breeding (Ceccarelli et al. 2001) (see also below), in which selections are made, as far as possible, at the sites where the emerging varieties will be used. We examine some of the consequences of environmental interactions (G × E, G × S, etc.) in more detail below.

Genetic diversity is usually estimated at three levels. First, there is the total genetic diversity within a species. Different species may then partition their overall diversity in different ways, either more amongst populations (the second level) or within populations (the third level) (Berg and Hamrick 1997). Amongst crops, a large proportion of genetic diversity is present in traditional varieties, often called landraces (Carolina et al. 2005).

Genetic diversity also extends above the species level. One reason is that in many cases there are no hard boundaries between plant species. Crossing is often possible between different but closely related species. Thus, in searching for new sources of genetic variation, crop breeding can often turn to the wild relative species of crops to attempt introgression of desirable traits into the crop (Ross 1966; Placido et al. 2013). Although this approach is often dependent on techniques that are not compliant with the principles of organic plant breeding (Wilbois et al. 2012), it is useful to expand the consideration of genetic diversity beyond intraspecific diversity. One approach that has been tried in the context of organic systems is the recreation of hexaploid wheat by crossing tetraploid durum wheat (*Triticum durum* or *Triticum turgidum* subsp. *durum*) with derivatives of goat grass (*Aegilops tauschii*) (van Ginkel and Ogbonnaya, 2007); a similar crossing scheme involved durum and emmer (*T. turgidum* subsp. *dicoccon*) with einkorn (*Triticum monococcum*) (Megyeri et al. 2011).

13.2.2 Species and Ecosystem Diversity

In addition to genetic diversity, biodiversity at the species level (Peet 1974; Jurasinski et al. 2009) and at the ecosystem level (McNeely et al. 1990; Lapin and Barnes 1995) is of central importance for organic agriculture in general (van Elsen 2000; Bengtsson et al. 2005; Fuller et al. 2005; Hole et al. 2005; Crowder et al. 2010) and organic plant breeding in particular. A measure often used to assess diversity at the species level is species richness, which is the number of species present in a defined unit of study, although there are several other methods for measuring species diversity (Magurran 2004; Tuomisto 2010). Generally, there are three ways in which species diversity is linked to organic plant breeding.

1) Plant breeding and variety selection for organic systems aims at conditions that are characterised by higher species richness than in conventional systems. For example, crops in organic systems often interact with a larger diversity of weeds (Gabriel et al. 2010) and, depending on management, may also face a more biodiverse soil in the target environment.
2) Selection can take place for organic systems with increased crop species diversity, such as co-breeding for intercropping (Döring et al. 2011). Here, crop genotypes (e.g. of wheat) are

selected on the basis of their compatibility with a companion crop (e.g. a grain legume such as faba bean, *Vicia faba*).

3) There is an economic trade-off between diversity *amongst* crop species versus genetic diversity *within* species. Breeding efforts are diluted if they are spread over many species, but concentrating breeding investment in only few species inevitably neglects diversity at species level. With limited funds, it is therefore difficult to increase investments in both areas simultaneously. This dilemma highlights the urgent need to increase overall research and implementation of organic crop breeding, and the importance of revising the organisation of the entire plant breeding sector towards increasing efforts on currently neglected crop species, such as minor crops and orphan crops. For many plant species, there is also a vicious circle of low investment into breeding and seed production, low usage by farmers and low financial return for breeders and seed producers (Döring et al. 2012). This situation leads to a strong bias against the use of neglected but potentially valuable plant species in agriculture.

The importance of crop species diversity is illustrated by the example of cotton in India. In 1947, more than 90% of the cotton area in India was planted with traditional Indian Desi cotton. This term refers to two species, *Gossypium arboreum* and *Gossypium herbaceum*, which have good resistance to drought and other stresses. With the intensification of cotton production, however, the area grown with Desi cotton shrank dramatically, as it was being replaced by *Gossypium hirsutum* hybrids. By 2012, the proportion of *Gossypium arboreum* and *G. herbaceum* had each shrunk to well below 5% (Bharud 2014). At the same time, commercial breeding efforts in *G. hirsutum* hybrids were dedicated mainly to genetically modified (GM) insect-resistant *Bt*-cotton. As a consequence, the non-GM cotton seed market was eroded, with the loss of locally adapted Desi cotton varieties (Messmer et al. 2014). Participatory plant breeding projects have therefore been set up to re-establish the supply of GM-free seed and to support organic and low-input cotton farmers.

Unfortunately, there is currently no evidence-based strategy in the (organic) plant breeding sector that would inform the allocation of resources on within-species diversity versus amongst-species diversity. This key question of optimal partitioning of diversity should be decided by studying the degree to which diversity amongst species versus diversity within species can deal best with the challenges and stresses of varying unpredictability (Fridley and Grime 2010). In Section 13.3, we will explore the relationship between crop genetics and such challenges and stresses in detail.

Ecosystem diversity plays a role for organic agriculture in terms of crop rotations, the presence and quality of semi-natural habitats on the farm, and in agroforestry systems. However, so far, relationships of ecosystem diversity with crop breeding have not been extensively studied.

13.2.3 Effects of Crop Diversity: Types of Mechanisms

With respect to the ways in which plant diversity interacts with the services provided by agro-ecosystems, there are four broad mechanisms: capacity, complementation, compensation and co-operation (Döring et al. 2011; Creissen et al. 2013).

The first type of mechanism, *capacity*, is more or less equivalent to the category of functional identity as discussed by Costanzo and Bàrberi (2014) and does not necessarily involve any mixing (i.e. co-presence) of different genotypes or species in the field. Capacity refers to the genetic potential to deal with some issue; it relates to the presence of traits entailing the provision of given ecosystem services. For example, a single variety may have none, one or many known genes for resistance to common bunt (*Tilletia caries* syn. *Tilletia tritici*). Such variation can only be expressed amongst varieties rather than within a variety. Plant diversity affects cultivar choice indirectly through the availability of different capacities, such as whether cultivars with the desired traits actually exist. In the case of bunt resistance, the diversity of cultivars from which organic farmers can choose is currently extremely narrow, because conventional wheat breeding has neglected resistance to common bunt for decades (Matanguihan et al. 2011). Organic wheat breeding therefore aims to increase the diversity of cultivars from which farmers can choose varieties in order to provide the relevant capacity, that is, resistance to this disease (Borgen 2014).

The second type of mechanism, *complementation*, broadly equates to the concept of functional composition as described by Costanzo and Bàrberi (2014) and refers to the complementary effect of different plant traits. Here, different elements (e.g. genotypes or species) co-occur in the field and each element in the mixture can contribute to the same or a different set of agro-ecosystem services. For complementation to take effect, the elements that are co-present in the mixture do not necessarily interact with each other; that is, their behaviour is not necessarily affected by the presence of the other partners in the mix. An example is the use of two different functional groups, namely deep-rooting and shallow-rooting species in a cover crop mixture. By mixing these two functionally different species, resource use and productivity are increased through the exploitation of different spatial niches, thereby reducing niche overlap. Another well-known example is the mixture of N_2-fixing legumes and non-legumes (Tilman et al. 2001). As both elements have complementary functions, overall service provision of the mixture is higher than when the component elements are grown as monocultures.

The third category, *compensation*, refers to the situation in which one element is able to compensate for the failure of another one. For example, consider a pathosystem of wheat and yellow rust. When a mixture of two cultivars (A and B) is grown in which each cultivar is resistant to a different set of pathogen races (Finckh and Mundt 1992), the presence of one race of the rust fungus may reduce the growth of one of the two cultivars (A), but the other (B) is then able to compensate through increased growth facilitated by the reduced competition from A. Importantly, this effect can also be expected for the opposite situation, when B is infected and A is able to compensate for B's reduced growth.

Finally, the mechanism of *co-operation*, often also called facilitation, involves the direct benefiting of one partner from the presence of the other one. Examples include the nitrogen transfer from a legume partner to a non-legume partner in a species mixture and the induction of acquired resistance to pathogens through contact with avirulent pathogens. Co-operation is usually thought of as mutualism, where all partners benefit, although commensalism, where only one, major, partner benefits, may also be valuable.

We now turn to the question of how crop diversity behaves in complex and dynamic environments.

13.3 Crop Genetics in Complex and Dynamic Environments

13.3.1 The Organic Principle of Ecology

The IFOAM principle of ecology (IFOAM 2005) states that organic farming should 'be based on living ecological systems and cycles, work with them, emulate them and help sustain them'. This involves four more specific objectives: (i) to base production on ecological processes and recycling; (ii) to operate in a site-specific manner, that is, to adapt management to local conditions, ecology, culture and scale; (iii) to reduce inputs; and (iv) to maintain genetic and agricultural diversity. All four areas intersect with plant breeding and plant genetics. Here we concentrate on the second objective, local adaptation, which in plant breeding research is explored under the heading of genotype × environment (G × E) interactions.

13.3.2 The Ecology of G × E Interactions

In 2010, Zeller and co-workers published a study on experiments with transgenic spring wheat that had been transformed to improve resistance to powdery mildew (Zeller et al. 2010). Four GM lines and their four corresponding non-GM lines were grown under different environmental conditions in the glasshouse, by varying soil nutrient conditions and fungicide treatment. Similar experiments were performed in the field. However, the results were far from what had been expected. Although only one gene had reportedly been inserted using genetic transformation techniques, the differences between GM and non-GM lines were dramatically dependent on the environmental conditions under which the lines were grown. In the greenhouse yield performance without fungicidal treatment was doubled in the GM lines compared with control lines. In the field, as the authors candidly state, 'these results were reversed'. In fact, half of the GM lines showed up to 56% yield reduction compared with their control lines in the field experiment. In addition, when grown in the field, the GM lines were significantly more severely infected by a different disease, namely ergot (*Claviceps purpurea*), than the control lines.

Essentially, the study provides a striking example of significant G × E interactions, that is, how different genotypes can respond in contrasting ways when subjected to different environments (Annicchiarico 2002). If there were no interactions, the performance of any tested genotype (G) in any tested environment (E) could be calculated by adding up the effect of each genotype (averaged over all environments) and the effect of each environment (averaged over all genotypes). In the case of G × E interactions, however, effects are non-additive, and the observed performance of at least one genotype in at least one environment deviates significantly from the sum of effects of G and E. Generally, significant G × E interactions will occur if generalist and specialist genotypes are tested together in different environments, or if genotypes with different specialist adaptations to environments are tested together.

The E part of G × E experiments can be deconstructed into various factors; these could be location, year, and different management measures such as application of organic fertilisers, tillage operations or seed density. In many cases, however, complex ecological interactions amongst environmental factors make it difficult to disentangle these factors any further when evaluating genotypes in different environments. Also, cropping management responds dynamically to environmental changes, for example as sowing depth and plant density are adapted to meet soil conditions found in a particular site and year. Therefore, it is more useful to employ a systemic view of plant breeding and variety evaluation by evaluating genotypes in clusters of agro-ecological conditions, but even if sensibly clustered agro-environments are identified, the large number of environments then poses a tremendous challenge for evaluation of genotypes.

In view of the complexity of environments and the ubiquity of G×E interactions, the call for thorough testing of genotypes in many different environments is faced with a (statistically based) trade-off: with a given experimental effort (e.g. a fixed number of plots), increasing the number of environments will lead to a reduction in either the number of genotypes that can be tested or the number of replicates. Reducing the number of genotypes curtails the diversity subjected to the test, while reducing the number of replications increases uncertainty of the information gained in each environment. In addition, both mean that the power for testing significance of G×E interactions is diminished.

This dilemma is particularly problematic for organic systems, which exhibit more environmental variation (e.g. amongst farms) than is observed for conventional systems. It is further complicated when higher-level interactions are involved (e.g. sowing density × tillage × genotype). In the case of higher level interactions, it is useful to employ advanced statistical tools such as Additive Main effects and Multiplicative Interaction (AMMI) which are designed to deal with this complexity (Sadeghi et al. 2011). Part of the problem lies in the fact that traditional statistics rely on arbitrary but inflexible significance thresholds (Garamszegi et al. 2009). But even with novel and more flexible statistical tools, the fundamental trade-off between broad picture (across many environments) and fine resolution (accurate information per environment) will remain. A structural solution lies in the decentralisation of breeding and testing, as often realised in evolutionary (Murphy et al. 2005; Döring et al. 2011) and participatory plant breeding projects (Ceccarelli et al. 2001; Dawson et al. 2008; Messmer et al. 2014).

13.3.3 Implications of G×E Interactions for Testing Varieties for Organic Agriculture

G×E interactions play a central role for plant genetics in organic agriculture. Specifically, there are two inter-related questions concerning plant varieties that are targeted at organic systems. First, should varieties be evaluated under organic conditions? And second, should plant breeding take place under organic conditions? With regard to the first question, if the environmental factor 'management system' S (organic versus non-organic) significantly interacts with the factor 'genotype', that is, if there is a significant G×S interaction, it is necessary to test varieties under organic conditions. Otherwise, variety test results obtained under conventional management would be sufficiently informative for organic farmers.

Over the last few decades, some evidence has been found for significant G×S interactions, for different behaviours of crop varieties in conventional versus organic systems, including agronomic performance of wheat (Murphy et al. 2007), dry beans (Heilig and Kelly 2012) and lentils (Vlachostergios and Roupakias 2008), as well as chlorophyll content of maize leaves (Goldstein et al. 2012) and tuber quality in potatoes (Maggio et al. 2008). For maize grain yield, it was found that genotypic correlation between organic and non-organic trials was only 0.54, suggesting that achievable selection gains can only be realised with organic testing (Goldstein et al. 2012).

However, in other cases no significant G × S interactions were observed, for example in wheat (Hildermann et al. 2009) and oats (Saastamoinen et al. 2008). In a study on spring wheat, G × S interactions did occur but 'the rank of the highest yielding cultivar was unchanged across the 14 paired comparisons' (Carr et al. 2006). Similarly, in the study on dry beans, genotypes with poor performance under organic management also performed poorly when grown conventionally (Heilig and Kelly 2012). An Italian study on soft wheat varieties concluded that correlations between the systems were high and that genotype × location interactions were much more important than G × S interactions (Annicchiarico et al. 2010). The same conclusion was reached in a similar study on lucerne (*Medicago sativa*) (Annicchiarico and Pecetti 2010). Recently, Wortman et al. (2013) evaluated varieties of maize, soybean and wheat under organic and conventional management and did not find consistent G × S interactions.

Underlying all of these observations, however, is the question of the form of management used in the systems compared. For example, 'extreme' organic compared with 'extreme' conventional is much more likely to generate G × S interactions than a comparison involving marginal organic and conventional systems. The balance of evidence for or against will thus depend, first, on the directions taken in the future development of the two systems (will they tend to merge or separate?), and second, on the impacts of external factors such as the price of oil and the severity of climate instability.

An important argument in the discussion on the necessity of organic variety trials is that normally conventional variety trials are run 'anyway' and that it would not be a realistic decision-making strategy for an organic farmer to completely ignore the results from conventional trials (Przystalski et al. 2008). The question then becomes a slightly different one: is the *additional* information gained through running organic variety trials along with the conventional trials worth the effort? According to this approach, the maximum yield difference observed for various cereal species in Europe was 0.63 dt/ha (Przystalski et al. 2008), so this would be the maximal yield loss if the organic farmer's decisions was based on conventional trials results only. Again, this figure is dependent on the background of the trials analysed; it is also subject to the question of future changes in management systems and the external environment.

In summary, however, the results from G × S studies for variety testing remain inconclusive. There are several potential reasons for this. First, one factor that might generate significant G × S interactions, differences in nutrient use efficiency (NUE) amongst varieties, might not vary sufficiently amongst tested varieties, simply because NUE has not been the focus of mainstream plant breeding in the past. Two further reasons refer to the effect of plant diseases, which is another factor that could potentially lead to G × S interactions. Theoretically, yield responses of varieties to pathogens can be expected to differ when they are either conventionally treated or organically managed. In the trials, however, there might not be strong differential response of varieties to systems, because experimental plots are usually small so that the most susceptible varieties affect the less susceptible varieties, masking their potential advantages because of the relatively high and continuous spore load that they generate. Also, the plant disease spectrum usually varies between years. So while specific resistance interactions with pathogens can lead to significant G × S interactions in any single year, these might be lost in the average of several years. However, what is relevant is the *risk* of choosing the wrong (i.e. susceptible) variety. This risk is better captured by the variance than by the mean, but the evaluation of G × S interactions is usually based on the mean only.

Finally, while conventional systems stabilise environmental conditions for the crop by chemical inputs, comparatively similar growing conditions for the tested genotypes might be achieved through agronomic measures in organic systems, such as rotation design or organic matter management. Methods to optimise nutrient supply and to reduce pest and disease pressure may differ strongly between systems but in well-designed organic cropping systems, the resulting growing conditions may not be so different from conventional systems.

Apart from evaluating varieties under different management systems, a second question is whether *breeding* under organic conditions leads to better performance of varieties in organic systems. The breeding stage refers to the early selection phases, which usually leads to the removal of a large proportion of genotypes from further testing. Because the bulk of plant genetic variability is removed by the breeder during this stage, it can be argued that this phase may be an additional reason why only inconclusive G × S interactions are observed at later stages of testing, as genotypes with differential responses in organic and non-organic systems may have already been deselected from the testing procedure at the early stages. In other words, there is the potential that adaptation to organic systems can be observed only if a larger pool of genetic diversity is subjected to testing.

As for variety testing, however, the evidence for G × S interactions at the selection stage is relatively inconclusive. In a study on wheat cultivars in the long-term DOK trial in Switzerland, it was found that 'in the organic systems, the organically bred cultivars could not outperform the conventionally bred cultivars in grain yield and NUE parameters' (Hildermann et al. 2010). Similarly, for maize it was suggested that varieties for organic systems can be developed by screening conventional varieties for their performance under organic conditions (Lorenzana and Bernardo 2008), which was also indicated for spring wheat (Carr et al. 2006). On the other hand, a study on selecting lines from spring wheat populations in Canada showed that separate selection on organic land was necessary (Reid et al. 2009). Similarly, selecting maize genotypes for organic systems was found to be more efficient if it was carried out in the target environment (Rodrigues de Oliveira et al. 2011).

Thus, for both early and later stages of genotype evaluation, there is evidence for, as well as against, a differential response of genotypes to organic management. However, even if there were no G × S interactions at all, organic variety trials would fulfil an important function in providing confirmation and information independently from the conventional community. This can be seen as an important motivation in a relationship between the systems that is characterised by asymmetrical distribution of power and resources; especially when plant breeding is concentrated in few hands, it is difficult for organic farmers to influence which selection criteria are applied by breeders (Osman et al. 2008).

So how should an organic variety trialling programme be set up? Generally, the small number of small-scale replicated field trials typically used for evaluation of genotypes is not able to replicate the complexity of the environment on working farms (Klepatzki et al. 2014). For organic farming, this is partly a structural problem because organic agriculture, though growing, is still a niche sector, and economic constraints do not allow a strong expansion of replicated small-plot field trials. A potential solution could be the massive expansion of decentralised on-farm trialling. This could also contribute to the on-farm conservation of plant genetic resources, especially for minor crops, including the potential for crop improvement through local adaptation over time (Horneburg and Becker 2008).

Since geographic location alone is a poor predictor of ecological conditions, this could be run within the framework of 'virtual ecological villages', in which each 'village' represents a cluster of similar ecological and management conditions. To be useful, such an approach would require the ability to measure genotype performance as well as tight informational connectivity amongst farms with data to be made available about site conditions, allowing feedback of accumulated and condensed information within the network.

13.3.4   Genetic Properties of Crops for Suitability in Organic Systems

The importance of traits relevant for selection differs between organic and low-input management (Löschenberger et al. 2008). The traits currently targeted in organic plant breeding activities include weed suppression and competitiveness of the crop plant, resistance to pests and diseases, nitrogen use efficiency, end-use quality and nutritional value (Arterburn et al. 2012). More specifically, the traits linked to improved weed control include tall plant stature in cereals, a high ability to cover the ground quickly, and an ability to regenerate after mechanical measures of weed control.

One complex that has received particular attention in plant breeding is resource use efficiency (Messmer et al. 2012b). This can refer to various resources, including essential crop nutrients and water, and can be defined generally as the yield of grain (or harvested product) obtained per unit of the resource available to the crop. Plant breeding and variety selection, however, are only two of many strategies to increase resource use efficiency. A German study on organically grown wheat varieties concluded that breeding more nitrogen-efficient varieties alone will 'help only little' to increase nitrogen utilisation by the crop (Baresel et al. 2008). Also, because traits associated with efficient nitrogen use vary between extensive and more favourable conditions of nitrogen supply, selection for high nitrogen use efficiency is difficult to achieve.

Since currently available cultivars have been selected mainly under high-input conditions, it could be expected that older varieties might be better able to use available nitrogen. In fact, it was found that old wheat varieties were outyielded by modern ones even under conditions of zero mineral nitrogen input (Guarda et al. 2004). Through selection for increased grain yield, wheat breeding may have led indirectly to improved nitrogen uptake and use. Also in a different study, NUE increased with the year of release of cultivars (Hildermann et al. 2010). Even under conditions of limited nitrogen supply, the best choice appears to be modern cultivars rather than the old landraces if NUE is the main selection criterion.

An important aspect of resource use efficiency that goes beyond plant physiological and morphological traits is the role of micro-organisms in the rhizosphere for nutrient uptake by plants. A promising pathway in organic crop breeding is therefore selection for optimised interactions between plants and microbes, for example symbioses with nitrogen-fixing rhizobia or with arbuscular mycorrhizal fungi (AMF) that can increase phosphorus uptake. In the latter case, modern plant breeding has reduced the plant's dependence on mycorrhizal symbiosis (Hetrick et al. 1992; Zhu et al. 2001). Currently underexplored are genetically based variations in the crop's ability to form beneficial (or least detrimental) relationships with endophytes and free-living rhizobacteria. This is particularly important for perennial crops, another area largely neglected by plant breeding.

Apart from resource use efficiency, a further vital target for breeding and variety evaluation is high yield stability. This is associated with several traits including the compensatory ability of successive yield components (plasticity) and broad resistance to pests and pathogens. Yield stability is of particular importance for organic farming insofar as the variability of environmental conditions is greater on organic than on conventional farms. However, stability is difficult to breed for because of its low heritability (Link et al. 1994). Also, assessing yield stability requires datasets of high quality, where genotypes are tested over many environments. Again, decentralised networks of farmers collecting relevant (yield) data could be a low-cost solution to this challenge, as could the use of populations in cereals, which we discuss below.

13.3.5 Crop Genetics for Ecological Cropping Systems Design

In some cases, the selection of suitable crop genotypes may also change the design of organic cropping systems. Here we briefly discuss three cases with various degrees of system change: soybean (*Glycine max*), oil seed rape (canola, *Brassica napus*) and pea (*Pisum sativum*).

In soybean breeding for central European climates, a relatively simple suite of traits determines the suitability of this crop, including the time to maturity (Hüsing et al. 2011; Messmer et al. 2012a), which is closely linked to cold tolerance. In central and northern Europe, soybean can be integrated into existing organic rotations only when genotypes with sufficiently low temperature requirements for reaching maturity have been identified. If so, soybean simply replaces other spring-sown grain legumes such as pea or faba bean. With its often superior nutritional quality, soybean can then be used for improved supply of home-grown high-quality feed for organic livestock.

A counter-example is oil seed rape for organic systems. From a cropping system perspective, this is an excellent crop species, with multiple benefits in the rotation. However, there are several problems that currently block wider adoption of this crop in organic farming, including numerous insect pests (e.g. the pollen beetle *Meligethes aeneus*), foliar and root diseases, occasional weed problems, and, probably the most important reason, comparatively high nitrogen demand (Böhm et al. 2011, 2013). Breeding oil seed rape to achieve improvements in only one of these areas would itself be challenging, so the simultaneous introduction of multiple resistances plus better nitrogen use efficiency and improved competitiveness against weeds is (currently) an unrealistic target for (organic) oil seed rape breeding.

One case in which breeding and variety selection did achieve a system change, which has made it possible to design entirely new cropping systems, is the rediscovery and development of winter peas in Germany (Urbatzka 2010; Haase et al. 2013). In this region, the dominant form of grain legume cropping has been spring-sown forms of peas, faba beans, lentils or lupins to produce protein-rich feed and to provide nitrogen to the following crop. Prior to the introduction of winter pea genotypes with sufficient frost tolerance, winter peas had been grown mostly in regions with milder winters, such as France. After frost-tolerant winter pea genotypes had been identified it became possible to extend the growing season of grain legumes within the rotation, thereby increasing the amount of N_2 fixed (Urbatzka et al. 2011b) and reducing weed problems. In addition, the use of intercropping systems with grain legumes could now be extended to include winter-sown cereals (Urbatzka et al. 2012; Gronle and Böhm 2014) or winter oil seed rape (Urbatzka et al. 2011a). Also, optimised rotations for energy crops were developed, by using a winter pea cereal intercrop as a pre-crop for energy maize (Graß and Scheffer 2005; Graß et al. 2013). Rather than just replacing another crop species in the same rotation, winter peas therefore made it possible to completely rethink organic crop rotations in the region. As with winter pea, innovations in organic crop rotations are also being studied with winter faba beans (Roth 2010; Neuhoff and Range 2012) following efforts to increase winter hardiness of this species (Link et al. 2010).

With these examples in mind, it seems likely that one of the key plant traits with system-changing potential is the genetically determined requirement for time to maturity or, more generally, the typical length of the growing season for a particular genotype. As demonstrated above, extending (or shortening) the growing season of a crop can have far-reaching consequences for the design of the rotation. An extreme case is the effort to render annual crops perennial, as attempted for wheat and rye (Jaikumar et al. 2012). A true game-changer, perenniality affects every aspect of the cropping system, from tillage to disease control, from rotation planning to harvesting technology.

13.3.6   Limitations of Crop Genetics and the Role of Plant Genetic Diversity

What is the best way to deal with the large range of environmental conditions found on organic farms? Do we just need different varieties which are better adapted to each of these conditions in order to overcome the yield gap of organic crop production? Or will there always be a mismatch between genotype and environment? Based on the evidence presented above, we argue that although breeding and selection for organic agriculture definitely has a role to play in optimising organic crop performance, the potential of this genetic pathway might be limited.

One reason is that genetic effects within any one crop species may be small relative to higher level genetic effects, that is, differences amongst crop species. An example is the integration of agroforestry into organic farming in temperate systems (Smith et al. 2012). When designing a silvoarable system, with tree and arable components, species selection of the tree component (e.g. trees for fruit or nut production, for energy or wood production) will largely determine the structural, ecological and economic properties of the system. Variety selection within the chosen species will then be secondary, as it is used for fine-tuning the system.

Because of the often much larger differences between species than between genotypes within a species, it may be more promising to target crop production problems by a wider consideration for the choice of crop species. Promotion of organic crop breeding should therefore not be at the expense of the number of crop species available for organic agriculture. Further, improvement of crop genetics should not lead to neglect of agronomic solutions. Environmental conditions usually play a much greater role in determining yield (or other performance parameters) than genotype effects or $G \times E$ interactions. Therefore, ecological manipulation of the site conditions, through mulching, appropriate tillage, organic matter management, landscaping or selection of suitable sites, is likely to provide greater benefits than variety choice and breeding. Also, as demonstrated above, genetic variation in crop varieties and breeding material is difficult to assign to the right environments, if the environmental conditions are changing, fluctuating and increasingly losing predictability.

13.4   Crop Genetics for Health

13.4.1   The Organic Principle of Health

The principle of health plays a central role organic agriculture. It is often referred to as the most important of the four principles, possibly because the history of organic agriculture is strongly linked to the issue of health. In particular, Lady Eve Balfour, one of the founding figures of the organic agriculture movement in Europe, stated in her book *The Living Soil* (Balfour 1943) that the health of soil, plant, animal and man is 'one and indivisible'. Several decades of her life were dedicated to testing this fundamental hypothesis.

With a nod to Eve Balfour, the organic principle of health therefore states that organic agriculture should 'sustain and enhance the health of soil, plant, animal, human and planet as one and indivisible' (IFOAM 2005). The link amongst the different domains of health is further explained in the principle in that healthy crops depend on healthy soils, whereas the health of animals and people is in turn dependent on healthy food. More recently, it has been shown that links between soil health, plant health, animal health and human health are widespread but also much more complex in terms of mechanisms than originally hypothesised by Eve Balfour (Vieweger and Döring 2014a).

13.4.2   What is Health?

Health, according to the understanding expressed in the organic principle, is the:

> … wholeness and integrity of living systems. It is not simply the absence of illness, but the maintenance of physical, mental, social, and ecological well-being (IFOAM 2005).

This is of course not the only understanding of health. In fact, health is a complex, fuzzy, dynamic and controversial term. Over the last few decades, there has been intensive debate on different concepts of health, especially with regard to humans (Gimmler 2002; Nordenfelt 2006; Huber et al. 2012).

In the past, the debates on health in agricultural contexts have largely been separate in the domains of soil, plants, animals and humans, but they have recently been brought together (Vieweger and Döring 2012; Döring et al. 2014; Vieweger and Döring 2014a). This revealed that the concept of resilience can serve as a common denominator, as a universal criterion of health, applicable across all domains affected by agriculture (Döring et al. 2014). Although resilience is itself a fuzzy and dynamic term, its core concept is recognised by ecology, soil science, plant pathology, veterinary science and human medicine as the ability to bounce back after disturbance or, put differently, to recover functionality after a stress event. With a focus on resilience, the concept of health assumes a more dynamic perspective, and recent attempts to reform the understanding of human health amongst medical scientists and professionals (Huber et al. 2011) are reflected in it.

13.4.3   Connections Between Crop Genetics and Health

Crop genetics affect the health of crops, but also soil health and the health of farm animals and humans (Figure 13.3). We first discuss the most obvious case – the relationships between plant genetics and plant health. In the literature, the expression that the plant is healthy is often just shorthand to describe that the plant is free from a particular disease (Döring et al. 2012a). In this simple case, crop genetics and breeding already play a paramount role for health through genetically inherited resistance to pests and pathogens (Döring et al. 2012b). There are several examples of pathosystems for which resistance breeding is considered to be highly important in maintaining or improving crop productivity in organic agriculture (Table 13.1).

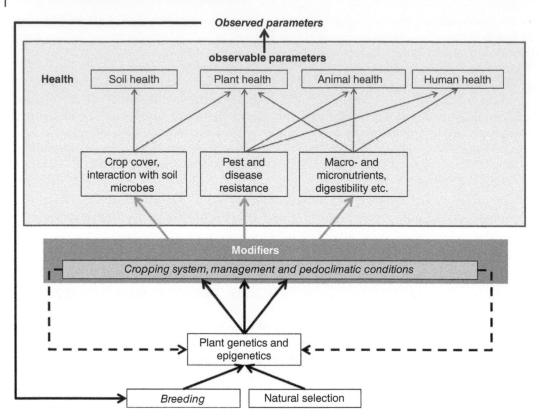

Figure 13.3 Crop breeding for health in organic systems, illustrating the concept of modifiers; these modifiers change the way in which plant genetics determine or influence observable parameters of health; human activities are shown in italics.

However, beyond individual plant diseases and genetically conferred specific resistance, there are other aspects to plant health, especially when it is seen as a wider, more complex concept (Döring et al. 2012a). An example illustrating an indirect link between crop genetics and plant health is the genetically variable suitability to be grown in a mixture or as a bi-crop (Lammerts van Bueren et al. 2002). Here, timing of maturity, plant morphological traits and the ability to compete for light, nutrients and water with the partner crop in a balanced way are the primary selection criteria. In a pea-barley bi-crop for example, semi-leafless pea genotypes that are not entirely outcompeted by the barley would be preferred over leafless types. The disease-reducing effects of this intercrop, through barrier or dilution effects, would then constitute a secondary aspect of plant health, as the health of both intercropping partners are indirectly affected by the pea genotype.

A further example of the relationship between crop genotype and plant health, though without any link to plant diseases, is the compensatory ability of crops in response to abiotic stresses via the plasticity of yield components. A stress event (e.g. drought or flooding) may reduce crop performance with regard to one yield component (e.g. number of tillers per unit area) but some genotypes may be able to compensate for this effect, responding to the reduction in tiller

Table 13.1 Examples of pathosystems for which resistance breeding is considered as a potential tool for crop improvement in organic farming

Crop	Disease or pest and causal agent	References
Cereals		
Wheat	Common bunt (*Tilletia caries*)	Matanguihan et al. (2011)
	Loose smut (*Ustilago tritici*)	Kassa et al. (2014)
	Yellow rust (*Puccinia striiformis*)	Lammerts van Bueren et al. (2002)
Barley	Barley leaf stripe (*Pyrenophora graminae*)	Müller (2011)
Rice	Various insect pests	Vanaja et al. (2013)
Legumes		
Lupin	Anthracnose (*Colletotrichum lupini*)	Nirenberg and Feiler (2003)
Soybean	Soybean aphid (*Aphis glycines*)	Brace and Fehr (2012)
Faba bean	Chocolate spot (*Botrytis fabae, Botrytis cinerea*)	Taylor and Cormack (2002)
Red clover	*Fusarium* root rot (*Fusarium* ssp.)	Wallenhammar et al. (2006)
Other crops		
Potato	Late blight (*Phytophthora infestans*)	Lammerts van Bueren et al. (2014)
Tomato	Late blight (*P. infestans*)	Horneburg and Myers (2012)
Brassicas	Club root (*Plasmodiophora brassicae*)	Myers et al. (2012)
Onions	Downy mildew (*Peronospora destructor*)	Scholten et al. (2007)
Carrots	Carrot leaf blight (*Alternaria dauci*)	Lammerts van Bueren et al. (2002)

number by (strongly) increasing the number of grains per tiller (Sadras and Slafer 2012). Such resilience against environmental stresses, including climatic factors, may be seen as a component of plant health. In a comparative study on genetically diverse composite cross-populations, variety mixtures and corresponding monocultures of winter wheat, it was found that the yield advantage of diversity is strongest in the most stressful environments (Figure 13.4). This has been interpreted as an indicator of increased resilience in the more diverse material (Döring et al. 2010).

A final example with particular relevance for organic agriculture is the genetically variable ability to penalise non-co-operative strains of bacteria in the (generally symbiotic) relationship between legumes and rhizobia (Kiers et al. 2007). Here and in many other contexts, it is necessary to employ an evolutionary and system understanding when using genetics for promoting plant health (McDonald and Linde 2002; Moury et al. 2011).

Particularly in organic farming systems, genotypic effects on plant health are strongly modified by environment and crop management. Soil organic matter management, such as through application of farmyard manure, or the position in a crop rotation determine how plant genetic properties generate a phenotype that then has an effect on plant health (see Figure 13.3).

As with plant health, the health of soil has no universally acknowledged definition. Ideas often associated with soil health include the function of the soil to support crop productivity and the sustainability of this function over time (Vieweger and Döring 2014b). In the view of

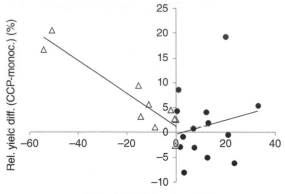

Figure 13.4 Relationship between a measure of between-environment variability (B(E), *x*-axis) and relative yield difference of composite cross-populations (CCP) compared to monocultures (*y*-axis). Open symbols for negative values, filled symbols for positive values. When the average yield of an environment was *below* the system's average (B(E) < 0), the yield effect of the populations was negatively correlated with B(E), i.e. the lower the yield of the environment was below the expectation, the higher was the advantage of the populations (adjusted R^2 = 0.84, P < 0.001, df = 8). For positive values of B, i.e. above-average yielding environments, the CCPs did not show any significant correlations with B(E).

Kibblewhite et al. (2008), a healthy soil is 'capable of supporting both an adequate production of food and fibre and also the continued delivery of other essential ecosystem services'. These authors further stress that an important aspect of soil health is its ability to withstand disturbance (resistance) or to recover after disturbance (resilience), in particular with reference to agricultural interventions, and that soil health depends on the maintenance of carbon transformations, nutrient cycles, soil structure and the regulation of pests and diseases.

Based on these and similar views of soil health (Doran and Zeiss 2000), several connections between the health of soils and crop genetics can be identified. One example is the use of plant genetic variability to select and breed new cover crops that can protect the soil (Reents and Baresel 2008). Other genetically governed crop traits with an impact on soil health include (i) the way in which beneficial relationships between plants and soil microbes are formed, for example between onions and mycorrhiza (Scholten and Kuyper 2012); (ii) rooting depth and root architecture for improved soil structure and infiltrability; (iii) early vigour as a property that helps protect soil from erosion; and (iv) the amount and chemical composition of above- and below-ground plant residues affecting soil biology and soil structure.

A good example of this last trait is harvest index of cereals. With the wider dissemination of dwarf genotypes during the Green Revolution, plant breeders have managed to substantially increase the harvest index in all major cereals, thus increasing the allocation of biomass to grain. As a consequence, the amount of straw that remains in the field, or that is eventually returned to the soil after use by livestock, is lower with modern than with older genotypes. In organic systems, however, there are good arguments for increased plant height of cereals, including lower lodging propensity due to moderate nitrogen levels, decreased incidence of disease infection of the ear and higher competitive ability of tall crops against weeds. With regard to soil health, larger amounts of straw associated with tall crops also mean that more carbon is ultimately returned to the soil, thereby promoting microbial activity and improving soil structure. However, this case clearly illustrates the inevitable trade-off between allocation to grain yield and allocation to straw. There is no easy solution to reconcile these competing functions.

A further example of how soil health is indirectly affected by crop genetics is illustrated in the efforts to design reduced tillage systems for organic agriculture. Here, reduced intensity of tillage is known to be beneficial to the soil, but crop genotypes are needed that allow the use of reduced tillage through improved abilities to scavenge nutrients and compete against weeds.

Crop genetics are also linked to human health and animal health through many pathways, such as through nutritional quality of crop products with respect to macronutrients, micronutrients or secondary plant metabolites. For example, studies on nitrate contents in lettuce varieties grown under organic or biodynamic conditions (Lucarini et al. 2012) or in conventional growing systems (Burns et al. 2011) found large differences amongst genotypes in nitrate accumulation in the leaves, though earlier assessments of nitrates as harmful for human health have recently been called into question (Hord et al. 2009; Weitzberg and Lundberg 2013). Genotype has also been found to be a major factor influencing levels of some antioxidants in wheat (Roose et al. 2008; Hussain et al. 2012b), compounds generally considered to be beneficial for human health. More comprehensive studies have identified genotypic variation of several bioactive substances in wheat such as the dietary fibre arabinoxylan (Shewry et al. 2012) and heavy metals (Hussain et al. 2012a).

These examples highlight the fact that plant breeding and genotype selection by farmers do affect the levels of many plant compounds which may have subsequent effects on human health. But can benefits to human health via improved nutritional quality of crops be targeted directly by plant breeding? Although there may be some strong cases for plant breeding aiming at health effects, for example by selecting genotypes with low tendency to accumulate heavy metals, several issues are likely to limit the realisation of this general approach. First, many of the genotypic effects on composition of raw food products will depend largely on environmental factors and agronomic management, rather than genotype only (see Figure 13.3). As the strength and direction of these factors and their interactions with genotypes will often be unknown, the nutritional composition of crops may often be difficult to control by genotype selection alone. Second, even if differences in chemical composition amongst genotypes are robust at the production level, food processing may alter significantly the quality and quantity of active compounds, so that potential health effects may be lost through inappropriate processing. In wheat production for bread making, for example, fibre content and fibre quality may be enhanced by wheat breeding, but these effects are lost if bread is produced ultimately only from white flour. Also, in terms of quality traits, food processors may often focus on quality for easy industrial processing, long shelf-life and visual appearance of the product rather than its nutritional quality for health benefits.

One example is provided by the role of grain protein content in wheat for the quality of bread. In a review on wheat quality in organic agriculture, it was stated that 'Most Northwestern European consumers, including those preferring organic food, like their daily bread to be voluminous and easy to chew' (Osman et al. 2012). Under standard, industrial baking procedures, loaf volume is known to be largely influenced by protein content and protein composition. It was therefore suggested by Osman et al. that organic wheat breeding needs to concentrate on increasing the proportion of certain protein fractions, namely glutenins and gliadins, that are known to increase loaf volume. However, this view disregards any effects that the quantity and quality of grain protein may have on human health, for example through digestibility of the product. This is particularly relevant in relation to coeliac disease (gluten intolerance). Over the past few decades there has been a notable rise in the incidence of this disease (White et al. 2013), and although this increase is not necessarily a result of higher gluten content in wheat through breeding (Kasarda 2013), it has been linked to an increased gluten intake at the population level. To honour the principle of health, organic plant breeding should therefore redefine wheat quality with a closer connection to actual health effects and take into account alternative ways of bread making such as artisanal baking with longer dough fermentation.

But even if plant breeding manages to influence human health positively through the improved quality of a particular processed food such as bread, the breadth of the human diet further limits the potential of using plant breeding for health improvement. The composition of human diets is usually highly diverse with respect to the intake of different plant species and compositional differences amongst crop species are likely to dwarf those within species. This is complicated by the fact that different compounds from various components of the diet may interact physiologically with each other, but health effects of such interactions are largely unknown.

Beyond approaches focusing on individual plant compounds, there are also alternatives which may potentially have stronger effects on human health. Such an indirect approach to promoting human health through plant breeding is the genetic improvement of grain legumes for human consumption in order to replace animal protein sources and thereby reduce meat consumption. Here, breeding may target agronomic traits and resistance to diseases to make grain legumes more attractive to farmers, or taste and digestibility for a better uptake by consumers. Thus, crop breeding can contribute to a greater diversity of plants for human consumption.

13.4.4 The Role of Plant Genetic Diversity for Health

There is evidence that plant genetic diversity has beneficial effects on health in all domains – soils, plants, animals, humans and ecosystems. The ways in which health is affected by diversity through capacity, such as the availability and choice of appropriate genotypes, have been outlined above (see Section 13.4.3). Here we discuss diversity effects based on mechanisms that involve mixing genotypes or species in space or over time.

With regard to plant health, increased diversity in mixed compared to homogeneous crop stands has often been shown to reduce the incidence and severity of plant diseases, for example in intercropping systems (Boudreau 2013) and cultivar mixtures (Wolfe 1985; Finckh and Mundt 1992; Wolfe 2000; Zhu et al. 2000). Mechanisms underlying the disease-suppressing effects of plant diversity in the field include induced resistance, barrier effects and dilution effects. Further, because of reduced selection pressure on pathogens in diverse crop stands, durable resistance against plant diseases is more likely to be achieved with higher diversity in the field. However, undesired effects of increased plant diversity have also been observed in relation to plant diseases. For example, in agroforestry systems a moister microclimate next to tree rows can promote fungal diseases of the arable component (Jung et al. 2013). Plant diseases are also known to be affected by crop rotation, that is, plant diversity over time. Here, more diverse rotations promote plant health, especially in the case of soil-borne pests and pathogens.

Further, increased biological diversity has been found to promote various parameters associated with soil health (Stefanowicz et al. 2012). In a seven-year static field experiment with grassland-savanna species, it was found that microbial biomass, respiration and fungal abundance increased with plant diversity (Zak et al. 2003). These effects resulted mainly from higher plant productivity in the treatments with higher plant species richness rather than from diversity *per se*. However, it was suggested that the process of microbial mineralisation of nitrogen in this experiment was positively affected by plant diversity rather than increased productivity. Finally, plant diversity also affects human health (Fanzo et al. 2013) as well as animal health (Yoshihara et al. 2013) through the composition of diets.

The generally beneficial effects of diversity for health in various contexts raise the question about potential dose–effect relationship between diversity and health. For instance, how many species or genotypes should be mixed for obtaining optimal effects on health? How diverse do rotations need to be for detecting positive effects on health? For several parameters, it has been found that effects tend to saturate; that is, the benefits of adding further diversity decrease with increasing diversity level (Zak et al. 2003). However, little information exists about how to design the proportions of elements in a plant-diverse mix for optimal health effects.

13.5 Socioeconomics, Policies and Regulations

13.5.1 The Organic Principle of Fairness

The principle of fairness emphasises the need to strive for 'equity, respect, justice and steward-ship of the shared world' (IFOAM 2005). The principle refers to fair and just distribution of life opportunities and resources of the common environment. It touches all people involved in organic agriculture, at all levels of the food supply chain –producers, processors, distributors, traders and consumers – and also refers to the needs of future generations. It is stressed that agricultural and food systems need to account for actual environmental as well as social costs.

A central tenet of the principle of fairness is summarised in the term *food sovereignty*, a dynamic concept with different connotations according to geographic and political context (Jarosz 2014). Food sovereignty can broadly be defined as 'the right of nations and peoples to control their own food systems' (Wittman et al. 2010, p. 2). As a concept, it was developed by peasant movements in critical response to neoliberal globalisation, free trade and corporate-led food systems. Its origins marked the renunciation of the concept of food security, which was seen to focus too much on the maximisation of food production. Although food access opportunities are incorporated in the concept of food security, too little emphasis was seen to be placed on the question of 'how, where and by whom food is produced' (Wittman et al. 2010, p. 3).

In the discourses on both food security and food sovereignty, seeds and plant genetics play a vital role (Bezner Kerr 2013), and the access to plant genetic resources and the sharing of the associated costs and benefits are subject to intensive debates at global scale. With respect to organic agriculture, these issues have recently been reviewed elsewhere (Chable et al. 2012);

building on this extensive review, we briefly discuss selected areas of conflict with particular relevance to crop genetic diversity.

13.5.2   Traditional Landraces and the Protection of Plant Genetic Resources

Agriculture is experiencing the continued loss of genetic diversity of cultivated plant species around the globe. Already described more than two decades ago by Fowler and Mooney (1990), this dramatic trend of genetic erosion is still ongoing, as landraces are being replaced by modern cultivars (Carolina et al. 2005; van de Wouw et al. 2009), although when the analysis of global crop genetic diversity trends is restricted to modern cultivars, no loss of diversity is evident (van de Wouw et al. 2010). Still, many regions of the world lose plant genetic resources at an alarming rate (Ahn et al. 2011).

However, as shown above, productivity of organic systems depends strongly on high genetic diversity of cultivated plants and on the availability and maintenance of suitable germplasm. This is also because the loss of plant genetic diversity mainly affects germplasm that is (or would be) especially suitable for low-input or organic conditions, as it stems from traditional forms of agriculture that do not use synthetic inputs for plant protection or plant nutrition. Because of this dependency of organic crop performance on plant genetic diversity, and because of the need to preserve cultural heritage associated with plant genetic resources, there is a large overlap between the interests of organic agriculture, on the one hand, and the conservation of plant genetic resources, in particular of traditional landraces, on the other. However, there are also potential conflicts between the two (Döring 2011); these conflicts are primarily issues of fair distribution of plant genetic resources.

The first potential conflict lies in the hidden use of unfairly obtained germplasm – sometimes termed 'gene grab' or biopiracy. Landraces in traditional agriculture may already offer a degree of preadaptation to conditions of (certified) organic agriculture. It is therefore desirable that their traits are used in breeding programmes for organic and low-input farming. However, it is crucial that the collection and use of these valuable genetic resources are based on fairness and the principles of access and benefit sharing. Farmers will find it difficult to check if this is the case, since information on the source of genetic material used in the breeding process is usually not easily accessible. Thus, as demand for varieties with such traits grows, regulations should prevent varieties from entering the market that are ecologically and agronomically suitable for organic farming but do not comply with the principle of fairness.

A second conflict is the success of organic farming itself, as it comes under increasing economic pressure to reduce production costs and competition intensifies within the sector. As bigger markets demand larger quantities of more uniform produce, there is the risk that organic farming practice moves towards genetically simplified systems by shedding 'messy' but extremely valuable bits of plant genetic diversity. In view of these dynamic developments, benefits and costs of the maintenance of landraces and other plant genetic resources need to be shared fairly within the organic community.

Thus, organic agriculture does not provide an automatic guarantee for preserving plant genetic resources. Vigilance will therefore be necessary to prevent organic agriculture from contributing, if only in unintended ways, to the further erosion of plant genetic resources. Organic farming needs to engage at all organisational levels in the on-farm conservation of plant genetic resources.

13.5.3 Sharing the Costs and Benefits of Plant Breeding

A critical question is who ultimately pays for the necessary investments in plant breeding. There are real dilemmas for which fair solutions are not easy to find. On one hand, breeders' investments need to be protected. If there is no return on a variety, for example because its seed is completely freely available to all, without restrictions, incentives are low for developing new varieties. On the other hand, farmers' independence from breeders, by saving seed, is also valuable. However, such independence is difficult to achieve if the seed sector is controlled mainly by a few large organisations. Indeed, in the conventional seed and plant breeding sector, there is currently an unprecedented level of market concentration. Three-quarters of the international seed market is shared by the top 10 seed companies and only six corporations control 75% of the entire private sector plant breeding research (ETC Group 2013).

In the organic seed sector, concentration has also been increasing but smaller enterprises are recognised for their vital contributions. However, despite the known demand for varieties bred for organic systems, organic plant breeders are currently finding it difficult to cover the costs of their work as they come under increasing economic pressure from rising fees for registering new organic varieties. This is relevant for the development of the whole organic crop production chain because there is currently not enough choice of varieties for organic cropping conditions. A related question is the use of conventionally produced seed in organic agriculture (Döring et al. 2012). Organic agriculture regulations prescribe the use of organic seed, but for many crops seed is not available in organic quality. Regulations therefore allow the use of conventional seed under defined circumstances. Although such derogations are temporarily justifiable to ensure access of organic farmers to a large range of varieties, they also thwart present and future investment in the organic seed sector and organic plant breeding.

An alternative system that does not separate farmers and breeders is participatory plant breeding. This system is more amenable to the fair distribution of resources and power in the seed and breeding sector, and also more conducive to the maintenance of crop diversity which requires a more distributed model of access and ownership of germplasm, less centralised than formal seed systems currently prevailing in industrialised countries. Insofar as people recognise that protection of diversity necessitates non-central structures, it can be expected that plant genetic diversity itself may help to promote fairness amongst farmers, between breeders, seed producers, farmers, traders and consumers.

13.5.4 Hybrid Varieties in Organic Farming

A question that has been strongly debated within the organic community is the use of hybrids. Over the last few decades, the proportion of hybrids has massively increased in many crop species, including leeks, lettuce, tomato, cucumber and rye. For cereals, it has been found that hybrids show greater yield stability than inbred lines (Mühleisen et al. 2014); also, hybrids often show greater mean yields than inbred lines.

However, the use of hybrids has also been criticised, mainly because of the increased control over genetic resources. In contrast to non-hybrid varieties, it is not legally required to make germplasm of hybrids, or their parents, available to the public in gene banks. Further, as seed saving in hybrids is not feasible for farmers, the expansion of hybrids shifts power and seed sovereignty further away from farmers and towards breeding companies. As non-hybrids disappear from the seed catalogues, attempts have therefore been made to save the associated germplasm from being lost from public access by dehybridisation. There is also an increasing push, particularly from the biodynamic sector, for concentration on 'open-pollinated varieties' (OPVs), which will maintain higher levels of diversity in vegetable production systems.

13.6 Indeterminism and Crop Genetics

13.6.1 The Organic Principle of Care

The organic principle of care is based on the view that agriculture and food systems are dynamic, and that understanding of these systems is inevitably incomplete (IFOAM 2005). The consequences of this indeterministic perspective affect the ways in which knowledge about organic systems should be obtained and how new and existing technologies need to be assessed.

Organic agriculture builds on scientific knowledge to assess whether practices are healthy, safe and ecologically sound. Apart from the role of science, however, it also emphasises the need to integrate practical experience and indigenous knowledge. With regard to the assessment of (new) technologies, care, responsibility, precaution and the prevention of significant risks provide the main guidelines in organic agriculture. The principle of care further stresses that the processes of technology assessment should be based on transparency and participation of all potentially affected parties.

13.6.2 Implications of the Care Principle for Crop Breeding

A direct reference to plant breeding in the organic principle of care is the explicit rejection of genetic engineering. This is grounded in the view that genetic transformation of crops is a technology with particularly unpredictable consequences. An example of such unpredictability is provided by the transgenic spring wheat (Zeller et al. 2010) discussed above (see Section 13.3.2). However, unpredictability is only one amongst many arguments (Antoniou et al. 2014) that the organic sector has put forward against the use of genetic transformation of crops. Also, genetic engineering is only one of many new breeding technologies (Messmer 2014). In response to the galloping development of biotechnological techniques, the organic breeding community has defined the limit of acceptable interference at the cell level. Thus, breeding techniques violating the integrity of the plant's cell or genome (Lammerts van Bueren et al. 2003) are not considered to be conforming to organic breeding standards (IFOAM 2014, p. 43).

A further aspect of plant genetics touched by the principle of care is plant breeding in a changing climate. As highlighted in the introduction to this chapter, a major problem of climate change is that not only are the means of temperature and precipitation shifting (with uncertain pace), but their variance is also increasing, thereby reducing the predictability of climatic parameters at the regional and local level.

Finally, the principle of care puts a strong emphasis on the complementary role of science and farmers' experience and indigenous knowledge. Thus it provides a mandate for participatory plant breeding, in which the development of selection criteria, selection itself and even crossing is done by farmers, based on their needs and values (Almekinders and Hardon 2006; Chable et al. 2014; Messmer et al. 2014).

13.6.3   The Role of Plant Genetic Diversity for the Care Principle

As pointed out above, crop diversity provides insurance against instability and ecological surprises. In particular through compensatory mechanisms, crop diversity within and amongst species helps to hedge farmers' bets. Moreover, the care principle provides an important argument for the maintenance and conservation of plant genetic diversity: once a genotype is lost, it is lost forever. Therefore, if you think you do not need a particular genotype at this point in time, you should, because of precaution, still retain it and not throw it away.

However, there is also an important trade-off between (crop) diversity and the principle of care, and this has to do with the time necessary to assess relevant potentials and risks. Even genotypes created with traditional breeding techniques may not necessarily be without risk. An example is the undetected susceptibility to fungal pathogens that lead to high levels of mycotoxins. However, thoroughly assessing genotypes takes time, as there is a trade-off between thoroughness of assessment in multiple environments and manageable crop diversity. Especially because of the dynamic nature of agricultural systems, sufficient time needs to be spent to observe how things unfold.

Thus, the principle of care simultaneously demands the conservation and maintenance of many genotypes and the thorough and careful assessment of each. Given limited resources, a solution to this challenge is again provided by crop diversity. When risks of an agricultural technique (in this case, a particular genotype) are likely to be small, but principally unknown, the idea is to keep the size of its potential impact low by growing it only on small areas. This approach requires – and promotes – crop diversity. Decentralised gathering of knowledge such as in participatory plant breeding may then help in assessing the risks and potentials in effective ways.

13.7   Conclusion

How can plant breeding contribute to achieving the organic principles? In the sections above we have provided several examples of how crop diversity within and amongst species can help organic farming practice to honour its principles of ecology, health, fairness and care. However, the maintenance of diversity requires appropriate structures and investment. Privatisation and concentration of the plant breeding and seed sector have led to simplification, loss of diversity and unfair distribution of power and resources. Changes in the legal system are therefore necessary to prevent the loss of genetic diversity and encourage its maintenance. Intensified cooperation and networking within the organic community at the global level are necessary to deal with this challenge.

In addition, networking at the local and regional level is expected to help in maintaining and developing crop diversity, for example through farmer-led breeding. Stronger networking amongst farmers will also be necessary to bring relevant information about variety performance together for more effective variety assessment and understanding of underlying potentials and constraints.

Further, genetic diversity within species is critically important but more use should be made of the tremendous functional diversity present at species level since a lot of valuable germplasm has been entirely left behind in the conventional breeding process. It will be difficult to decide how we should distribute resources amongst the vast number of species. Again, however, we believe that diversity itself will provide an answer.

Acknowledgements

Writing this chapter would not have been possible without the support for ongoing organic plant breeding research provided by various projects including SOLIBAM (funded by the European Commission through FP7, grant no. 245058), Wheat Breeding LINK (LK 0999, funded by Defra), and COBRA (EU ERANET CORE Organic II). We would also like to thank Riccardo Bocci, James Brown, Henry Creissen, Nick Fradgley, Robbie Girling, Sally Howlett, Hannah Jones, Jack Kloppenburg, Samuel Knapp, Monika Messmer, Karl-Josef Müller, Bruce Pearce, John Snape, Andrew Whitley and Lawrence Woodward for discussing matters of organic plant breeding.

References

Ahn, W.S., Yoon, S.H., Kim, S.G., Byeon, H.D., Ahn, C.H., 2011. Collecting and research on native crops in Geosan-gun, Korea. 17th IFOAM Organic World Congress: Organic Seed Preconference. Heuksalim Institute, Cheongwon-gun, Korea, pp. 162–166.

Alfred, J., Dangl, J.L., Kamoun, S., and McCouch, S.R. (2014). New horizons for plant translational research. *PLoS Biol.* 12: e1001880.

Almekinders, C. and Hardon, J. (2006). *Bringing Farmers Back into Breeding. Experiences with Participatory Plant Breeding and Challenges for Institutionalisation*, Agromisa Special 5, 135. Wageningen, Netherlands: Agromisa.

Annicchiarico, P. (2002). *Genotype x Environment Interactions: Challenges and Opportunities for Plant Breeding and Cultivar Recommendations*, vol. 174. Rome: Food & Agriculture Organization.

Annicchiarico, P. and Pecetti, L. (2010). Forage and seed yield response of lucerne cultivars to chemically weeded and non-weeded managements and implications for germplasm choice in organic farming. *Eur. J. Agron.* 33: 74–80.

Annicchiarico, P., Chiapparino, E., and Perenzin, M. (2010). Response of common wheat varieties to organic and conventional production systems across Italian locations, and implications for selection. *Field Crop Res.* 116: 230–238.

Antoniou, M., Robinson, C., and Fagan, J. (2014). *GMO Myths and Truths – An Evidence-Based Examination of the Claims Made for the Safety and Efficacy of Genetically Modified Crops.* London: Earth Open Source.

Arnell, N.W. (2003). Relative effects of multi-decadal climatic variability and changes in the mean and variability of climate due to global warming: future streamflows in Britain. *J. Hydrol.* 270: 195–213.

Arterburn, M., Murphy, K., and Jones, S.S. (2012). Wheat: breeding for organic farming systems. In: *Organic Crop Breeding* (ed. E.T. Lammerts van Bueren and J.R. Myers), 163–174. Hoboken: Wiley-Blackwell.

Balfour, E.B. (1943). *The Living Soil.* London: Faber and Faber.

Baresel, J., Zimmermann, G., and Reents, H. (2008). Effects of genotype and environment on N uptake and N partition in organically grown winter wheat (*Triticum aestivum* L.) in Germany. *Euphytica* 163: 347–354.

Bengtsson, J., Ahnstrom, J., and Weibull, A.C. (2005). The effects of organic agriculture on biodiversity and abundance: a meta-analysis. *J. Appl. Ecol.* 42: 261–269.

Berg, E.E. and Hamrick, J.L. (1997). Quantification of genetic diversity at allozyme loci. *Can. J. For. Res.* 27: 415–424.

Bezner Kerr, R. (2013). Seed struggles and food sovereignty in northern Malawi. *J. Peasant Stud.* 40: 867–897.

Bharud, R.W. (2014). *Cotton Developments in India*. Rahuri-Maharashtra: Mahatma Phule Agricultural University.

Blair, M.W. (2013). Breeding approaches to increasing nutrient-use efficiency – examples from common beans. In: *Improving Water and Nutrient-Use Efficiency in Food Production Systems* (ed. Z. Rengel), 161–176. Oxford: Wiley-Blackwell.

Böhm, H., Kühne, S., Ludwig, T., et al. 2013. Stickstoff-und Schwefelversorgung von Rapsbeständen im Ökologischen Landbau. In: Neuhoff, D., Stumm, C., Ziegler, S., Rahmann, G., Hamm, U., Köpke, U., eds. 12. Wissenschaftstagung Ökologischer Landbau. Ideal und Wirklichkeit – Perspektiven Ökologischer Landbewirtschaftung, 5-8 March. Verlag Dr Köster, Berlin.

Böhm, H., Sauermann, W., Alpers, G., Aulrich, K., 2011. Sortenwahl und Schädlingsdruck im ökologischen Rapsanbau. In: Leithold, G., Becker, K., Brock, C., et al., eds. 11. Wissenschaftstagung Ökologischer Landbau: Es geht ums Ganze: Forschen im Dialog von Wissenschaft und Praxis, Justus-Liebig-Universität Gießen, 15–18 March.

Borgen, A., 2014. Virulence pattern in Danish races of common bunt (*Tilletia caries*). 7th ISTA Seed Health Symposium, 12–14 June, Edinburgh. http://orgprints.org/26891.

Boudreau, M.A. (2013). Diseases in intercropping systems. *Annu. Rev. Phytopathol.* 51: 499–519.

Brace, R.C. and Fehr, W.R. (2012). Impact of combining the *Rag1* and and *Rag2* alleles for aphid resistance on agronomic and seed traits of soybean. *Crop Sci.* 52: 2070–2074.

Burns, I.G., Zhang, K., Turner, M.K. et al. (2011). Screening for genotype and environment effects on nitrate accumulation in 24 species of young lettuce. *J. Sci. Food Agric.* 91: 553–562.

Carolina, T., Villa, C., Maxteda, N. et al. (2005). Defining and identifying crop landraces. *Plant Genet. Resour. Charact. Util.* 3: 373–384.

Carr, P.M., Kandel, H.J., Porter, P.M. et al. (2006). Wheat cultivar performance on certified organic fields in Minnesota and North Dakota. *Crop Sci.* 46: 1963–1971.

Ceccarelli, S., Grando, S., Bailey, E. et al. (2001). Farmer participation in barley breeding in Syria, Morocco and Tunisia. *Euphytica* 122: 521–536.

Chable, V., Dawson, J., Bocci, R., and Goldringer, I. (2014). Seeds for organic agriculture: development of participatory plant breeding and farmers' networks in France. In: *Organic Farming, Prototype for Sustainable Agricultures* (ed. S. Bellon and S. Penvern), 383–400. Berlin: Springer.

Chable, V., Louwaars, N.P., Hubbard, K. et al. (2012). Plant breeding, variety release, and seed commercialization: laws and policies applied to the organic sector. In: *Organic Crop Breeding* (ed. E.T. Lammerts van Bueren and J.R. Myers), 139–159. Oxford: Wiley.

Cordell, D., Drangert, J.-O., and White, S. (2009). The story of phosphorus: global food security and food for thought. *Glob. Environ. Chang.* 19: 292–305.

Costanzo, A. and Bàrberi, P. (2014). Functional agrobiodiversity and agroecosystem services in sustainable wheat production. A review. *Agron. Sustain. Dev.* 34: 327–348.

Creissen, H.E., Jorgensen, T.H., and Brown, J.K. (2013). Stabilization of yield in plant genotype mixtures through compensation rather than complementation. *Ann. Bot.* 112: 1439–1447.

Crowder, D.W., Northfield, T.D., Strand, M.R., and Snyder, W.E. (2010). Organic agriculture promotes evenness and natural pest control. *Nature* 466: 109–113.

Dawson, J.C., Murphy, K., and Jones, S.S. (2008). Decentralized selection and participatory approaches in plant breeding for low-input systems. *Euphytica* 160: 143–154.

Denison, R.F. (2012). *Darwinian Agriculture – How Understanding Evolution Can Improve Agriculture*. Princeton: Princeton University Press.

Doran, J.W. and Zeiss, M.R. (2000). Soil health and sustainability: managing the biotic component of soil quality. *Appl. Soil Ecol.* 15: 3–11.

Döring, T.F., 2011. Organic seed and the conservation of plant genetic resources – a call for vigilance. 17th IFOAM Organic World Congress: Organic Seed Preconference. Heuksalim Institute, Cheongwon-gun, Korea, pp. 195–197.

Döring, T.F., Bocci, R., Hitchings, R. et al. (2012). The organic seed regulations framework in Europe – current status and recommendations for future development. *Org. Agric.* 2 (3–4): 173–183.

Döring, T.F., Kovacs, G., Wolfe, M.S., and Murphy, K. (2011). Evolutionary plant breeding in cereals – into a new era. *Sustainability* 3: 1944–1971.

Döring, T.F., Pautasso, M., Finckh, M.R., and Wolfe, M.S. (2012a). Concepts of plant health – reviewing and challenging the foundations of plant protection. *Plant Pathol.* 61: 1–15.

Döring, T.F., Pautasso, M., Finckh, M.R., and Wolfe, M.S. (2012b). Pest and disease management in organic farming: implications and inspirations for plant breeding. In: *Organic Crop Breeding* (ed. E.T. Lammerts van Bueren and J.R. Myers), 39–59. Oxford: Wiley-Blackwell.

Döring, T.F., Vieweger, A., Pautasso, M. et al. (2015). Resilience as a universal criterion of health. *J. Sci. Food Agric.* 95: 455–465. https://doi.org/10.1002/jsfa.6539.

Döring, T.F., Wolfe, M., Jones, H., Pearce, H., Zhan, J., 2010. Breeding for resilience in wheat – nature's choice. In: Goldringer I, Dawson J, Rey F, Vettoretti A, eds. Breeding for Resilience: A Strategy for Organic and Low-Input Farming Sytems? Eucarpia 2nd Conference of the Organic and Low-Input Agriculture Section. 1–3 December, Paris, pp. 45–48. http://orgprints.org/18171/1/Breeding_for_resilience-Book_of_abstracts.pdf

DoVale, J.C., Oliveira DeLima, R., and Fritsche-Neto, R. (2012). Breeding for nitrogen use efficiency. In: *Plant Breeding for Abiotic Stress Tolerance* (ed. R. Fritsche-Neto and A. Borém), 53–65. Berlin: Springer.

Easterling, D.R., Meehl, G.A., Parmesan, C. et al. (2000). Climate extremes: observations, modeling, and impacts. *Science* 289: 2068–2074.

ETC Group, 2013. Putting the cartel before the horse... and farm, seeds, soil, peasants, etc. ETC Group, Ottawa.

European Commission, 2007. Council Regulation (EC) No 834/2007 of 28 June 2007 on organic production and labelling of organic products and repealing Regulation (EEC) No 2092/91. http://eur-lex.europa.eu/LexUriServ/LexUriServ.do?uri=OJ:L:2007:189:0001:0023:EN:PDF, p. 23.

Fanzo, J., Hunter, D., Borelli, T., and Mattei, F. (2013). *Diversifying Food and Diets: Using Agricultural Biodiversity to Improve Nutrition and Health*. Routledge, London.

Finckh, M.R. and Mundt, C.C. (1992). Plant competition and disease in genetically diverse wheat populations. *Oecology* 91: 82–92.

Fowler, C. and Mooney, P. (1990). *Shattering: Food, Politics, and the Loss of Genetic Diversity*. University of Arizona Press, Tucson.

Fridley, J.D. and Grime, J.P. (2010). Community and ecosystem effects of intraspecific genetic diversity in grassland microcosms of varying species diversity. *Ecology* 91: 2272–2283.

Fritsche-Neto, R. and DoVale, J.C. (2012). Breeding for stress-tolerance or resource-use efficiency? In: *Plant Breeding for Abiotic Stress Tolerance* (ed. R. Fritsche-Neto and A. Borém), 13–19. Berlin: Springer.

Fuller, R.J., Norton, L.R., Feber, R.E. et al. (2005). Benefits of organic farming to biodiversity vary among taxa. *Biol. Lett.* 1: 431–434.

Gabriel, D., Sait, S.M., Hodgson, J.A. et al. (2010). Scale matters: the impact of organic farming on biodiversity at different spatial scales. *Ecol. Lett.* 13: 858–869.

Garamszegi, L.Z., Calhim, S., Dochtermann, N. et al. (2009). Changing philosophies and tools for statistical inferences in behavioral ecology. *Behav. Ecol.* 20: 1363–1375.

Gimmler, A. (2002). The concept of health and its normative implications – a pragmatic approach. In: *Health and Quality of Life – Philosophical, Medical, and Cultural Aspects* (ed. A. Gimmler, C. Lenk and G. Aumüller), 69–79. Münster: Lit Verlag.

Goldstein, W.A., Schmidt, W., Burger, H. et al. (2012). Maize: breeding and field testing for organic farmers. In: *Organic Crop Breeding* (ed. E.T. Lammerts van Bueren and J.R. Myers), 175–189. Oxford: Wiley-Blackwell.

Graß, R., Heuser, F., Stülpnagel, R. et al. (2013). Energy crop production in double-cropping systems: results from an experiment at seven sites. *Eur. J. Agron.* 51: 120–129.

Graß, R. and Scheffer, K. (2005). Alternative Anbaumethoden: Das Zweikulturnutzungssystem. *Natur und Landschaft* 9/10: 435–439.

Gronle, A. and Böhm, H. (2014). Untersuchungen zur Unkrautunterdrückung in Rein-und Mischfruchtbeständen von Wintererbsen unterschiedlichen Wuchstyps. *Julius-Kühn-Archiv* 443: 431–440.

Grundbacher, F.J. (1963). The physiological function of the cereal awn. *Bot. Rev.* 29: 366–381.

Guarda, G., Padovan, S., and Delogu, G. (2004). Grain yield, nitrogen-use efficiency and baking quality of old and modern Italian bread-wheat cultivars grown at different nitrogen levels. *Eur. J. Agron.* 21: 181–192.

Haase, T., Quendt, U., Mindermann, A., Müller, K.-J., Heß, J., 2013. Prüfung von Wintererbsengenotypen auf ihre Winterhärte. In: Neuhoff, D., Stumm, C., Ziegler, S., Rahmann, G., Hamm, U., Köpke, U., eds. 12. Wissenschaftstagung Ökologischer Landbau. Verlag Dr Köster, Berlin.

Hawksworth, D.L. (1995). *Biodiversity: Measurement and Estimation*. London: Chapman & Hall and The Royal Society.

Heilig, J.A. and Kelly, J.D. (2012). Performance of dry bean genotypes grown under organic and conventional production systems in Michigan. *Agron. J.* 104: 1485–1492.

Hetrick, B., Wilson, G., and Cox, T. (1992). Mycorrhizal dependence of modern wheat varieties, landraces, and ancestors. *Can. J. Bot.* 70: 2032–2040.

Heyden, B., 2004. Bedeutung von Regionalsorten im Getreidebau – Abschlussbericht zum Forschungsprojekt Nr. 02OE494. Keyserlingk-Institut, Salem.

Hildermann, I., Messmer, M., Dubois, D. et al. (2010). Nutrient use efficiency and arbuscular mycorrhizal root colonisation of winter wheat cultivars in different farming systems of the DOK long-term trial. *J. Sci. Food Agric.* 90: 2027–2038.

Hildermann, I., Thommen, A., Dubois, D. et al. (2009). Yield and baking quality of winter wheat cultivars in different farming systems of the DOK long-term trial. *J. Sci. Food Agric.* 89: 2477–2491.

Hoad, S.P., Bertholdsson, N.O., Neuhoff, D., and Köpke, U. (2012). Approaches to breed for improved weed suppression in organically grown cereals. In: *Organic Crop Breeding* (ed. E.T. Lammerts van Bueren and J.H. Myers), 61–76. Oxford: Wiley-Blackwell.

Hole, D.G., Perkins, A.J., Wilson, J.D. et al. (2005). Does organic farming benefit biodiversity? *Biol. Conserv.* 122: 113–130.

Hord, N.G., Tang, Y., and Bryan, N.S. (2009). Food sources of nitrates and nitrites: the physiologic context for potential health benefits. *Am. J. Clin. Nutr.* 90: 1–10.

Horneburg, B. and Becker, H.C. (2008). Crop adaptation in on-farm management by natural and conscious selection: a case study with lentil. *Crop Sci.* 48: 203–212.

Horneburg, B. and Myers, J.R. (2012). Tomato: breeding for improved disease resistance in fresh market and home garden varieties. In: *Organic Crop Breeding* (ed. E.T. Lammerts van Bueren and J.R. Myers), 239–249. Oxford: Wiley-Blackwell.

Huber, M., Bakker, M.H., Dijk, W. et al. (2012). The challenge of evaluating health effects of organic food; operationalisation of a dynamic concept of health. *J. Sci. Food Agric.* 92: 2766–2773.

Huber, M., Knottnerus, J.A., Green, L. et al. (2011). How should we define health. *BMJ.* 343: d4163.

Hughes, R.A., Inouye, B.D., Johnson, M.T.J. et al. (2008). Ecological consequences of genetic diversity. *Ecol. Lett.* 11: 609–623.

Hüsing, B., Haase, T., Trautz, D., Heß, J., Schliephake, U., 2011. Sortenprüfung frühabreifender Sojabohnensorten im ökologischen Landbau. In: Leithold, G., Becker, K., Brock, C., et al., eds. 11. Wissenschaftstagung Ökologischer Landbau. 15–18 March. Verlag Dr Köster, Berlin. pp. 298–301.

Hussain, A., Larsson, H., Kuktaite, R., and Johansson, E. (2012a). Healthy food from organic wheat: choice of genotypes for production and breeding. *J. Sci. Food Agric.* 92: 2826–2832.

Hussain, A., Larsson, H., Olsson, M.E. et al. (2012b). Is organically produced wheat a source of tocopherols and tocotrienols for health food? *Food Chem.* 132: 1789–1795.

IFOAM, 2005. The Principles of Organic Agriculture. Available at: www.ifoam.bio/sites/default/files/poa_english_web.pdf

IFOAM, 2014. The IFOAM Norms for Organic Production and Processing – Version July 2014. International Federation of Organic Agriculture Movements, Bonn.

Jaikumar, N., Snapp, S., Murphy, K., and Jones, S. (2012). Agronomic assessment of perennial wheat and perennial rye as cereal crops. *Agron. J.* 104: 1716–1726.

Jarosz, L. (2014). Comparing food security and food sovereignty discourses. *Dialogues Hum. Geogr.* 4: 168–181.

Jung, L.S., Bärwolff, M., and Vetter, A. (2013). Einflüsse auf die Erntequalität von Ackerfrüchten in Agroforstsystemen. *Mitt. Ges. Pflanzenbauwiss* 25: 114–115.

Jurasinski, G., Retzer, V., and Beierkuhnlein, C. (2009). Inventory, differentiation, and proportional diversity: a consistent terminology for quantifying species diversity. *Oecology* 159: 15–26.

Kasarda, D.D. (2013). Can an increase in celiac disease be attributed to an increase in the gluten content of wheat as a consequence of wheat breeding? *J. Agric. Food Chem.* 61: 1155–1159.

Kassa, M.T., Menzies, J.G., and McCartney, C.A. (2014). Mapping of the loose smut resistance gene Ut6 in wheat (*Triticum aestivum* L.). *Mol. Breed.* 33: 569–576.

Kibblewhite, M.G., Ritz, K., and Swift, M.J. (2008). Soil health in agricultural systems. *Philos. Trans. R. Soc. B Biol. Sci* 363: 658–701.

Kiers, E.T., Hutton, M.G., and Denison, R.F. (2007). Human selection and the relaxation of legume defences against ineffective rhizobia. *Proc. R. Soc. B Biol. Sci.* 274: 3119–3126.

Klepatzki, J., Döring, T.F., Macholdt, J., and Ellmer, F. (2014). Comparing the reliability of maize variety data from on-farm trials and experimental stations. *J. Kult.* 66: 389–395.

Lammerts van Bueren, E.T., Engelen, C., Hutten, R.C.B., 2014. Participatory potato breeding model involving organic farmers and commercial breeding companies in the Netherlands. 7th Organic Seed Growers Conference, Organic Seed Alliance, Port Townsend, WA, pp. 69–73.

Lammerts van Bueren, E.T., Jones, S.S., Tamm, L. et al. (2010). The need to breed crop varieties suitable for organic farming, using wheat, tomato and broccoli as examples: a review. *J. Life Sci.* 58: 3–4.

Lammerts van Bueren, E., Struik, P., and Jacobsen, E. (2002). Ecological concepts in organic farming and their consequences for an organic crop ideotype. *J. Life Sci.* 50: 1–26.

Lammerts van Bueren, E.T., Struik, P.C., Tiemens-Hulscher, M., and Jacobsen, E. (2003). Concepts of intrinsic value and integrity of plants in organic plant breeding and propagation. *Crop Sci.* 43: 1922–1929.

Lampkin, N. (1994). *Organic Farming*. Ipswich: Farming Press.

Lapin, M. and Barnes, B.V. (1995). Using the landscape ecosystem approach to assess species and ecosystem diversity. *Conserv. Biol.* 9: 1148–1158.

Link, W., Balko, C., and Stoddard, F.L. (2010). Winter hardiness in faba bean: physiology and breeding. *Field Crop Res.* 115: 287–296.

Link, W., Stelling, D., and Ebmeyer, E. (1994). Yield stability in faba bean, *Vicia faba* L. 1. Variation among inbred lines. *Plant Breed.* 112: 24–29.

Lorenzana, R.E. and Bernardo, R. (2008). Genetic correlation between corn performance in organic and conventional production systems. *Crop Sci.* 48: 903–910.

Löschenberger, F., Fleck, A., Grausgruber, H. et al. (2008). Breeding for organic agriculture: the example of winter wheat in Austria. *Euphytica* 163: 469–480.

Lotter, D.W. (2003). Organic agriculture. *J. Sustain. Agric.* 21: 59–128.

Lucarini, M., d'Evoli, L., Tufi, S. et al. (2012). Influence of growing system on nitrate accumulation in two varieties of lettuce and red radicchio of Treviso. *J. Sci. Food Agric.* 92: 2796–2799.

Luttikholt, L.W.M. (2007). Principles of organic agriculture as formulated by the international federation of organic agriculture movements. *J. Life Sci.* 54: 347–360.

Maggio, A., Carillo, P., Bulmetti, G.S. et al. (2008). Potato yield and metabolic profiling under conventional and organic farming. *Eur. J. Agron.* 28: 343–350.

Magurran, A.E. (2004). *Measuring Biological Diversity*. Oxford: Blackwell.

Matanguihan, J.B., Murphy, K.M., and Jones, S.S. (2011). Control of common bunt in organic wheat. *Plant Dis.* 95: 92–103.

McDonald, B. (2010). How can we achieve durable disease resistance in agricultural ecosystems? *New Phytol.* 185: 3–5.

McDonald, B.A. and Linde, C. (2002). Pathogen population genetics, evolutionary potential, and durable resistance. *Annu. Rev. Phytopathol.* 40: 349–379.

McNeely, J.A., Miller, K.R., Reid, W.V. et al. (1990). *Conserving the World's Biological Diversity*. Washington, DC: IUCN, World Resources Institute, Conservation International, WWF-US and the World Bank.

Megyeri, M., Mikó, P., Molnár, I., and Kovács, G. (2011). Development of synthetic amphiploids based on *Triticum turgidum* × *T. monococcum* crosses to improve the adaptability of cereals. *Acta Agron. Hung.* 59: 267–274.

Messmer, M. (2014). *Evaluation of New Plant Breeding Techniques for Organic Agriculture.* Brussels: IFOAM EU Group Meeting.

Messmer, M., Berset, E., Zimmer, S., et al. 2012a. Breeding for improved soybean-Bradyrhizobia symbiosis for cool growing conditions in Central Europe. 10th European Nitrogen Fixation Conference (ENFC), 2–5 September, Munich, Germany. http://orgprints.org/21797.

Messmer, M., Hildermann, I., Thorup-Kristensen, K., and Rengel, Z. (2012b). Nutrient management in organic farming and consequences for direct and indirect selection strategies. In: *Organic Crop Breeding* (ed. E.T. Lammerts van Bueren and J.R. Myers), 15–38. Oxford: Wiley-Blackwell.

Messmer, M.M., Shivas, Y., Verma, R. et al. (2014). Participatory cotton breeding and cultivar evaluation for organic smallholders in India. *Build. Org. Bridges* 2: 671–674.

Mohammadi, S. and Prasanna, B. (2003). Analysis of genetic diversity in crop plants – salient statistical tools and considerations. *Crop Sci.* 43: 1235–1248.

Moury, B., Fereres, A., García-Arenal, F., and Lecoq, H. (2011). Sustainable management of plant resistance to viruses. In: *Recent Advances in Plant Virology* (ed. C. Caranta, M.A. Aranda, M. Tepfer and J.J. Lopez-Moya), 219–336. Norfolk: Caister Academic Press.

Mühleisen, J., Piepho, H.-P., Maurer, H.P. et al. (2014). Yield stability of hybrids versus lines in wheat, barley, and triticale. *Theor. Appl. Genet.* 127: 309–316.

Müller, K.J. (2011). New hulless spring barley 'Pirona' and how it was developed under organic farming. In: *Organic Plant Breeding: What Makes the Difference?* 30. Frick: European Consortium for Organic Plant Breeding.

Murphy, K.M., Campbell, K.G., Lyon, S.R., and Jones, S.S. (2007). Evidence of varietal adaptation to organic farming systems. *Field Crop Res.* 102: 172–177.

Murphy, K., Lammer, D., Lyon, S. et al. (2005). Breeding for organic and low-input farming systems: an evolutionary-participatory breeding method for inbred cereal grains. *Renewable Agric. Food Syst.* 20: 48–55.

Myers, J.R., McKenzie, L., and Voorrips, R.E. (2012). Brassicas: breeding cole crops fro organic agriculture. In: *Organic Crop Breeding* (ed. E.T. Lammerts van Bueren and J.R. Myers), 251–262. Oxford: Wiley-Blackwell.

Neuhoff, D. and Range, J. (2012). Unkrautkontrolle durch Zwischenfruchtmulch von Sonnenblume (*Helianthus annuus*) und Buchweizen (*Fagopyrum esculentum*) im ökologischen Anbau von Winter-Ackerbohnen. *Journal fur Kulturpflanzen-Journal of Cultivated Plants* 64: 229–236.

Nirenberg, H.I. and Feiler, U. (2003). *Eindämmung des Erregers der Anthraknose der Lupine im ökologischen Landbau durch Anbau weniger anfälliger Lupinensorten der drei landwirtschaftlich wichtigen Lupinenarten*, 37. Braunschweig: Bundesforschungsinstitut für Kulturpflanzen.

Nordenfelt, L. (2006). *Animal and Human Health and Welfare: A Comparative Philosophical Analysis.* Wallingford: CABI.

Osman, A.M., Almekinders, C.J.M., Struik, P.C., and Lammerts van Bueren, E.T. (2008). Can conventional breeding programmes provide onion varieties that are suitable for organic farming in the Netherlands? *Euphytica* 163: 511–522.

Osman, A., Struik, P.C., and Lammerts van Bueren, E. (2012). Perspectives to breed for improved baking quality wheat varieties adapted to organic growing conditions. *J. Sci. Food Agric.* 92: 207–215.

Parentoni, S.N., Ferreira Mendes, F., and Moreira Guimarães, L.J. (2012). Breeding for phosphorus use efficiency. In: *Plant Breeding for Abiotic Stress Tolerance* (ed. R. Fritsche-Neto and A. Borém), 67–85. Berlin: Springer.

Parlevliet, J.E. (2002). Durability of resistance against fungal, bacterial and viral pathogens; present situation. *Euphytica* 124: 147–156.

Peet, R.K. (1974). The measurement of species diversity. *Annu. Rev. Ecol. Syst.* 5: 285–307.

Placido, D.F., Campbell, M.T., Jin, J. et al. Introgression of novel traits from a wild wheat relative improves drought adaptation in wheat (*Triticum aestivum*). *Plant Physiol.* 161 (4): 1806–1819.

Przystalski, M., Osman, A., Thiemt, E.M. et al. (2008). Comparing the performance of cereal varieties in organic and non-organic cropping systems in different European countries. *Euphytica* 163: 417–433.

Reents, H.J., Baresel, J.P., 2008. Annual clovers and medics in living mulch systems: competition and effect on N supply and soil fertility. Cultivating the Future Based on Science: 2nd Conference of the International Society of Organic Agriculture Research. ISOFAR, Modena, 18–20 June.

Reid, T.A., Yang, R.C., Salmon, D.F., and Spaner, D. (2009). Should spring wheat breeding for organically managed systems be conducted on organically managed land? *Euphytica* 169: 239–252.

Reusch, T.B.H., Ehlers, A., Hämmerli, A., and Worm, B. (2005). Ecosystem recovery after climatic extremes enhanced by genotypic diversity. *Proc. Natl. Acad. Sci.* 102: 2826–2831.

Richter, G.M. and Semenov, M.A. (2005). Modelling impacts of climate change on wheat yields in England and Wales – assessing drought risks. *Agric. Syst.* 84: 77–97.

Rodrigues de Oliveira, L., Glauco Vieira, M., Oliveira DeLima, R. et al. (2011). Combining ability of tropical maize cultivars in organic and conventional production systems. *Ciênc. Rural* 41: 739–745.

Roose, M., Kahl, J., and Ploeger, A. (2008). Influence of the farming system on the xanthophyll content of soft and hard wheat. *J. Agric. Food Chem.* 57: 182–188.

Rosenzweig, C., Iglesius, A., Yang, X.B. et al. (2001). Climate change and extreme weather events – implications for food production, plant diseases, and pests. *Glob. Chang. Hum. Health* 2: 90–104.

Ross, H. (1966). The use of wild *Solanum* species in German potato breeding of the past and today. *Am. J. Potato Res.* 43: 63–80.

Roth, F., 2010. Evaluierung von Winterackerbohnen als Zwischenfrucht für eine ökologische Biogasproduktion. University of Göttingen, Göttingen, p. 88.

Saastamoinen, M., Hietaniemi, V., and Pihlava, J.M. (2008). ß-Glucan contents of groats of different oat cultivars in official variety, in organic cultivation, and in nitrogen fertilization trials in Finland. *Agric. Food Sci.* 13: 68–79.

Sadeghi, S.M., Samizadeh, H., Amiri, E., and Ashouri, M. (2011). Additive main effects and multiplicative interactions (AMMI) analysis of dry leaf yield in tobacco hybrids across environments. *Afr. J. Biotechnol.* 10: 4358–4364.

Sadras, V.O. and Slafer, G.A. (2012). Environmental modulation of yield components in cereals: heritabilities reveal a hierarchy of phenotypic plasticities. *Field Crop Res.* 127: 215–224.

Schär, C., Vidale, P., Lüthi, D. et al. (2004). The role of increasing temperature variability in European summer heatwaves. *Nature* 427: 332–336.

Scholten, O.E. and Kuyper, T.W. (2012). Onions: breeding onions for low-input and organic agriculture. In: *Organic Crop Breeding* (ed. E.T. Lammerts van Bueren and J.R. Myers), 263–272. Oxford: Wiley-Blackwell.

Scholten, O.E., van Heusden, A., Khrustaleva, L. et al. (2007). The long and winding road leading to the successful introgression of downy mildew resistance into onion. *Euphytica* 156: 345–353.

Semenov, M.A. (2007). Development of high-resolution UKCIP02-based climate change scenarios in the UK. *Agric. For. Meteorol.* 144: 127–138.

Shewry, P.R., Charmet, G., Branlard, G. et al. (2012). Developing new types of wheat with enhanced health benefits. *Trends Food Sci. Technol.* 25: 70–77.

Smith, J., Pearce, B.D., and Wolfe, M.S. (2012). A European perspective for developing modern multifunctional agroforestry systems for sustainable intensification. *Renew. Agric. Food Syst.* 27: 323–332.

Sourdille, P., Cadalen, T., Gay, G. et al. (2002). Molecular and physical mapping of genes affecting awning in wheat. *Plant Breed.* 121: 320–324.

Stefanowicz, A.M., Kapusta, P., Szarek-Łukaszewska, G. et al. (2012). Soil fertility and plant diversity enhance microbial performance in metal-polluted soils. *Sci. Total Environ.* 439: 211–219.

Taylor, B.R., Cormack, W.F., 2002. Choice of cereal and pulse species and varities. In: Younie, D., Taylor, B.R., Welch, J.P., Wilkinson, J.M., eds. Organic Cereals and Pulses. Papers presented at conferences held at Heriot-Watt University, Edinburgh, and at Cranfield University Silsoe Campus, Bedfordshire, 6 and 9 November 2001. Chalcombe Publications, Southampton, pp. 9–28.

Tilman, D., Reich, P.B., Knops, J. et al. (2001). Diversity and productivity in a long-term grassland experiment. *Science* 294: 843–845.

Tuomisto, H. (2010). A consistent terminology for quantifying species diversity? Yes, it does exist. *Oecology* 164: 853–860.

Urban, D., Roberts, M.J., Schlenker, W., and Lobell, D.B. (2012). Projected temperature changes indicate significant increase in interannual variability of U.S. maize yields. *Clim. Chang.* 112: 525–533.

Urbatzka, P., 2010. Anbauwürdigkeit von Wintererbsen – Ein Vergleich zu Sommererbsen in Rein- und Gemengesaat unter den Bedingungen des Ökologischen Landbaus. Universität Kassel, Fachgebiet Ökologischer Land- und Pflanzenbau.

Urbatzka, P., Graß, R., Haase, T. et al. (2011a). Mischanbau von Winterraps und Wintererbse zur Erhöhung der Ressourcenausnutzung. In: *Wissenschaftstagung Ökologischer Landbau – Es geht ums Ganze: Forschen im Dialog von Wissenschaft und Praxis* (ed. G. Leithold, K. Becker, C. Brock, et al.), 193–194. Berlin: Verlag Dr Köster.

Urbatzka, P., Graß, R., Haase, T. et al. (2012). Influence of different sowing dates of winter pea genotypes on winter hardiness and productivity as either winter catch crop or seed legume. *Eur. J. Agron.* 40: 112–119.

Urbatzka, P., Graß, R., Haase, T. et al. (2011b). The level of N_2-fixation of different genotypes of winter pea in comparison to spring pea in pure and mixed stands. *J. Kult.* 63: 374–386.

van Bueren, E.T.L., Thorup-Kristensen, K., Leifert, C. et al. (2014). Breeding for nitrogen efficiency: concepts, methods, and case studies. *Euphytica* 199: 1–2.

van de Wouw, M., Kik, C., van Hintum, T. et al. (2009). Genetic erosion in crops: concept, research results and challenges. *Plant Genet. Resour. Charact. Util.* 8: 1–15.

van de Wouw, M., van Hintum, T., Kik, C. et al. (2010). Genetic diversity trends in twentieth century crop cultivars: a meta analysis. *Theor. Appl. Genet.* 120: 1241–1252.

van Elsen, T. (2000). Species diversity as a task for organic agriculture in Europe. *Agric. Ecosyst. Environ.* 77: 101–109.

van Ginkel, M. and Ogbonnaya, F. (2007). Novel genetic diversity from synthetic wheats in breeding cultivars for changing production conditions. *Field Crop Res.* 104: 86–94.

Vanaja, T., Mammootty, K., and Govindan, M. (2013). Development of organic indica rice cultivar (*Oryza sativa* L.) for the wetlands of Kerala, India through new concepts and strategies of crop improvement. *J. Org. Sys.* 8: 18–28.

Vieweger, A. and Döring, T.F. (2012). Resilience: linking health in soils, plants, animals and people. *Org. Res. Cent. Bull.* 111: 11.

Vieweger, A. and Döring, T.F. (2014a). Assessing health in agriculture – towards a common research framework for soils, plants, animals, humans and ecosystems. *J. Sci. Food Agric.* 95: 438–446.

Vieweger, A. and Döring, T.F. (2014b). The meaning of 'health' in the organic principle of health. *Build. Org. Bridges* 4: 1135–1138.

Vlachostergios, D.N. and Roupakias, D.G. (2008). Response to conventional and organic environment of thirty-six lentil (*Lens culinaris* Medik.) varieties. *Euphytica* 163: 449–457.

Wallenhammar, A., Adolfsson, E., Engström, M., et al. 2006. Field surveys of Fusarium root rot in organic red clover leys. Sustainable Grassland Productivity: Proceedings of the 21st General Meeting of the European Grassland Federation, Badajoz, Spain, 3–6 April. Sociedad Española para el Estudio de los Pastos (SEEP), Madrid, pp. 369–371.

Weitzberg, E. and Lundberg, J.O. (2013). Novel aspects of dietary nitrate and human health. *Annu. Rev. Nutr.* 33: 129–159.

White, L.E., Merrick, V.M., Bannerman, E. et al. (2013). The rising incidence of celiac disease in Scotland. *Pediatrics* 132: e924–e931.

Wignall, K., 2012. Untitled cartoon. Farmers' Guardian, 3rd August, p. 9.

Wilbois, K.P., Baker, B., Raaijmakers, J.M., and Lammerts van Bueren, E.T. (2012). Values and principles in organic farming and consequences for breeding approaches and techniques. In: *Organic Crop Breeding* (ed. E.T. Lammerts van Bueren and J.R. Myers), 125–138. Oxford: Wiley-Blackwell.

Wittman, H., Desmarais, A., and Wiebe, N. (2010). The origins and potential of food sovereignty. In: *Food Sovereignty: Reconnecting Food, Nature and Community* (ed. H. Wittman, A. Desmarais and N. Wiebe), 1–14. Oakland: Food First.

Wolfe, M.S. (1985). The current status and prospects of multiline cultivars and variety mixtures for disease resistance. *Annu. Rev. Phytopathol.* 23: 251–273.

Wolfe, M.S. (2000). Crop strength through diversity. *Nature* 406: 681–982.

Worthington, M. and Reberg-Horton, C. (2013). Breeding cereal crops for enhanced weed suppression: optimizing allelopathy and competitive ability. *J. Chem. Ecol.* 39: 213–231.

Wortman, S.E., Francis, C.A., Galusha, T.D. et al. (2013). Evaluating cultivars for organic farming: maize, soybean, and wheat genotype by system interactions in Eastern Nebraska. *Agroecol. Sust. Food Syst.* 37: 915–932.

Yoshihara, Y., Mizuno, H., Yasue, H. et al. (2013). Nomadic grazing improves the mineral balance of livestock through the intake of diverse plant species. *Anim. Feed Sci. Technol.* 184: 80–85.

Zak, D.R., Holmes, W.E., White, D.C. et al. (2003). Plant diversity, soil microbial communities, and ecosystem function: are there any links? *Ecology* 84: 2042–2050.

Zeller, S.L., Kalinina, O., Brunner, S. et al. (2010). Transgene environment interactions in genetically modified wheat. *PLoS One* 5: e11405.

Zhu, Y.Y., Chen, H.R., Fan, J.H. et al. (2000). Genetic diversity and disease control in rice. *Nature* 406: 718–722.

Zhu, Y.-G., Smith, S.E., Barritt, A.R., and Smith, F.A. (2001). Phosphorus (P) efficiencies and mycorrhizal responsiveness of old and modern wheat cultivars. *Plant Soil* 237: 249–255.

14

Exploring the Systems Concept in Contemporary Organic Farming Research

Christine A. Watson[1,2] and Bruce D. Pearce[3]

[1] *SRUC, Aberdeen, Scotland, UK*
[2] *Swedish University of Agricultural Sciences (SLU), Uppsala, Sweden*
[3] *ORC, Newbury, UK*

14.1   Introduction

The title we have given to this chapter poses a series of inherent questions. What do we mean by systems in this context? How does contemporary thinking influence how we view systems and any future developments? Are there issues concerning research on organic farming which might be different from other types of farming research?

Wikipedia defines a system as:

> a regularly interacting or interdependent group of items forming a unified whole. Every system is delineated by its spatial and temporal boundaries, surrounded and influenced by its environment, described by its structure and purpose and expressed in its functioning.

Systems can clearly be either open or closed depending upon whether they exchange both matter and energy with their surroundings. All farming systems are open systems, but they will vary in terms of the nature and the amounts of exchange and it is here that organic systems will differ from most other approaches to agriculture. As exemplified by earlier chapters in this volume, these differences relate principally to the involvement of people, a policy of restricting inputs to the system and the intention to optimise natural processes. This intention gives a clear link to ecology where the ecosystem was identified as the most important concept (Cherrett 1989), probably because of its link to holism, and explains why the principle of ecology is one of the four overarching principles of organic agriculture (IFOAM 2005).

Waring (1989) suggested that in an ecological context, ecosystems were bound to the circulation, accumulation and transformation of energy and matter through the medium of living things and their activities. He also suggested that it was dimensionally undefined because the systems boundary depended on the problem and time scales being addressed with many key issues operating at scales which cross commonly identified boundaries and go beyond a single

The Science Beneath Organic Production, First Edition. Edited by David Atkinson and Christine A. Watson.
© 2020 John Wiley & Sons Ltd. Published 2020 by John Wiley & Sons Ltd.

human generation. Despite this, ecosystem studies in general, and also as applied to farming, have moved from a descriptive to a predictive phase which requires adequate understanding so that models can be developed which are based on a clear understanding of basic principles. Paul (1989) suggested that soil was both an ecosystem component and a controller of ecosystem processes. This results in its being at the core of organic farming's approach to food production.

In differentiating organic systems from others, the issue of the systems boundary is important. In the introduction to this volume, it was stressed that one significant difference between organic and other forms of production was its emphasis on the need to achieve a series of equally important objectives rather than having a predominant focus on production. As a result, the boundary of organic production includes a range of issues traditionally linked to the social rather than to the natural sciences, some of which are discussed earlier in this volume. As transformation of matter, through controlling soil processes, is at the heart of systems management, the time scales linked to such processes will lead to different temporal boundaries. Together, these differences suggest why the research needs of organic production may, in some cases, be distinct from other forms of farming at a systems level.

14.2 The Importance of the Systems Concept in Organic Farming

The history of the organic movement has been told in many different ways (Conford 2001; Heckman 2006, Rahmann et al. 2017). However, all accounts recognise that the health of soil, plant, animal and man are 'one and indivisible' (Balfour 1943). Sir Albert Howard described a holistic approach to agriculture in 1940 as 'nature's farming', based on his experience in India (Howard 1940), but Walter Northbourne first used the term *organic* to apply to farming systems 'having a complex but necessary interrelationship of parts, similar to that in living things' (Northbourne 1940). In other words, there is a recognition that complex relationships exist between different system components and that the sustainability of the system is dependent upon the functioning of a whole integrated and inter-related system (Atkinson and Watson 2000).

A major difference between organic and conventional approaches is that conventional agriculture often uses targeted approaches to control a problem, such as a fungicide to control a disease outbreak. In contrast, organic systems aim at prevention rather than control by using systems-level approaches such as rotation. Vogt (2000) expresses this as organic farming relying on 'biological' (living organisms) and 'ecological' tools (agricultural management of the ecosystem, habitat diversity) rather than external inputs to maintain and enhance productivity.

The concept of 'eco-functional intensification' (Schmid et al. 2009) is increasingly being considered and developed within the organic movement and is understood as the replacement of external inputs with knowledge of how the system works and inter-relates, that is, 'more knowledge, information and organisation per hectare'. The various practices used in organic farming, such as the use of legumes or avoidance of manufactured fertiliser, do not in themselves make a farming system 'organic' as any farmer can use these approaches. It is the combination of the different practices brought together within the framework of the organic principles that make up the organic system.

Agricultural practices, worldview and values all contribute to the distinction between organic and conventional agriculture (Watson et al. 2006). Organic agriculture follows an explicit set of principles, regulations and standards for production and processing set out in law in Europe (EC 2007), the USA (www.ams.usda.gov/about-ams/programs-offices/national-organic-program) and elsewhere. These regulations and the resulting standards certify the process rather than the final product itself and often contain lists of actions that are prohibited rather than those that are permitted. Even with processes being certified, standards have dealt with the more easily controlled and monitored elements of the system, such as raw materials, livestock numbers, etc., and have tended to avoid more complex issues related to such things as energy and environment (Woodward 1998).

14.3   How are Systems Reflected in Regulation?

The concept of 'systems' and 'systems research' is integral to organic farming and research. Organic farming has developed around the concept of 'the farm as an organism', an idea discussed by the pioneers of both organic and biodynamic farming. In other words, the farm is a living and dynamic entity or 'system'. This basic concept of an agricultural system that works with ecological processes rather than external inputs has given rise to the four principles of organic farming (IFOAM 2005) on which current organic standards are based.

- The principle of health.
- The principle of ecology.
- The principle of fairness.
- The principle of care.

The systems concept is particularly strongly expressed in the principle of ecology: 'Organic Agriculture should be based on living ecological systems and cycles, work with them, emulate them and help sustain them'. The principle of health also encompasses the health of the system: 'Organic Agriculture should sustain and enhance the health of soil, plant, animal, human and planet as one and indivisible'. Padel et al. (2007) describe systems thinking as an integrative value linking the four principles. Thus, the ecology principle is about working in compliance with the ecosystem, and the standards focus principally on the farm.

Organic principles set the ideal for the organic system while regulations and standards interpret these and implement them into a real-world situation that deals with the changing environment, culture and markets in different countries, regions, etc. Arguably, standards are also about unifying consumer perceptions with actual organic systems management (Rahmann et al. 2017). Standards should not be seen as a target or endpoint but as a pathway to continual improvement, that will change and develop over time, moving organic farming practices along the sustainability pathway one step at a time.

14.4   Applying the Systems Concept to Organic Production

Spedding (1988) defines a system as:

> a group of interacting components, capable of reacting as a whole to external stimuli applied to one or more components and having a specified boundary based on the inclusion of all significant feedbacks.

This requires an understanding that what happens to one part of a system will affect other parts to a greater or lesser extent, although those effects may not be seen for some time. Thus, defining both the boundaries and the significant components of the system is important for management (Tow et al. 2011). Of course, it is rarely possible to study the whole system, and scientists frequently study subsystems or single components. However, without looking at the overall system to identify which components critically affect others and their relative importance when acting together, it is difficult to improve the performance of the overall system. Bawden (1991) suggests that a systems thinking approach is the key to sustainable agriculture and Ikerd (1993) differentiates between industrialised and sustainable agriculture through the concept that in a sustainable system, people are seen not only as components of the system but as deriving their well-being from it.

It is helpful to explore some of the properties of systems and thus how they are reflected in organic agriculture. Dillon (1984) described farm systems as being both dynamic and purposeful. Dynamic as in responding to external or internal influence, and purposeful in having specific aims or goals. The idea of organic food and farming systems embodies the concept of systems nested within other systems and while we are primarily dealing here with biological aspects of systems, we cannot divorce the production of food completely from consumption as marketing channels influence the supply of nutrients and energy within the wider system.

While organic consumers often prefer the idea of short food chains, direct marketing, etc. (Wägeli and Hamm 2016), large conventional companies also play an important part in organic food retailing (Howard 2013). Many supermarkets have introduced schemes for selling local organic produce as a dynamic response to consumer demand and government intervention (Seyfang 2008) but global trade and global players are still important. While within the EU organic regulations, animal feed must be imported from within the region, organic fruit and vegetables for human consumption come from other parts of the globe. It is simply not feasible to recycle waste over such distances and hence there is a risk of depleting soil fertility in organic systems in the producing country. Hence, as Bawden (1991) points out, changes to one system can lead to changes in other systems which are not necessarily considered as connected. There is increasing recognition of the need to close the nutrient loop at a small scale where in practice agreements between producers are leading to the exchange of materials between farms to prevent mining one system to feed another (Nowak et al. 2013) but at the scale of global trade, this is unlikely and probably impossible.

Organic agriculture is undoubtedly purposeful in its aims, as described earlier, and these are governed by regulation. At a grassroots level, there is a desire to ensure that organic production meets its given aims and both producers and researchers are striving to ensure that organic regulations and standards are developed and implemented to meet these aims (Løes et al. 2016; Niggli et al. 2016).

It is interesting to consider whether organic producers consider the farm system in a different way to conventional producers, although there has been little research exploring this perspective. However, a number of studies exist which compare differences in the way in which organic and conventional farmers approach management challenges within the farm system. Several studies show that organic and ex-organic farmers express greater environmental awareness than conventional farmers (Burton et al. 2003; Flaten et al. 2006). Läpple and Kelley (2013) surveyed farmers converting to organic in their attitudes to the environment but none of their statements embodied the idea of the farm or organic farming as a system. It is interesting to speculate whether researchers designing surveys embody the holistic systems concept. In a small study of organic and conventional farmers in Sweden, Chongtham et al. (2016) recently showed that farmers who had been organic for longer periods were more likely to follow strict crop rotations than recently converted or conventional farmers. The explanation for this was that they had experienced challenges such as weed management that could only be overcome in organic farming by using a systemic solution such as changing the rotation.

In the current environment where agriculture has become increasingly specialised and crops and livestock increasingly separated, there has been an ongoing debate about the reliance of organic crop production on nutrients from within or outside organic farming (Oelofse et al. 2013). The idea of linking crop and livestock farms in a given region to create a 'mixed farming system' by moving the boundaries beyond the farmgate is one option for embodying the organic principles of a system into contemporary specialised agriculture.

A study by Nowak et al. (2013) set out to explore interfarm nutrient exchanges in three regions of France to assess the extent to which organic systems rely on nutrients from conventional farming. Perhaps unsurprisingly, the reliance on nutrients from conventional farm sources was lower in mixed farming districts than in areas of farm specialisation. Moraine et al. (2016) and Ryschawy et al. (2017) have studied scenarios to develop exchanges between specialised crop farms and livestock farms and thus reintegrate crops and livestock. All the farmers were interested in crop–manure exchanges, including investment in collective equipment. While the scenarios increased impact and had positive environmental benefits, there were logistical and social aspects that limited uptake in practice and thus need further exploration. Trust between farmers was a critical issue and these authors suggest that smaller collective size seems more favourable for developing strong co-operation and taking these ideas forward. Proximity of farms and issues around distance of transport have also been studied by Asai et al. (2014) and de Wit et al. (2006) in the context of northern Europe.

This group of studies shows the importance of transdisciplinary research and the need to integrate social and natural science in order to solve problems in a way that is socially, technically and economically acceptable. Rahmann et al. (2017) point out the value and challenges of collaboration and co-operation for improving the efficiency of the wider organic food and farming system in relation to the supply chain but also the need for co-operation between stakeholders and researchers.

14.5 How is the Systems Concept Reflected in Organic Farming Research?

Earlier we discussed how the concept of the system is embedded in the aims and regulations that govern organic farming systems but what do we mean by 'systems concept' in the context of organic research? We do not mean 'farming systems research' which is a specific approach characterised by on-farm experimentation with a management orientation developed to assist agriculture in less developed parts of the world (Pemberton 1987). Any meaningful research on organic agriculture should work with the principles of organic agriculture, or at least with a significant understanding and respect of those principles (which, as we have discussed above, accepts the concept of the system within them). An excellent example of this is Barberi (2002) who explores why weed control in organic farming should be tackled from a systems perspective.

In addition, agricultural research cannot be entirely independent of human values as agriculture is about the human management of nature using a combination of ecological and biological understanding, technology and social science. The importance of the human as an integral part of the system has historically been more widely recognised amongst the organic research community than the conventional one. Padel et al. (2010) argue that the human needs to be at the centre of organic research in that new innovation is produced by the application of existing, often tacit, knowledge, such as management practices building on understanding of ecological principles, and that it is these producers who will need to apply the outcomes in their farming systems. Putting the human at the centre may not be just about the involvement of farmers in research but also the willingness to 'think outside the box' in terms of interpreting results. For example, Martini et al. (2004) questioned the conclusions of an agronomic experiment that yield changes following conversion to organic farming related to soil health and suggested instead that they resulted from improved skill on the part of the farmer.

Organic research has often preached the value of holistic approaches where the whole is seen to exceed the sum of the parts, although it is questionable to what extent it actually achieves this. Holistic research is a very challenging ambition due to the complexity and integrated nature of whole systems. This reflects the difficulty of interpreting results which represent the net effect of management practices on a system rather than the impact of changes in individual management practices. Reductionist, single-discipline approaches are therefore regarded as easier to publish in the scientific literature. Reductionist approaches can, however, sometimes lead to misunderstanding in terms of knowledge exchange from research into policy and practice. In other words, there is a need to move on from what Darnhofer et al. (2010) call 'exploitative innovations' or the focus amongst the agricultural research community on improving specific farm management practices.

There is now evidence of a wider acceptance by the whole research community of the need for integrated approaches. For example, McIntyre et al. (2009) in the IAASTD report *Agriculture at a Crossroads* talk about systemic innovation – reorienting agricultural science and technology towards more holistic approaches. This reflects the grand global challenges of food security and climate change which cannot be solved in a reductionist manner (Godfray et al. 2010; Ingram 2011). Of course, use of a systems-orientated approach is not to imply that disciplinary research does not have a very important place, but it does suggest that researchers need to take into account the consequences of reductionism. Reductionist approaches are vital in agricultural research as they allows us to isolate cause and effect and these have been integral in improving both the agronomic and the environmental performance of agriculture over the last 200 years. However, even with powerful inputs of fertiliser, biocides and genetics, the complexity of nature and the farming system means that the whole is greater than the sum of the parts or, to put it simply, what we see in the laboratory or greenhouse often does not necessarily result in the same outcome in the field.

The simplest example of this is the increased productivity we see in polycultural systems where productivity can be multiples of the same area in monoculture. A clear example of the problem of translation of research from glasshouse to field is a study on GM wheat crop. Zeller et al. (2010) studied transgenic bread wheat lines expressing a wheat gene against the fungus powdery mildew *Blumeria graminis* f.sp. *tritici*. Comparison was made between the transgenic line and its corresponding non-GM control line in both glasshouse and field trials. The results showed that, depending on the insertion event, a particular transgene can have large effects on the entire phenotype of a plant and that these effects can sometimes be reversed when plants are moved from the glasshouse to the field. In the glasshouse, GM lines had increased vegetative biomass and seed number and a twofold yield compared with control lines but in the field these results were reversed. Additionally, isolated farming practices or small groups of practices may deliver specific ecosystem services well but have negligible or negative impacts on others.

In an EU document looking back at a decade of organic research, Lutzeyer and Kova (2012) suggested that there is a need for research to:

> assess the degree to which organic agriculture complies with the principles and – in a wider perspective – delivers on the promises regarding important societal goals (e.g. reducing externalities).

Accepting, then, that the concept of a system is central to the understanding of organic farming, to what extent does the concept of a system feature in organic farming research? How relevant is non-systems research to organic farming and how does published research help us to understand how organic farming systems function and thus how we can improve them? Many studies of research in organic farming have concluded that appropriate holistic and reductionist approaches both have a place in aiding our understanding of systems (Rahman et al. 2017; Watson et al. 2008). However, research design and interpretation are fundamental in determining how effectively the results of research can be used to improve organic systems. Lutzeyer and Kova (2012) set out some appropriate methodologies for organic research which are given in Box 14.1, illustrating the need for a variety of research approaches and the involvement of appropriate stakeholders. Lutzeyer and Kova (2012) also highlight the need for site-specific solutions. Apart from the obvious constraints of climatic and edaphic factors, several studies have now shown that the landscape context of a given site can be important. For example, the impact of specific management practices or even the entirety of an organic system on biodiversity can differ depending on the complexity of the landscape which can still have a greater influence on above- and below-ground biodiversity than the management practices on an individual farm in that landscape (Flohre et al. 2011).

Box 14.1 Appropriate methodologies for organic research (Lutzeyer and Kova 2012)

Organic farming research methods include the following.

- Research which generates general and communicable knowledge.
- Whole-systems, multidisciplinary and interdisciplinary approaches should be used (rescaling continually the focus of research, e.g. cell, plant, field, farm, region).
- Both short- and long-term impact on agro-ecosystems should be considered – this includes models that allow amplification of environmental cost or benefits of a technology.
- Views of stakeholders (e.g. farmer, processors, consumers, environmentalists) should be integrated (participatory/action research).
- A specific analysis of stakeholders' expectations, since, according to organic farming objectives and principles, any situation is specific; therefore, paradigms and technical solutions cannot be implemented in the same way everywhere.

It is notable that organic farming researchers have been using participatory approaches for many years, even in developed countries (Ponzio et al. 2013), and recognise the role of farmer innovation in problem solving (Disler et al. 2013; Gunnarsson et al. 2013), perhaps because 'Organic farming with its stringent rules on external input use has to be even more innovative to solve production problems, sometimes opening up new avenues' (McIntyre et al. 2009). Padel et al. (2015) highlight the issue that innovation in agriculture is not just about technology but also about knowledge and application of knowledge.

We are not attempting a comprehensive review of organic farming research or of research approaches but a discussion highlighting where published research does and does not embrace the systems concepts so highly valued in organic farming and how important this is in defining the use of the research. In the section below, we set out some examples to highlight where published research has and has not embraced the system concept as well as some illustrations of where organic farming has specific research needs. We have selected (i) systems comparisons because they are fraught with controversy in the scientific literature; (ii) food quality as a topical area where the product quality and the production system are inextricably linked; (iii) weed control as an example of an issue that can limit crop yield in organic production but can be managed through the system or by direct intervention; and (iv) approaches to organic plant breeding that address a more variable and less controllable environment than in non-organic systems.

14.5.1   Example 1. Comparison of Production Systems

There are many examples of field experiments which set out to compare organic and conventional systems. Comparisons between systems are fraught with difficulty but useful comparisons can be made (Spedding 1988) where the basis of the comparison is open and fair, in particular the definition of starting points, boundaries and time-scale. Watson and Atkinson (2002) pointed out the need to separate aspects of the system that need to be assessed at the whole-systems level (i.e. those which are dominated by interactions or large-scale ecological processes, and those which can be compared using small plots). Kirchmann et al. (2016) recently set out three boundary conditions for field experiments comparing organic and conventional farming. Clearly, purpose is important and there are many different approaches on the ground, some of which explicitly set out to compare different organic rotations or systems with a view to improving their management (e.g. Taylor et al. 2006) and others which compare organic with biodynamic and/or conventional approaches (e.g. Mäder et al. 2002).

True systems comparisons are rare due to cost, issues of funding continuity and design. However, there are a number of such experiments in Europe that aim to compare fully functioning systems, including Scheyern in Germany (Lin et al. 2016) and Lögarden in Sweden (Stenberg et al. 2012). Others have taken a small farmlet approach where the systems are big enough to include grazing livestock, such as Apelsvoll (Eltun 1994). A more comprehensive guide to long-term experiments is available in Raupp et al. (2006).

However, the vast majority of comparisons are small plot rotational experiments and incomplete from a systems perspective. This is often due to the design and use of small plots such as commonly used in variety trialling. In other cases, the experiment has been designed in the manner of a traditional replicated field experiment and the plots representing different courses of the rotation are not contiguous. Plots are often too small, and costs too high, to maintain actual grazing livestock and returns from grazing are frequently simulated by mowing, missing the inherent spatial variability of dung and urine returns as well as the natural actions of trampling and defoliation. This kind of approach can provide very useful information from a reductionist perspective related to, for example, the value of different kinds of organic inputs. However, it is more difficult to scale up such results to quantify what happens in organic farms on the ground as the interactions with landscape are not present. In real farm situations, practical issues such as distance of fields from buildings are more likely to determine the use of manure on crops than prescriptions of nutrient requirements. There are still relatively few long-term studies of perennial systems although these have increased in recent years due to undertakings such as the H2020 AgFORWARD project (https://www.agforward.eu/index.php/en).

An alternative approach to systems comparison for arable farms was undertaken in the UK within the Scale Project (Gabriel et al. 2009). Comparator farms were selected using a statistical model which collated data from 30 variables which, through further analysis, were reduced to six. This showed that organic farms were spatially aggregated at the regional and neighbourhood scales and that their presence in a 10×10 km grid square can be predicted from the farm size/type. The Scale Project used this approach to select farms to study a range of agronomic, social and ecological aspects of farming (Sutherland et al. 2012).

14.5.2 Example 2. Food Quality and Its Relation to Production Systems

The growing market and globalisation of organic food have led to an emerging area of science which tests the authenticity of organic food (Capuano et al. 2013; Mie et al. 2014). It aims to find a method of chemical analysis which can provide definitive answers on whether food is organically produced. This begs the question about what defines organic and what unique properties of the organic system are likely to be measurable in the product. It would be reasonable to expect that nutritional composition could be influenced by cropping history (rotation) and fertilisation (use of legumes and manures) as well as variety and location. It would also seem that samples collected for research purposes should be clearly traceable in terms of place of origin, certification and farming history. Brandt et al. (2011) illustrate the criteria used by seven different authors or groups of authors to decide whether data on organic food quality is valid to be used within meta-analyses. Some have used organic certification as a criterion while others have required an indication of inputs used and conversion period. Lester and Saftner (2011) specify an extensive list of preharvest, harvest and postharvest criteria for valid comparisons compiled from a variety of other studies. These actually reflect that previous cropping has an impact on product quality which is rarely mentioned in food quality comparisons.

Meeting these criteria is a major challenge for future studies. Brandt et al. (2011) demand that the word 'organic' is used within a legally certified context, compared to Hunter et al. (2011) who required only the use of the word 'organic' or 'biodynamic' or utilised 'organic manure' without synthetic inputs. Hunter et al. (2011) did, however, control for other variables such as variety (which raises issues of appropriateness of any given variety across farming systems). These sets of rules are quite different and inevitably result in different data being used in the meta-analysis. This kind of issue with data selection pervades the literature and affects comparison of a range of 'products' from organic farming whether they are market goods or other products of the system, such as carbon sequestration. There are also examples in the literature of comparison of products taken from supermarket shelves without any knowledge of the history of the product other than certification (e.g. Angood et al. 2008). While this may provide information of value to the consumer, it does not help us to understand and develop the link between production systems and food quality.

14.5.3 Example 3. Weed Control

In organic farming, the design of the crop rotation plays a very strong role in weed management. In organic systems, the aim of weed control is not to totally eliminate weeds but to find a balance between managing their damaging effects and the biodiversity and other benefits that weeds can bring (Niggli et al. 2016). The use of bi- or multiannual crops in the rotation has been shown to control 71–98% of weeds compared to a rotation containing only annual crops (Lundkvist et al. 2011). In ley/arable systems, perennial leys with appropriate cutting or grazing treatments control annual weeds effectively. Perennial weeds remain a major challenge in all organic systems, especially stockless and no-till (Niggli et al. 2016). Research is ongoing to improve the control of weeds in organic systems using both systems-based approaches and more interventionist approaches. Participatory methods have been very successfully used to harness knowledge and innovation from the farming community (e.g. Turner et al. 2007).

There is an increasing interest in precision approaches to weed control, for persistent perennials such as docks (*Rumex obtusifolius*) (Norremark et al. 2009). Where such approaches are being developed, organic farming will need to consider the ethical and energetic aspects of their use. For example, a technique has been developed to control docks using jets of hot water (Latsch and Sauter 2014) but from a systems perspective it will be important to consider what impact this has on non-target organisms, such as soil micro- and macrofauna. Similarly, while mechanical weed control approaches can utilise more energy in the field than chemical approaches, there is a need to take into account the wider systems perspective and account for higher indirect energy use in conventional systems (Cormack and Metcalfe 2000).

14.5.4 Example 4. Plant Breeding

Approaches to breeding crop varieties for organic production need to take into account the way organic systems are managed. Modern crop varieties have generally been bred to perform best in conventional systems where much of the biotic pressures can be mitigated through the use of inputs such as nitrogen fertilisers, fungicides, pesticides and herbicides. Very few of these inputs are available with an organic system so there is a need to design a plant breeding system that is better suited to the organic system and the organic farmer's needs.

Two approaches have been pioneered within organic research: participatory plant breeding (PPB) and the development of composite cross-populations (CCP). Although not unique to organic systems, PPB has been used by farmers and scientists in both less developed and developed areas; it can be a powerful and cost-effective approach for developing locally adapted cultivars for a range of crops (Ceccarelli 2015) and some have been shown to outyield modern F1 varieties (Campanelli et al. 2015). CCPs are a way of addressing the biotic and abiotic stresses in organic systems with diversity itself. The concept is that by increasing the genetic diversity within a cultivar, it has the genetic potential to deal with the vagaries of changing growing conditions through:

- capacity – more phenotypic and genotypic variation
- complementation – the different genotypes within the CCP complement each other
- compensation – if some plants fail, others take their place
- change – evolutionary shifts in response to selection.

Competition between the different genotypes may also have negative effect. There is evidence that this approach works, with increases in yield and competitiveness being seen (Döring et al. 2015).

14.6 Cautionary Tales

From a systems perspective, there are many possible pitfalls in the analysis and descriptions of research results and here we highlight a few issues that have come to our attention while writing this chapter. We suggest that researchers working in organic systems need to be familiar with both the systems concept and the principles and regulatory frameworks that affect the possible design of research studies and the interpretation of results.

When it comes to the big questions such as whether organic agriculture can feed the world, research must be careful to be honest and open about the data being used. There has been much controversial discussion in the scientific literature which is beyond the scope of this chapter but, as Connor (2013) puts it, 'organic crops do not a cropping system make'. System productivity cannot be inferred from the yield of a single crop and any comparison of productivity should look at the output/input relationships as well as the sustainability of the inputs (not to mention the wider public goods outputs of a farming system). Researchers also need to be aware of the difficulties of interpreting research carried out under different certification schemes. We are referring here to differences between continents as opposed to individual certification bodies within a country. What is acceptable in the USA under NOP is not necessarily acceptable under EU Regulations. Regulations also change over time and there are instances in the literature of historical examples of farming practice being quoted which are outside the current regulations.

Terminology is also very important here and a phrase which continues to appear regularly in the literature is 'organic soil', meaning 'organically farmed' or 'organically managed soils' (e.g. Wang et al. 2013; Yossa et al. 2010). The term *organic soil* is incorrect in this context as it has a quite different meaning in soil science. This could reflect that researchers are focused on one component of the system rather than the system itself.

14.7 Are the Research Needs of Organic Farming Different from Conventional Farming?

The answer to this question must be 'yes' and 'no', depending on the context. Research around standards, whether it is meeting the current standards or the development of new standards, must be focused on organic systems. However, it is clear that much of the research carried out in, and for, organic systems is of relevance to non-organic production. Changes in regulation regarding use of agrochemicals will undoubtedly cause non-organic farmers to look for more cultural or systemic methods of weed, pest and disease control. The recent focus on soil in the past few years has also seen conventional farmers utilising cover crops and rotations that would have been unheard of a decade ago. A general move towards reduced reliance of agriculture on fossil fuels and a greater focus on recycling of nutrients in society will also stimulate interest in techniques and systems used by organic farmers. Knowledge exchange is clearly not a one-way route between organic and non-organic farming and much conventional research is clearly applicable to organic systems as long as the solutions provided fall within the approach of organic principles and certification.

References

Angood, K.M., Wood, J.D., Nute, G.R. et al. (2008). A comparison between organic and conventional-produced lamb purchased from three major UK supermarkets: price, eating quality and fatty acid composition. *Meat Science* 78: 176–184.

Asai, M., Langer, V., Frederiksen, P., and Jacobsen, B.H. (2014). Livestock farmer perceptions of successful collaborative arrangements for manure exchange: a study in Denmark. *Agricultural Systems* 128: 55–65.

Atkinson D, Watson CA. (2000). The research needs of organic farming: distinct or just part of agricultural research? Proceedings of the BCPC Conference – Pests & Diseases 2000, pp. 151–158. Alton, British Crop Protection Council.

Balfour, E. (1943). *The Living Soil*. London: Faber and Faber.

Barberi, P. (2002). Weed management in organic agriculture: are we addressing the right issues? *Weed Research* 42: 177–193.

Bawden, R.J. (1991). Systems thinking and practices in agriculture. *Journal of Dairy Science* 74: 2362–2373.

Brandt, K., Leifert, C., Sanderson, R., and Seal, C.J. (2011). Agroecosystem management and nutritional quality of plant foods: the case of organic fruits and vegetables. *Critical Reviews in Plant Sciences* 30: 177–197.

Burton, M., Rigby, D., and Young, T. (2003). Modelling the adoption of organic horticultural technology in the UK using duration analysis. *Australian Journal of Agricultural and Resource Economics* 47 (1): 29–54.

Campanelli, G., Acciarri, N., Campion, B. et al. (2015). Participatory tomato breeding for organic conditions in Italy. *Euphytica* 204 (1): 179–197.

Capuano, E., Boerrigter-Eenling, R., van der Veer, G., and van Ruth, S.M. (2013). Analytical authentication of organic products: an overview of markers. *Journal of the Science of Food and Agriculture* 93 (1): 12–28.

Ceccarelli, S. (2015). Efficiency of plant breeding. *Crop Science* 55 (1): 87–97.

Cherrett, J.M. (1989). *Ecological Concepts: The Contribution of Ecology to an Understanding of the Natural World*. Oxford: Blackwell.

Chongtham, I.R., Bergkvist, G., Watson, C. et al. (2016). Factors influencing crop rotation strategies on organic farms with different time periods since conversion to organic production. *Biological Agriculture & Horticulture* 33: 14–27.

Conford, P. (2001). *The Origins of the Organic Movement*. Edinburgh: Floris Books.

Connor, D.J. (2013). Organically grown crops do not a cropping system make and nor can organic agriculture nearly feed the world. *Field Crops Research* 144 (20): 145–147.

Cormack W, Metcalfe P. 2000. Energy use in organic farming systems. In: Defra Final Project Report, 2000. Defra, London.

Darnhofer, I., Lindenthal, T., Bartel-Kratochvil, R., and Zollitsch, W. (2010). Conventionalisation of organic farming practices: from structural criteria towards an assessment based on organic principles. A review. *Agronomy for Sustainable Development* 30: 67–81.

Dillon, J.L. (1992). *The Farm as a Purposeful System*. Armidale: Department of Agricultural Economics & Business Management.

Disler, M., Schmid, K., Ivemeyer, S. et al. (2013). Traditional homemade herbal remedies used by farmers of northern Switzerland to treat skin alterations and wounds in livestock. *Planta Medica* 79: PL24.

Döring, T.F., Annicchiarico, P., Clarke, S. et al. (2015). Comparative analysis of performance and stability among composite cross populations, variety mixtures and pure lines of winter wheat in organic and conventional cropping systems. *Field Crops Research* 183: 235–245.

EC (2007) Council Regulation (EC) No 834/2007 of 28 June 2007 on organic production and labelling of organic products and repealing Regulation (EEC) No 2092/91. Official Journal of the European Union, L189 (2007), pp. 1–23 (20.7.2007)

Eltun, R. (1994). The Apelsvoll cropping system experiment. I. Background, objectives and methods. *Norwegian Journal of Agricultural Sciences* 8: 301–315.

Flaten, O., Lien, G., Ebbesvik, M. et al. (2006). Do the new organic producers differ from the 'old guard'? Empirical results from Norwegian dairy farming. *Renewable Agriculture and Food Systems* 21 (3): 174–182.

Flohre, A., Rudnick, M., Traser, G. et al. (2011). Does soil biota benefit from organic farming in complex vs. simple landscapes? *Agriculture Ecosystems and Environment* 141: 210–214.

Gabriel, D., Carver, S.J., Durham, H. et al. (2009). The spatial aggregation of organic farming in England and its underlying environmental correlates. *Journal of Applied Ecology* 46: 323–333.

Godfray, H.C.J., Beddington, J.R., Crute, J.I. et al. (2010). Food security: the challenge of feeding 9 billion people. *Science* 327: 812–818.

Gunnarsson, S., Fredriksson, P., Hoffmann, R. et al. (2013). Knowledge synthesis and dissemination in organic research in Sweden: integrating ethics. In: *The Ethics of Consumption* (ed. H. Röcklinsberg and P. Sandin), 494–498. Wageningen: Wageningen Academic Publishers.

Heckman, J. (2006). A history of organic farming: transitions from Sir Albert Howard's War in the Soil to the USDA National Organic Programme. *Renewable Agriculture and Food Systems* 21: 143–150.

Howard, A. (1940). *An Agricultural Testament*. London: Oxford University Press.

Howard, P. (2013). Organic industry structure. *Journal of the New Media Caucus* 5 (3): http://median.newmediacaucus.org/winter-2009-v-05-n-03-agriart-companion-planting-for-social-and-biological-systems-organic-industry-structure.

Hunter, D., Foster, M., McArthur, J.O. et al. (2011). Evaluation of the micronutrient composition of plant foods produced by organic and conventional agricultural methods. *Critical Reviews in Food Science and Nutrition* 51: 571–582.

IFOAM (2005). *Principles of Organic Agriculture*. Bonn: International Federation of Organic Agriculture Movements.

Ikerd, J.E. (1993). The need for a systems approach to sustainable agriculture. *Agriculture Ecosystems and Environment* 46: 147–160.

Ingram, J. (2011). A food systems approach to researching food security and its interactions with global environmental change. *Food Security* 3: 417–431.

Kirchmann, H., Kätterer, T., Bergström, L. et al. (2016). Flaws and criteria for design and evaluation of comparative organic and conventional cropping systems. *Field Crops Research* 186: 99–106.

Läpple, D. and Kelley, H. (2013). Understanding the uptake of organic farming: Accounting for heterogeneities among Irish farmers. *Ecological Economics* 88: 11–19.

Latsch, R. and Sauter, J. (2014). Optimisation of hot-water application technology for the control of broad-leaved dock (*Rumex obtusifolius*). *Journal of Agricultural Engineering* 45: 137–145.

Lester, G.E. and Saftner, R.A. (2011). Organically versus conventionally grown produce: common production inputs, nutritional quality, and nitrogen delivery between the two systems. *Journal of Agricultural and Food Chemistry* 59: 10401–10406.

Lin, H.C., Huber, J.A., Gerl, G., and Hülsbergen, K.J. (2016). Nitrogen balances and nitrogen-use efficiency of different organic and conventional farming systems. *Nutrient Cycling in Agroecosystems* 105 (1): 1–23.

Løes, A.K., Bünemann, E.K., Cooper, J. et al. (2016). Nutrient supply to organic agriculture as governed by EU regulations and standards in six European countries. *Organic Agriculture* 7: 395–418.

Lundkvist A, Fogelfors H, Ericson L, Verwijst T. (2011). The effects of crop rotation and short fallow on the abundance of perennial sow-thistle (Sonchus arvensis L.). Proceedings of 24th NJF Congress, Food, Feed, Fuel and Fun – Nordic Light on Future Land Use and Rural Development, p. 76, Uppsala, Sweden.

Lutzeyer, H.-J. and Kova, B. (eds.) (2012). *A Decade of EU-Funded, Low-Input and Organic Agriculture Research (2000–2012)*. Luxembourg: Publications Office of the European Union.

Mäder, P., Fliessbach, A., Dubois, D. et al. (2002). Soil fertility and biodiversity in organic farming. *Science* 296: 1694–1697.

Martini, E.A., Buyer, J.S., Bryant, D.C. et al. (2004). Yield increases during the organic transition: improving soil quality or increasing experience? *Field Crops Research* 86: 255–266.

McIntyre, B.D., Herren, H.R., Wakhungu, J., and Watson, R.T. (eds.) (2009). *Agriculture at a Crossroads. Global Report by the International Assessment of Agricultural Knowledge, Science and Technology for Development (IAASTD): Synthesis Report*. Washington DC: International Assessment of Agricultural Knowledge, Science and Technology for Development (IAASTD).

Mie, A., Laursen, K.H., Åberg, M. et al. (2014). Discrimination of conventional and organic white cabbage from a long-term field trial study using untargeted LC-MS-based metabolomics. *Analytical and Bioanalytical Chemistry* 406: 2885–2897.

Moraine, M., Grimaldi, J., Murgue, C. et al. (2016). Co-design and assessment of cropping systems for developing crop–livestock integration at the territory level. *Agricultural Systems* 147: 87–97.

Niggli U, Willer H, Baker BP. (2016). A Global Vision and Strategy for Organic Farming Research. TIPI - Technology Innovation Platform of IFOAM – Organics International. Research Institute of Organic Agriculture (FiBL), Frick.

Norremark M, Swain KC, Melander B. (2009). Advanced Non-Chemical and Close to Plant Weed Control System for Organic Agriculture. Proceedings of the 10th International Agricultural Engineering Conference, Bangkok, Thailand, 7–10 December. www.researchgate.net/publication/267765746_Advanced_Non-Chemical_and_Close_to_Plant_Weed_Control_system_for_Organic_Agriculture

Northbourne, L. (1940). *Look to the Land*. London: J.M. Dent.

Nowak, B., Nesme, T., David, C., and Pellerin, S. (2013). To what extent does organic farming rely on nutrient inflows from conventional farming? *Environmental Research Letters* 8: 044045.

Oelofse, M., Jensen, L.S., and Magid, J. (2013). The implications of phasing out conventional nutrient supply in organic agriculture: Denmark as a case. *Organic Agriculture* 3: 41–55.

Padel, S., Niggli, U., Pearce, B. et al. (2010). *Implementation action plan for organic food and farming.research*. Brussels: IFOAM-EU Group.

Padel P, Röcklinsberg H, Verhoog H, et al. (2007). Balancing and integrating basic values in the development of organic regulations and standards: proposal for a procedure using case studies of conflicting areas (D2.3). Report from the Organic Revision Project. http://orgprints.org/10940

Padel, S., Vaarst, M., and Zaralis, K. (2015). Supporting innovation in organic agriculture: a European perspective using experience from the SOLID project. *Sustainable Agriculture Research* 4: 32–41.

Paul, E. A. 1989 Soils as components and controllers of ecosystem processes. Toward a More Exact Ecology: 30th Symposium of the British Ecological Society, p. 353. Cambridge University Press, Cambridge.

Pemberton, C.A. (1987). Improving the methodological approach to farming systems research. *Agricultural Administration and Extension* 26: 91–100.

Ponzio, C., Gangatharan, R., and Neri, D. (2013). The potential and limitations of farmer participatory research in organic agriculture: a review. *African Journal of Agricultural Research* 8: 4285–4292.

Rahmann, G., Ardakani, M.R., Bàrberi, P. et al. (2017). Organic Agriculture 3.0 is innovation with research. *Organic Agriculture* 7 (3): 169–197.

Raupp, J., Pekrun, C., Oltmanns, M., and Köpke, U. (eds.) (2006). *Long-term Field Experiments in Organic Farming*. Berlin: International Society of Organic Agriculture Research (ISOFAR).

Ryschawy, J., Martin, G., Moraine, M. et al. (2017). Designing crop–livestock integration at different levels: toward new agroecological models? *Nutrient Cycling in Agroecosystems* 108 (1): 5–20.

Schmid, O., Padel, S., Halberg, N. et al. (2009). *Strategic Research Agenda for Organic Food and Farming. Technology Platform Organics*. Brussels: IFOAM-EU Group.

Seyfang, G. (2008). Avoiding Asda? Exploring consumer motivations in local organic food networks. *Local Environment* 13: 187–201.

Spedding, C.R.W. (1988). *An Introduction to Agricultural Systems*. London: Elsevier.

Stenberg, M., Ulén, B., Söderström, M. et al. (2012). Tile drain losses of nitrogen and phosphorus from fields under integrated and organic crop rotations. A four-year study on a clay soil in southwest Sweden. *Science of the Total Environment* 434: 79–89.

Sutherland, L.A., Gabriel, D., Hathaway-Jenkins, L. et al. (2012). The 'Neighbourhood Effect': a multidisciplinary assessment of the case for farmer co-ordination in agri-environmental programmes. *Land Use Policy* 29: 502–512.

Taylor, B.R., Younie, D., Matheson, S. et al. (2006). Output and sustainability of organic ley/arable crop rotations at two sites in Northern Scotland. *Journal of Agricultural Science, Cambridge* 144: 435–447.

Tow, P., Cooper, I., Partridge, I. et al. (2011). Principles of a systems approach to agriculture some definitions and concepts. In: *Rainfed Farming Systems* (ed. P. Tow, I. Cooper, I. Partridge and C. Birch), 3–43. Berlin: Springer Science+Business Media B.V.

Turner, R.J., Davies, G., Moore, H. et al. (2007). Organic weed management: a review of the current UK farmer perspective. *Crop Protection* 26 (3): 377–382.

Vogt, G. (2000). *Entstehung und Entwicklung des ökologischen Landbaus im deutschsprachigen Raum*. Bad Duerkheim: Stiftung Ökologie und Landbau.

Wägeli, S. and Hamm, U. (2016). Consumers' perception and expectations of local organic food supply chains. *Organic Agriculture* 6: 215–224.

Wang, X.L., Ye, J., Gonzalez, P.P. et al. (2013). The impact of organic farming on the soluble organic nitrogen pool in horticultural soil under open field and greenhouse conditions: a case study. *Soil Science & Plant Nutrition* 59: 237–248.

Waring, R.H. (1989). Ecosystems: Fluxes of matter and energy. In: *Ecological Concepts* (ed. J.M. Cherrett), 17–41. Oxford: Blackwell.

Watson CA, Atkinson D. (2002). Organic farming – the appliance of science. Proceedings of the UK Organic Research 2002 Conference. pp. 13–17. Organic Centre Wales, Aberystwyth.

Watson, C.A., Kristensen, E.S., and Alroe, H.F. (2006). Research to support the development of organic food and farming. In: *Organic Agriculture: A Global Perspective* (ed. P. Kristiansen, A. Taji and J. Reganold), 361–383. Collingwood: CSIRO Publishing.

Watson, C.A., Walker, R.L., and Stockdale, E.A. (2008). Research in organic production systems – past, present and future. *Journal of Agricultural Science, Cambridge* 146: 1–19.

de Wit, J., Prins, U., and Baars, T. (2006). Partner farms: experiences with livestock farming systems research support intersectoral cooperation in the Netherlands. Livestock farming systems: product quality based on local resources leading to improved sustainability. *EAAP Publications* 118: 317–322.

Woodward, L. (1998). *Consumer Perceptions of Organic Food Quality*. Newbury: Elm Farm Research Centre.

Yossa, N., Patel, J., Miller, P., and Lo, M. (2010). Antimicrobial activity of essential oils against *Escherichia coli* O157:H7 in organic soil. *Food Control* 21: 1458–1465.

Zeller, S.L., Kalinina, O., Brunner, S. et al. (2010). Transgene6 environment interactions in genetically modified wheat. *PLoS One* 5 (7): e11405.

15

Science Base of Organic Agriculture

Some Conclusions

David Atkinson and Christine A. Watson

SRUC, Aberdeen, Scotland, UK

15.1   Introduction

The debate between the proponents of organic farming and those who practise other approaches to agriculture continues, frequently with a degree of virulence which is rarely seen in relation to most of the other divides in farming practice. Why this should be the case is important as we come to review how the science beneath organic farming differs from that which underpins the rest of agricultural practice and to suggest where there is a need for developments to inform the organic science base for the future. A number of the chapters in this volume have suggested that the aims of all those who farm are important to the nature of the division and so contribute to the passions involved in the debate.

The forms of agriculture characterised as conventional or mainstream at any point in time have ever been clearly linked to the currently prevailing business and social models. These tend to emphasise producing as much as is easily possible and for the minimum cost. They necessarily, within even a limited view of these objectives, require compliance with legislative requirements such as elements of environmental and employment law, the need to market and position the product within a market and working with a community. However, cost control tends to dominate or to be a driving element. This has led to the sector adopting new technologies wherever they seemed likely to either increase productivity, commonly through increasing yields, or reducing costs, commonly by reducing labour requirements. This business model has resulted in farm sizes increasing and in many becoming specialist in the production of arable crops, livestock or horticultural produce. The science base and social science/economic model which support such an approach share much in common with that underpinning many other businesses.

A comparison of this approach with the basics of organic production detailed by Lawrence Woodward in Chapter 2 shows the importance to organic production of very different values and both a science base which incorporates much of what is described as the social sciences, discussed in Chapter 3 by Pete Ritchie, and a different part of the more conventional science base; here synecology is particularly important. People and their skills tend not to be the most

The Science Beneath Organic Production, First Edition. Edited by David Atkinson and Christine A. Watson.

important of priorities. The scale of adverse environmental impact is commonly limited by legislative compliance rather than a desire to work with natural processes. Chapter 3 emphasises the very different view of food production which organic farming represents. Looking from a more societal perspective and a social science viewpoint, it questions whether more really is always better, disputes a view which ignores externalities which it sees as being part of production rather than someone else's problems and introduces global ethics and food justice as key issues for all systems of production. Social science asks difficult questions about the purpose of farming and what defines good farming As Chapter 3 concludes:

> This all takes us far from 8 tonnes a hectare or 10 tonnes, to questions about who we are, and what we are doing in the world. Such questions are just as much part of the organic conversation as discussion of yields and feeding the world. The task of organic philosophy is to set out a whole-system picture of what better farming looks like, and to use this as a guide to achievable and desirable change.

For over half a century, the conventional business model has resulted in 'conventional agriculture' becoming a major user of products from chemical industries. The dominance of the business model has been that much, perhaps even most, agricultural research over the recent past has focssed on either optimising the use of chemical inputs or developing systems designed to work with such inputs. Organic production, however, can learn and has learned from how these inputs work and the consequences of their effects.

The basic science which underpins all crop production is discussed in Chapter 8 and the measures which crops can take to increase nutrient uptake in Chapter 10. Both emphasise the importance of the link between an early supply of nitrogen and subsequent crop yields. Our current understanding of the importance of this link for later development is a good example of an area where fertiliser technology has provided knowledge helpful to setting goals for organic production. Similarly, work with herbicides and fungicides has shown just how damaging even moderate levels of infection or presence can be and so helped in the targeting of alternative measures. These goals have helped to develop the modifications to organic systems discussed in Chapter 6 on rotations and Chapter 12 on crop protection.

However, in general, little of this research has contributed to our core understanding of the most important areas for organic production. The most important aspect of all information relates to practical results about the workings of soil and other biological processes and the available options for managing these. Issues around soil health are discussed in Chapter 4 and soil microbiology in Chapter 11. The importance of soil biology would seem to be the reason why the largest of the UK's organic farming organisations is named the Soil Association. In addition, one of the oldest discussions about the distinctness of organic production spells out the reasons for the paramount importance of soil biology (Balfour 1943).

It is important to reflect on how we reached the current situation in respect of both conventional and organic production. The critical need for food production during World War II gave rise to a production agenda, at almost any cost. Production was maximised by increasing the land area under crops and by the use of fertilisers and other inputs which seemed likely to increase yields. For the next 50 years, this imperative of maximised production through the optimisation of inputs continued to dominate thinking and research expenditure. Feeding the developed world as cheaply as possible seemed to be a clear objective accentuated by free trade agreements within major trading blocks such as the EU or NAFTA. This shaped much of the research agenda not only of commercial companies but also as a result of funding models of publically funded research institutes and universities in the western world. Research focused on alternative methods of production was limited because these seemed unlikely to increase either production or efficiency as measured by cost and output per unit of labour. Studies comparing the productivity of organic systems with other systems were carried out but the underlying assumption that production was the key issue resulted in such studies suggesting the inefficiency of this means of production and, because of the design of the studies, to their giving limited information about how an organic system might best be run to meet its organic objectives.

The public agenda has changed and there is now real interest in recycling and in production focused on minimising the use of carbon-containing fuels. As the chapters in this volume indicate, there is huge potential to recycle organic waste materials from both agriculture and some other industries by using them as sources of mineral nutrients to power crop growth and of carbon to benefit soil structure. However, knowledge of soil microbial processes is limited and so we are currently far from being able to optimise our management of such activity. Rectifying this must be a major aim for all of agriculture.

Those who farm using organic principles are of course concerned with yields, because this is the basis of their income, but they have chosen to both produce food and achieve other objectives. It is the focus on other objectives which has largely resulted in organic production being reliant on a different science base, one which it shares with non-agricultural vegetation management. This distinctness in production methods and approaches and the lower yields commonly obtained have led to such food being labelled as 'organic' with the expectation that it will obtain a premium price in return for the embedded values (Atkinson et al. 2012) which the means of production have given to the product. The marketing associated with the organic label suggests that food produced in this way is different and perhaps even better. It is these inherent claims which tend to power the debates between those committed to different ways of farming. The ways in which organic produce differs from that produced by other systems are being increasingly documented. Whether organic produce is or is not better is open to a much wider debate and revolves around how the different foods have been produced, their environmental impact and their role in providing a fair return to individual producers and to communities.

15.2 Increasing the Contribution of Organic Agriculture to Global Food Production

A number of recent studies have assessed both future research needs (e.g. Hamm et al. 2017; Rahmann et al. 2017) and the ways in which organic farming research should be organised (e.g. Niggli et al. 2017, TPOrganics 2017). TPOrganics (2017) identified the importance of linking organic production to global needs for food and to the effects of general issues such as urbanisation and the approach to planetary limits and to sustainable development goals. They saw this as incorporating naratives linked to efficiency, consistency and sufficiency and pillars associated with redesigning the food system, ecological diversification and sustainable consumption.

Hamm et al. (2017) identified key targets as more research on plant breeding specifically targeted to organic production, plant–microbe and plant–plant interactions, field microrobots, alternative means of fungal disease control and the management of nutrients and soil fertility. New structures for plant breeding were seen as especially important as a consequence of the need for cultivars which capture and use nutrients efficiently and have significant ability to resist diseases. They identified that the current size of the organic market has led to commercial plant breeders being unwilling to invest in special cultivars for organic production and economically less important crops such as legumes with improved nitrogen fixation and significant changes to how plant breeding was organised internationally. They also identified the need for a broader vision for new plant breeding and an approach which recognised the holistic approach of organic production and, as a result, the need to make use of recent research on epigenetics and on interactions in the rhizosphere. Improved knowledge of micro-organisms was seen as important to improving both the nutrition and resistance of crops.

As targeted support and the use of ecological regulation mechanisms are at the core of organic production, making use of diverse cropping systems, digitisation, autonomous vehicles, drones and robots was seen as something to be explored. Fungal diseases have long been a major limit to production and have curtailed the overall environmental benefit of organic production. The improvement of resilience and health requires developments in epidemiological models of pathogens, diagnostic kits, breeding of resistant cultivars, optimising the use of acceptable crop protection materials and better co-ordination.

While nutrient storage is important, it links to organic matter accumulation and storage.

Both Hamm et al. (2017) and Rahmann et al. (2017) identified a need for research to be done in different ways such as focusing on efficient structures and additional funding. Many of the objectives are shared with conventional production, such as the need to produce sufficient food for an expanding world population, reducing greenhouse gas production, developing food chains and incorporating emerging ethics and lifestyle issues. Niggli et al. (2017) identified the need for a technology innovation platform which empowers rural areas, practises ecofunctional intensification and produces food for health (food products with identified health-sustaining outcomes). Again, they emphasised the importance of soil quality and healthy soil structure and the importance of soil microbes to both nutrient supply and the suppression of pests and diseases. They identified the need to explore the concept of yields – production per unit and profitability bases – relative to biodiversity resilience and stability.

In addition, the development of an organic approach to the production of food has been characterised as falling into three distinct eras or paradigms, identified as organic 1.0, organic 2.0 and organic 3.0 (Barabanova et al. 2015; IFOAM 2015; Rahmann et al. 2017; Arbenz et al. 2015). Organic 1.0 was the era of the founding visionaries beginning in the 1920s and included the development of microbiology and the identification of the importance of soil fertility, nitrogen fixation and the use of organic inputs. It was in this era that the principles which still underpin production were established. Organic 2.0 was the era of public recognition, standards and regulation, which began in the 1960s. This era was marked by an increase in the area of organically farmed land and food sales. This trend has continued, with the area of organically farmed land in Europe growing from 6.8 million ha in 2005 to 11.5 million ha in 2013 and retail sales of food by 138%. Organic 3.0 builds on this growth with the aim of becoming a global

Table 15.1 Future research needs aimed at improving the production of organic crops

Crop yield or quality objective (chapter where discussed)	Knowledge deficit/research target
Soil nitrogen availability (Chapters 8–10)	How soil nitrogen availability might be better geared to the potential photosynthesis made possible by available radiant energy Additional means of increasing heat flows into soils Better gearing of soil microbial activity to the release of nitrogen from organic materials in the soil Crop growth characteristics more closely geared to nitrogen supply rather than energy supply
Soil microbe impact on crop performance (Chapters 5, 7, 11)	The importance of signalling between crop plant and microbial communities in relation to nitrogen uptake and in response to pathogen attack The relationship between soil management through rotations and microbial functionality The importance of ecological diversity
Contribution of root system to crop performance (Chapter 10)	Better understanding of the joint and separate roles of root and fungal partners in nutrient uptake and under different conditions Scope for breeding for specific root system characters
Leaf area development and photosynthetic efficiency (Chapter 8)	Scope for adjusting the current relationships between light interception and temperature-linked growth
Maintenance of an effective leaf area (Chapter 12)	Means of controlling leaf pests and pathogens in the absence of pesticides
The contribution of soil structure to crop performance (Chapter 4)	The linkage between microbes and soil condition, especially the amount and make up of soil organic matter Mechanisms for the maintenance of soil organic matter and structure
The production of secondary metabolites and their contribution to health (Chapters 12 and 13)	The role of secondary metabolites such as those responsible for removing free radicals on energy capture The role of crop breeding in crop health

benchmark for sustainability through its emphasis on health, ecology, fairness and care. This requires an increase in funding from both private and public sources, especially in Europe and the USA. Its principal aim is to bring organic into the mainstream of world agriculture as a means of meeting the challenges for our societies.

This suggests that future research will be linked to both improving yields and documented quality criteria (Table 15.1) and to addressing key societal objectives and concerns (Table 15.2).

The impact of agriculture on soil structure, fertility and health has been an ongoing concern for many years. In the late 1960s, at a time when the use of mineral fertilisers had become established, developments in agricultural engineering had led to the use of larger and more powerful machines and crop production using pesticides was about to become the accepted norm, the UK government established a committee to assess the impact of modern farming on the soil. Its findings remain relevant to the current debate about the impact of agriculture

Table 15.2 Current societal objectives with the potential for organic farming to make a significant contribution

Societal objective	Potential contribution of organic agriculture
Mitigation of the impact and reduction of the magnitude of global climate change	Increased incorporation of atmospheric CO_2 into soil organic matter, reduced need for CO_2 generating inputs, increased recycling of wastes generated both by food production and by some non-agricultural industries
Food security	Emphasis on the management of natural processes, especially those in the soil, and working with prevailing ecological processes has potential to future proof food production against factors such as restrictions on the use of fossil fuels
Food quality	Organic food contains low/no residues of synthetic pesticides and as a consequence of varietal selection can have significant levels of polyphenols and other secondary metabolites with proven impacts on health
Carbon economy	Carbon offsetting has been an element of adjustment to climate change. The focus of organic production on building up organic reserves within the soil and eliminating many greenhouse gas-generating activities, such as the use of synthetic nitrogen fertilisers, should make organic production a key player in a more carbon-based economy
Business size and governance	Organic farming is labour intensive by its nature and so tends to sit predominantly within the small business sector. This gives rise to diversity, which is key to matching production to soil type and local production conditions. Governance is heavily influenced by organic standards and the fact that they are policed by a certifying body such as the UK's Soil Association
Employment in the digital era	Digital technologies are replacing the use of people in most sectors of the economy. Agriculture is not remote from this and embraces this in relation to cultivation and harvesting machinery. The detailed crop management upon which organic production depends will benefit in the future from remote sensing techniques but it is clear that this type of production will always require human input and perhaps to a greater extent than most industries

on the environment and upon soil condition and hence sustainability (Anon 1970). Having looked at the condition of soils in England and Wales, the committee came to the following conclusions.

- There were no concerns around the nutrient fertility of soils and that the replacement of ley farming and FYM by chemical fertilisers seemed not to have resulted in any loss of inherent fertility. They were, however, concerned with the potential impact of changes in farming practice on the availability of trace elements.
- There were concerns about effects on soil structure. On unstable soils the influence of organic matter is all important, making it important to know which soils were potentially unstable so that crop production might be sustained. There were significant concerns about the impact of heavy machinery and machinery use in unsuitable conditions. There was concern at the impact of modern systems on levels of soil organic matter. Organic matter is highest under permanent grass and lowest under arable rotations with ley systems being intermediate.

- Weeds were a cause of anxiety, especially the increase in grass weeds associated with continuous cereals. It is important to avoid unwise cropping sequences and to improve husbandry-based methods.
- There were economic difficulties in combining modern methods of farming and soil health. Methods of assessing the economic health of farm businesses can under-rate the importance of the effects of one year's cropping on subsequent crops because of the impact of farming methods on the structure and fertility of the soil. Comparisons of profitability need to take account of long-term effects on soil structure and organic matter.

All these issues continue to arise in current discussions of food production. They may have a marginally different context and they may impact in slightly different ways. They do raise the question of how long-term issues can be balanced against shorter-term financial returns. They suggest that assessing the merit of production methods solely or largely against the criterion of maximising yields is unwise.

Soil is critical to food production. Significant amounts are lost each year to urban developments such as housing, putting additional pressure on the remaining soil base. The organic carbon content remains important for all the reasons advanced in the above 1970 report but now additionally because of its importance both as a major reservoir of carbon and as one of the more easily manageable recipients of carbon dioxide released by both agricultural practice and other industries. It is imperative to develop agriculture to optimise incorporation of soil organic matter and minimise CO_2 generation, both at the site of food production and through its need for inputs. Current organic practices seem to have much to contribute to these current societal issues.

While farming and food criteria are important, agriculture is also a social, societal and economic activity and so it matters that we see it in the context of these wider issues and assess the opportunities for organic production to meet some current societal aspirations.

15.3 Challenges to Organic Production

Above we have focused on the potential contributions of organic production to food supply/ quality and to more general societal aims. We need to consider what future problems there might be and if there are elephants in the room. One of the major challenges to the organic movement in the 1990s was the advent of genetically modified crops (GMOs). While these became accepted in the USA and in many other countries, they were effectively banned in Europe and the UK. In view of the exit of the UK from the EU and potential closer trading links with the USA, it seems likely that such restrictions will be reconsidered in the UK at least. GMOs were never accepted by the organic movement in any country for reasons which were set out by IFOAM in 2001. GM crops were usually created by the insertion of genes not from that species, most commonly bacterial genes, into the crop genome. The most common means of insertion meant that the position of insertion into the genome was usually random. The source of the genes and the method of insertion raised questions about the naturalness of the process and its environmental impact. The changes produced were most commonly related to increasing the ease of crop production by making crops resistant to herbicides or to a limited number of pests or diseases. Few of the suggested changes affected things which could have made the ultimate food product more desirable to or more valued by consumers.

The earliest food GMO, a tomato with increased shelf-life, indicated the difficulty of continually modifying current varieties in situations of relatively high varietal turnover. At the same time, there were potential risks to the environment as a consequence of the movement of the introduced gene to native vegetation and to weed species. Initial concerns about the safety of GMO foods seem to have been unfounded but there was a feeling that all this was unnatural and public sentiment has remained opposed to the concept. In addition, few of the later developments which were supposedly in the pipeline, such as crops able to fix their own nitrogen or crops with advanced health properties, have been brought to market. It's clear that many of them were never realistic. The issues around the use of GMOs were reviewed comprehensively by Bruce and Bruce (1998).

The development of more targeted methods of gene editing in the past five or so years has now reopened this as an active subject of debate. Gene editing allows the genome to be modified by removing genes which are known to result in harmful effects, but also allows the accurate positioning of new genetic information found in other individuals of the same species. The process most commonly involves the use of a genetic construct known as CRISPR/Cas9 although there are also related technologies such as TALEN. These technologies have recently been reviewed by Doudna and Sternberg (2017). Unlike the situation with GMOs, the initial applications for gene editing technologies are likely to relate to farm animals and to humans rather than to crops. In this respect, they are more controllable and have a reduced environmental impact.

There seems to be a real likelihood of pigs being modified so as to be resistant to the viral disease porcine reproductive and respiratory syndrome (PRRS) and African swine fever in the foreseeable future and with significant benefit to pig welfare and, as a result of decreased piglet mortality, production costs. PRRS is one of the most economically important infectious diseases affecting pigs worldwide. The causative agent is PRRS virus. Infected pigs may present with a range of symptoms but the most devastating effects of infection are observed in young piglets and pregnant sows. In pregnant sows, full abortions or death of foetuses are observed and live-born piglets are often weak from an antenatal infection and display severe respiratory symptoms. The loss of pregnancies, death in young piglets and decreased growth rates result in major financial losses and compromise pig health and welfare. The virus has a very narrow host range, infecting only specific subsets of porcine macrophages. A range of gene editing approaches are being assessed which could result in either the removal of the host gene needed for viral infection or merely a small change to the receptor protein. By being much easier, genome editing has the effect of enabling more things to be tried, more often and in more places.

A recent paper confirmed IFOAM's opposition to GMO techniques and that this opposition included CRISPR and other new plant breeding techniques (NPBT) as genetic engineering was unacceptable in organic production (IFOAM 2015). During the GMO debates of the 1990s, it was possible for those opposed to the use of GMO technology to emphasis its potential adverse impacts on the environment and untested impacts on health. The use of CRISPR technology to increase farm animal welfare seems likely to be much harder to resist, especially if it results in less expensive food. The use of gene editing to improve pig welfare and the difficulties of separating pigs with this genetic transformation from the general pig-breeding population suggest that engaging popular support for not following this approach will be difficult. The IFOAM document centres its objections to GMOs on the basis that such techniques can lead to unpredictable side effects, that once released they cannot be recalled and that they seem likely to result in a reduction in biodiversity. All these claims may be harder to substantiate with gene-edited livestock, which enjoy better health and reduced mortality.

In addition, if gene editing techniques begin to be used to treat human diseases then this will help to substantiate the case for more general use. The organic movement seems likely to have to reassess the basis of its objection to this type of approach.

15.4   Conclusion

Organic farming is a decision to farm in a particular way and on the basis of clear principles. It is, however, also a societal construction in respect of the importance given to these principles rather than to other principles which form the basis of other approaches to production. In addition, it is bounded by a set of legal safeguards and regulatory protections which are subject to political processes. To remain sustainable, organic production will need to evolve to take account of the societal pressures which we identify above and the various technical issues which make up the substance of this volume. In opening this volume, we asked a series of questions. What is organic agriculture? What is distinctive about its science base? What are its links to ecology? Which bits of science matter? Which bits of science are very different? What is the current context of organic farming?

We began the introduction by defining our aims and said:

> Our objective is to demonstrate that a substantial body of science underpins organic food production. Much of this is shared with other systems of production but there are real and major differences in terms of the parts of the science base which are mission critical. The reliance of organic systems on natural soil processes means that it has much in common with the science base of natural ecosystems. Information derived from research on such systems has a greater significance for organic production than it has for some other systems of production.

In delivering this volume as it is, we have been unequivocal about the importance of science but have questioned whether society's current business model for western farming has a restricted view of which science matters and of how this sits with a wider social science-based take on the issues we face not just in producing food for now but on maintaining our ability to produce food for the populations of the future.

References

Anon 1970 Modern Farming and the Soil. Report of the Agricultural Advisory Council on Soil Structure and Soil Fertility. MAFF, Stationery Office, London.

Arbenz M, Gould D, Stopes C (2015) Organic 3.0. IFOAM Organics International, Bonn.

Atkinson, D., Harvey, W., Leech, C. et al. (2012). Food security: a Churches Together approach. *Rural Theology* 10: 27–42.

Balfour, E.B. (1943). *The Living Soil*. New York: Universal Books.

Barabanova, Y., Zanpli, R., Schluter, M., and Stopes, C. (2015). *Transforming Food and Farming: An Organic Vision for Europe in 2030*. Bonn: IFOAM Europe Group.

Bruce, D. and Bruce, A. (1998). *Engineering Genesis*. London: Earthscan.

Doudna, J. and Sternberg, S. (2017). *A Crack in Creation*. London: Bodley Head.

Hamm, U., Haring, A.M., Hulsbergen, K.-J. et al. (2017). Research strategy of the German Agricultural Research Alliance (DAFA) for the development of the organic farming and food sector in Germany. *Organic Agriculture* 7: 225–242.

IFOAM (2015). *IFOAM EU position paper on new plant breeding techniques*. Bonn: IFOAM Organics International.

Niggli, U., Andres, C., Willer, H., and Baker, B. (2017). Building a global platform for organic farming research, innovation and technology transfer. *Organic Agriculture* 7: 209–224.

Rahmann, G., Ardakani, M., Barberi, P. et al. (2017). Organic agriculture 3.0 is innovation with research. *Organic Agriculture* 7: 169–197.

TP Organics (2017). *Research and Innovation for Sustainable Food and Farming*. Brussels: IFOAM.

Index

The Science Beneath Organic Production, First Edition. Edited by David Atkinson and Christine A. Watson.
© 2020 John Wiley & Sons Ltd. Published 2020 by John Wiley & Sons Ltd.